GEL ELECTROPHORESIS OF PROTEINS

Gel Electrophoresis of Proteins

Edited by

Michael J Dunn

Lecturer Royal Postgraduate Medical School
Member of the Jerry Lewis Muscle Research Centre
Hammersmith Hospital
President of the International and the British Electrophoresis Society

WRIGHT

Bristol 1986

Published under the Wright Imprint by
IOP Publishing Ltd., Techno House, Redcliffe Way,
Bristol BS1 6NX, England.

British Library Cataloguing in Publication Data

Gel electrophoresis of proteins.
1. Proteins—Analysis 2. Electrophoresis
I. Dunn, M.J.
547.785046 QD431

ISBN 0 7236 0882 2

Typeset by
Mathematical Composition Setters Ltd. Salisbury, UK.

Printed in Great Britain by
J W Arrowsmith Ltd, Bristol

Preface

Methods of gel electrophoresis have been developed to such a state where in many situations they are the techniques of highest resolution available for protein analysis. This book is an attempt to give a comprehensive overview of the major techniques of analytical gel electrophoresis currently available. The first seven chapters give the theoretical basis of the major techniques of one-dimensional and two-dimensional gel electrophoresis, describe details of "state-of-the-art" methodologies and give examples of the ways in which these procedures can be applied to a variety of biochemical, biological and biomedical problems. There follows a chapter describing in detail the highly sensitive detection techniques now available for use in conjunction with the various electrophoretic procedures. The final chapter deals with qualitative and quantitative analysis of gel electrophoretograms, particularly with regard to those generated by high resolution two-dimensional methods. Each chapter has an extensive reference list, forming an excellent introduction to the literature for scientists unfamiliar with electrophoretic techniques who might be contemplating their use in a particular research project. For those already initiated into the mysteries of electrophoresis, it is hoped that the material contained within this volume can answer some questions, stimulate new ones and perhaps stimulate some advances in electrophoretic technology.

I would like to thank the authors for their willingness to contribute to this volume and for their punctuality in submitting manuscripts. I also express my gratitude to Dr P A Edge and the staff of John Wright for their patience and assistance in the completion of this book.

M J Dunn
April 1985

List of Contributors

A H M Burghes
Department of Genetics
The Hospital for Sick Children
555 University Avenue
Ontario
CANADA

M J Dunn
Jerry Lewis Muscle Research Centre
Roval Postgraduate Medical School
Ducane Road
LONDON
W12 OHS

K Gooderham
Bioseparations Research Department
LKB Produkter AB
Box 305
S-161 26 Bromma
SWEDEN

P M H Heegaard and T C Bøg-Hansen
The Protein Laboratory
University of Copenhagen
34 Sigurdsgade
DK-2200 Copenhagen N
DENMARK

C R Merril, M G Harasewych and M G Harrington
Section on Biochemical Genetics
Clinical Neurogenetics Branch
National Institute of Mental Health
Bethesda
Maryland 202051000
USA

P G Righetti, C Gelfi and E Gianazza
Faculty of Pharmacy and Department of
Biomedical Sciences and Technologies
University of Milano
Via Celoria 2
Milano 20133
ITALY

G M Rothe and W D Maurer
Institut für Allgemeine Botanik
Johannes-Gutenberg Universitat Mainz
Saarstrasse 21
6500 Mainz
FRG

C Schafer-Nielsen
The Protein Laboratory
University of Copenhagen
34 Sigurdsgade
DK-2200 Copenhagen N
DENMARK

S P Spragg, R Amess, M I Jones and R Ramasamy
Department of Chemistry
University of Birmingham
BIRMINGHAM
B15 2TT

Contents

List of Abbreviations

ACES	N-(2-acetamido)2-amino-ethane sulphonic acid
A/D	analogue/digital
Arg	arginine
Asn	asparagine
Asp	aspartic acid
b-Ala	beta-alanine
ATPase	adenosine 5′-triphosphatase
BAC	N,N′-bisacrylyl-cystamine
BEF	buffer isoelectric focusing
Bis	N,N′-methylene bisacrylamide
C	total g crosslinker per 100 ml
CA	carrier ampholyte
CCD	charge coupled device
CHAPS	3-[(cholamidopropyl)-dimethylammonio]-1-propane sulphonate
CMC	critical micelle concentration
Con A	concanavalin A
CRIE	crossed radio immunoelectrophoresis
CSF	cerebrospinal fluid
Cyt b5	cytochrome b5
Cyt c	cytochrome c
D/A	digital/analogue
DATD	N,N′-diallyltartardiamide
DDA	dodecyl alcohol
DDE	didodecyl ether
DDS	didodecyl sulphate
DHEBA	N,N′-(1,2-dihydroxyethylene) bisacrylamide
DMD	Duchenne muscular dystrophy
DNA	deoxyribonucleic acid
DNase	deoxyribonuclease
DPM	disintegrations per minute
EACA	epsilon amino caproic acid
E. coli	*Escherichia coli*
EDA	ethylene diacrylate
EDTA	ethylenediaminetetraacetic acid
EEO	electroendosmosis
ELISA	enzyme linked immunosorbent assay
FFT	fast Fourier transform
Gly	glycine
Hb	haemoglobin
HEPES	N-2-hydroxyethylpiperazine-N′-2-ethane-sulphonic acid

His	histidine
Hp	haptoglobin
HRP	horseradish peroxidase
HSA	human seum albumin
IEF	isoelectric focusing
IPG	immobilised pH gradients
ITP	isotachophoresis
K_d	dissociation constant
K_R	retardation coefficient
LPS	lipopolysaccharide
Lys	lysine
MDPF	2-methoxy-2,4-diphenyl-3(2H)-furanone
MES	morpholinoethane sulphonic acid
mol M	molecular mass
MTT	methyl thiazolyl tetrazolium
N_A	Avogadro's number, 6.022×10^3
NAA	neutron activation analysis
NAD	nicotinamide adenine dinucleotide
NADH	reduced form of NAD
NADP	nicotinamide adenine dinucleotide phosphate
NADPH	reduced form of NADP
NEPHGE	non-equilibrium pH gradient electrophoresis
PAA	polyacrylamide
PAG	polyacrylamide gel
PAGE	polyacrylamide gel electrophoresis
PAGGE	polyacrylamide gradient gel electrophoresis
PBS	phosphate buffered saline
PEG	polyethyleneglycol
pI	isoelectric point
PM	photomultiplier
PMS	phenazine methasulphate
PMSF	phenylmethanesulphonyl fluoride
RIA	radioimmunoassay
RNA	ribonucleic acid
RNase	ribonuclease
R_S	Stokes' radius
SB	sulphobetaine
SDS	sodium dodecyl sulphate
SNEP	snow electrophoresis
T	total g acrylamide + g Bis per 100 ml
TACT	N,N′,N′-triallylcitric triamide
TCA	trichloroacetic acid
TEMED	N,N,N′,N′-tetramethylethylenediamine
TEP	telescope electrophoresis

Tris	Tris hydroxymethyl aminomethane
Vh	volt hours
VT	vidicon camera
Wh	watt hours
2-D	two-dimensional

Chapter 1

Steady-state Gel Electrophoresis Systems

by

C. Schafer-Nielsen

1.1 Introduction

The term steady-state electrophoresis will in the present outline be used as a reference to electrophoresis characterised by a steady state, in which the electrolyte phases remain of constant ionic composition and in which there is no net transport of ion constituents by diffusion processes. The term is used in connection with isotachophoresis systems, moving boundary systems and isoelectric focusing systems.

Most workers in the field of protein analysis are acquainted with isotachophoresis through the daily use of discontinuous buffer systems in SDS-electrophoresis (sodium dodecyl sulphate). Here a sharp boundary between the electrolyte phases is employed for concentration of sample proteins in the early part of the run. Moving boundary systems are for the time being less widely used, but were in fact among the first systems applied in controlled electrophoresis of proteins and colloids by Tiselius in the first half of the century. A notable example of moving boundary electrophoresis is found in the early stage of isoelectric focusing with polyampholytes, i.e. before the final separation of ampholytes into discrete zones occupied by only one ampholyte species.

The aim of this article is to outline the various types of electrophoresis systems encountered in daily laboratory work and to provide the reader with an insight in the quantitative basis for calculation of their ionic composition. The reluctance with which the non-expert sets out to design electrophoresis systems is justified by the barrier presented by the somewhat complicated electrophoresis theory. However, a deeper understanding of the quantitative relations involved in the calculation of electrolyte phases for steady-state electrophoresis systems is not a pre-requisite for the worker who wants to design an electrophoresis system for his own needs. In fact, a qualitative understanding combined with simple calculation equipment is sufficient for this purpose in the majority of cases, especially where the task is to design a discontinuous electrophoresis system involving a few buffer constituents. At a later stage, if necessary, one can activate a computer program and obtain more precise knowledge of the quantitative composition of the systems. As will be demonstrated below, the main modest requirement for quantitative work often turns out to be programmable pocket calculator and a table of the pK and mobility values of the system constituents.

1.2 Historical Developments

The formation of sharp migrating boundaries between different electrolyte phases suspended in an electric field was observed independently by several workers in the nineteenth century. A quantitative theory accounting for the steady-state composition of the electrolyte phases was worked out by Friedrich Kohlrausch (1897). He realised that the ionic constituents were migrating independently with specific mobilities. In the first half of this century the phenomenon was extensively employed as a means for determination of ionic conductances and mobilities. However, although early experiments (Picton and Linder 1892) had been carried out on haemoglobin, the potential of the method, and indeed of electrophoresis in general, as a tool in biochemical investigations was not realised until Tiselius in the late 1920s carried out his studies on electrophoretic separation of proteins and colloids (Tiselius 1930). With the advent of the Tiselius apparatus, boundaries formed by electrophoretically migrating proteins in buffer solutions could be recorded optically and by light absorption and it became clear that the principle offered unique possibilities for the study of heterogenous biological fluids. The separation technique preceeded the application of useful chromatographic separations of proteins by more than a decade, and it was for a while the only useful method by which one could monitor the overall protein composition of, for example, blood serum. Tiselius himself did not formulate a quantitative theory for the moving boundary systems. A useful theory confined to strong electrolytes was published in 1945 by Dole and the theory was supported experimentally in an accompanying paper by Longsworth (1945), who carried out measurements in electrophoresis cells containing up to six monovalent ion species forming five independent boundaries. The moving boundary theory was extended to comprise weak electrolytes by Svensson (1948) and Alberty

(1950) and was later unified by the author in collaboration with P J Svendsen (Schafer-Nielsen and Svendsen 1981) to comprise any *n*-component steady-state electrophoresis system characterised by electrolyte phases of constant and self-adjusting composition.

The analytical device of Tiselius in which the sample ions are subjected to electrophoresis in free buffer, i.e. without an anticonvection medium, was supplemented in the late forties by the introduction of paper electrophoresis (Durrum 1950). The use of porous paper as an anti-convection medium led to a drastic reduction in the cost of equipment and at the same time a higher resolution was obtained, at least for some purposes. The paper strips were soon followed by gel based anti-convection media, e.g. starch (Smithies 1955), all of which are characterised by a low content although not complete lack of fixed net charges. The presence of fixed charges was found to interfere with electrophoresis, partly because of ionic binding of charged sample molecules and partly because fixed net charges give rise to electroendosmosis, i.e. to electrophoretic migration of water (Vanderhoff and Micale 1979). This intriguing phenomenon was a nuisance in steady-state electrophoresis systems, especially where local changes in the field gradient are accompanied by local deformation of the anticonvection medium due to changes in the electroendosmotic flow of water.

A major breakthrough in analytical electrophoresis followed from the advent of polyacrylamide gel (Raymond and Weintraub 1959), a hydrophilic non-ionic polymer with controllable protein permeability. In these gels one could obtain hitherto unprecedented resolution during electrophoresis of proteins, because of the low adsorption of the sample molecules and because the lack of electroendosmosis permitted the use of discontinuous steady-state electrophoresis systems. The first to employ the polyacrylamide gels in this context were Ornstein (1964) and Davis (1964) who introduced the technique under the name still used today: 'disc electrophoresis'. The word 'disc' refers both to the discontinuous electrolyte phases and to the discoid shape of the boundaries obtained in cylindrical electrophoresis tubes. Ornstein presented in 1964 a theoretical outline of the ionic composition of disc electrophoresis systems involving weak monovalent electrolytes and the theory was later expanded by others to comprise polyvalent electrolytes (Routs 1971, Jovin 1973, Everaerts 1976) and ampholytes (Schafer-Nielsen *et al* 1980).

With polyacrylamide as a nearly ideal anticonvection medium and disc electrophoresis as a perfect means for *in situ* concentration of applied sample proteins, the resolving power of electrophoresis was increased to an extent where up to about 20 protein species could be distinguished in a given run compared to the five to seven peaks obtainable with the Tiselius technique. However, two more refinements established analytical electrophoresis as an indispensable tool in biochemical work with protein mixtures.

The first of these was due to Svensson (1961), who, following on from results obtained by earlier workers (Williams and Waterman 1930, Kolin 1954, 1955), initiated studies around 1960 on the possibility for controlled use of ampholytes for electrophoretic generation of self-stabilising pH gradients. The basic consideration was, that if an aqueous solution of different ampholytes is subjected to an electric field in a closed system, the ampholytes will separate into discrete zones each occupied by one ampholyte species inducing a pH value close to its isoelectric point. In this way a stable pH gradient can be formed by electrophoresis of a mixture of ampholytes with different isoelectric points. Since proteins themselves are ampholytes, they can be separated by this process, and adequate spacing of the individual proteins can be obtained by addition of ampholytes with isoelectric points intermediate to those of the proteins. In his theoretical treatise on the principle (Svensson 1961) and the accompanying papers (Svensson 1962, Vesterberg 1966), Svensson pointed out that the number of known low molecular weight ampholytes with suitable isoelectric points is on the short side of what is needed for adequate spacing in isoelectric fractionation of complex protein mixtures. He further found that few of these ampholytes possess reasonable buffering capacity at their isoelectric points. The low buffering capacity is because most of the compounds (mainly amino acids and their derivatives) possess pK values two or more units from the isoelectric points. Svensson in addition pointed out that the low buffering capacity was accompanied by a low molar conductivity at the isoelectric points. The use of such ampholytes would thus lead to poorly defined pH gradients and the field strength required for their electrophoretic separation would be very high. The requirement then was for mixtures containing a large number of low molecular weight ampholytes with different isoelectric points and with closely spaced pK values. The riddle was solved when Vesterberg (1969) managed to work out a synthesis of highly diverse polyampholytes through a reaction of acrylic acid with oligoamines, the latter formed from polymerisation of ethylenediamine with dichloroethane. The product was made commercially available by LKB under the trade name 'Ampholine' and the type of electrophoresis became known as 'isoelectric focusing'.

The other major development was the introduction of sodium dodecyl sulphate (SDS) as the solubility agent for electrophoresis of proteins in polyacrylamide gels (Maizel 1969). This anionic detergent was known to be an extremely efficient solubiliser and denaturant of proteins. The efficiency is due to mutual electrostatic repulsion of the proteins following binding of large numbers of the negatively charged detergent molecules on their surface. In excess of the detergent, the amount bound to a given amount of protein differs little from one protein species to another, and the amount of sulphate groups in the protein–detergent complex is large compared to the number of charges of the native protein. The electrostatic repulsion of the sulphate groups is sufficient not only to dissociate protein aggregates, but also to change the tertiary structure of the individual protein molecules, so that they tend to assume a rod-shaped structure. Proteins treated in this way thus tend to have the same shape and charge density, and therefore the same electrophoretic mobility in a homogenous medium. In order to achieve an electrophoretic separation of SDS-treated proteins, it is necessary to perform the electrophoresis in an inhomogenous medium where separation parameters other than shape and charge can be exploited. Polyacrylamide gels provide an answer to this problem. With this material, it is possible to obtain a matrix in which the electrophoretic migration of macromolecules is restricted by partition between the polymer network and the aqueous phase. When SDS-treated proteins are subjected to electrophoresis in polyacrylamide, the migration velocity obtained shows an inverse linear correlation with the logarithm of the molecular weight. Although the exact mechanism of action is still open to theoretical interpretation (Bode 1980), it seems safe to conclude that the separation is due to the higher degree of interaction with the polymer network experienced by large as compared with small molecules, i.e. to the inhomogeneity obtained by the coexistence of a hydrated polymer and a free solvent. Molecular sieving of this type is observed in other types of gel as well, but due to its inertness and the unique possibility for controlled regulation of the sieving characteristics, polyacrylamide is still the material of choice.

The SDS–polyacrylamide gel electrophoresis was eventually refined by Laemmli (1970) through a combination with the disc electrophoresis principle of Ornstein and Davis. The resolving power was hereby increased to about 50 peptide species in a single analytical run and, because of the solubilising ability of dodecyl sulphate, almost any protein sample can be prepared for the purpose.

The next step forward in analytical electrophoresis came when O'Farrell (1975) designed a generally applicable combination of isoelectric focusing and SDS–polyacrylamide gel electrophoresis in which radio-labelled proteins could be separated analytically and detected by autoradiography with a resolving power of several hundred different peptides in a single sample.

Recent developments in analytical electrophoresis have mainly been centred on refinements and the extended use of the established techniques already mentioned. Thus Rosengren *et al* (1977) have introduced an elegant technique for obtaining protein titration curves by electrophoresis into isoelectrically separated ampholytes, and the technique has been extensively exploited by Righetti and Gianazza (1980) who have also been involved in the refinement of isoelectric focusing in polyacrylamide gels with pH gradients generated by acids and bases that are covalently coupled to the gel matrix (Gasparic *et al* 1975). This promising development was introduced by Righetti *et al* (1982) in collaboration with the LKB laboratories who have marketed the reactive electrolytes under the trade name 'Immobilines'. With immobilised pH gradients, the resolving power in isoelectric focusing of proteins approaches the highest conceivable level where proteins with isoelectric points differing by as little as 0.001 pH unit can be separated.

1.3 Fundamental Steady-state Electrophoresis Systems

As mentioned above, the steady-state electrophoresis systems usually involve boundaries that separate electrolyte phases with different ionic composition. The electrolyte phases are usually different in the qualitative sense, i.e. one or more ion species is present on only one side of the boundary. It is entirely possible, however, to design systems in which the same ions are present on both sides of the boundary with only their concentration ratios being different to those of the adjacent electrolyte phases. Figure 1.1 summarises the four types of electrophoresis systems most commonly employed in gel electrophoresis systems. In this figure, open circles represent ion constituents with similar sign of charge (positive or negative), whereas the closed circles represent the counter ions. Possible steady-state boundaries between different electrolyte phases are represented by the broken lines. In the zone electrophoresis system, the sample molecules (large open circles) migrate in a background electrolyte (the electrophoresis buffer) which is of uniform composition throughout the system.

In zone electrophoresis, the sample molecules are unrestricted from diffusion, and accordingly no sharp boundaries exist in such systems. The lack of stabilising boundaries excludes zone electrophoresis systems from the family of steady-state electrophoresis systems as defined in this outline.

The moving boundary systems (figure 1.1) are characterised by boundaries between electrolyte phases of different ionic composition in which at least one of the ion constituents that migrates in the same direction as a given boundary is present in the electrolyte phases on both sides of that

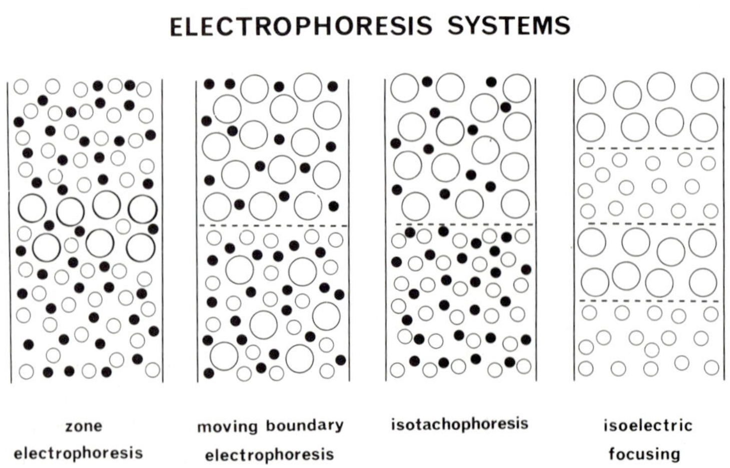

Figure 1.1 Schematic representation of the distribution of ion constituents in various types of electrophoresis systems. Open and closed circles represent ions carrying opposite net charge.

boundary. In the moving boundary system the large open circles represent that ion, whereas the small open circles represent an ion constituent that is present on only one side of the boundary.

The isotachophoresis systems are characterised by boundaries separating electrolyte phases which each contain one ion constituent that migrates in the same direction as the boundary and is only present in one of the phases.

The isoelectric focusing systems are characterised by a steady-state in which the constituents have separated in a sequence of zones according to their isoelectric points. In the ideal case, the separated constituents carry no net charge, and accordingly no counterions need to be present.

1.4 Fundamental Properties of Steady-state Electrophoresis Systems

The usefulness of steady-state electrophoresis systems in the separation of charged biomolecules is due to their ability to generate pH and/or conductivity gradients in which net diffusion of sample ions is abolished during electrophoresis. In the steady state, the sample ions are restricted to occupy discrete positions in the gradients, and their positions are dictated by their electrophoretic properties relative to the ion constituents defining the gradients. In most systems, steady-state gradients are generated by low molecular weight ion constituents which, because of different electrophoretic properties (pK values, isoelectric points, mobilities), arrange themselves in an ordered sequence during electrophoresis. These ion constituents are chosen so that their electrophoretic net mobilities in the system cover the net mobility range of the sample ions such that the latter become part of the gradient. The ion constituents carrying the same net charge sign as the sample ions are often referred to as 'spacer ions' or simply 'spacers', since their only function is to intercalate between the sample ions.

The key to defining the desirable properties of a steady-state electrophoresis system lies in the knowledge of the change in net mobility with pH in the separation medium for all ion constituents involved. A diagram showing such data for a set of different ion constituents is shown in figure 1.2. In this diagram, positive and negative net mobility of the constituents corresponds to migration towards the anode and the cathode respectively. The constituents A1–A3 are weak acids containing no amino groups, and accordingly they migrate towards the anode at all pH values. The constituents AM1 and AM2 contain both acidic and basic groups and are accordingly ampholytes characterised by their ability to migrate towards either the cathode or the anode depending on whether the pH is lower or higher than their isoelectric points. The constituent B is a weak base containing no acidic groups.

In the ideal case, such measurements of the net mobility as a function of pH in the separation medium should be available for all ion constituents in

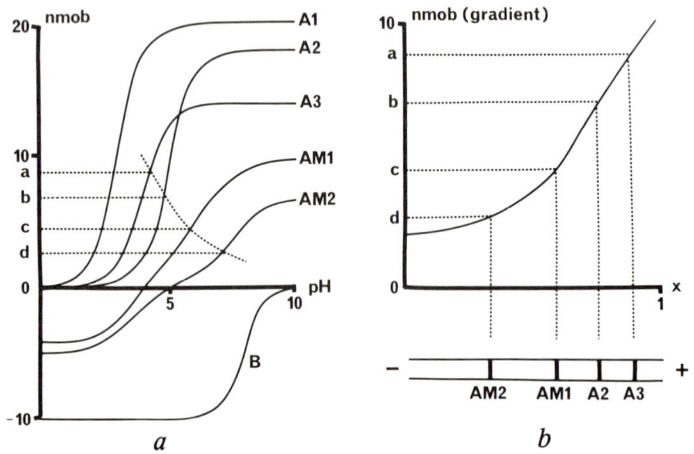

Figure 1.2 Diagrams relating the net mobility of ion constituents to their steady-state positions in an electrolyte gradient generated by a steady-state electrophoresis system. For details, see text.

a given electrophoresis system. This goes for both the sample ions and the ion constituents of the electrophoresis system proper. For the latter, the net mobility–pH relations can be calculated from their pK and mobility values (see later). However, for the sample ions which are often multicharged molecules with less predictable electrophoretic properties, one often has to rely on experimentally determined net mobility values. If the graphs in figure 1.2*a* represent data for the sample ions, a constructive way to predict their relative steady-state positions in a net mobility/pH gradient generated in a steady-state electrophoresis system may then be as follows:

Calculate for each point in the gradient its pH value and its steady-state migration velocity in a local voltage gradient of unit strength. This corresponds to a net mobility–pH data set for the gradient (rather than for its constituents), and the data can therefore be plotted together with the corresponding values for the sample constituents (as illustrated by the broken curve in figure 1.2*a*). The intercepts between this line and the graphs for the sample ions locate the steady-state positions of the sample ions with respect to two parameters, pH and net mobility, in the gradient. However, to get a picture of the spatial steady-state distribution of sample ions, the dimension of length must be included in the considerations. This can be done with the use of a graph depicting either the net mobility or the pH of the gradient as a function of its length (figure 1.2*b*). Since the relative positions of the sample molecules are fixed in the steady state and the absolute length of the gradient may increase as a function of time, it is convenient to represent only a unit gradient length in the diagram.

In figure 1.2*b*, the anodic end of the gradient corresponds to the value

1 and the cathodic end to the value 0 on the *x*-axis. The insert below the diagram is a projection onto the *x*-axis of the net mobilities at the interception points in figure 1.2*a*, and *b* is therefore an illustration of the relative positions of the sample ions in the steady-state gradient. From these premises, the sample ions must be confined to specific positions in the gradient, since they will migrate from all but one of these positions with either a higher or a lower velocity than the gradient, so as to be relocated to their steady-state positions. The question remains therefore, how a suitable net mobility–pH gradient can be generated with the available electrolyte components. The following sections provide information that may assist the reader in the formulation of an answer to this question.

1.5 Nomenclature and Definitions

In the following sections, reference is made to a given steady-state electrolyte phase by using the name of the ion constituent that defines that phase. The ion constituent defining an electrolyte phase is itself defined as an ion constituent that migrates in the same direction as the boundaries, and whose net mobility in the defined phase is intermediate to that of the ion constituents defining the adjacent phases. When only two phases are considered, the symbols L and T are employed, where L refers to the leading ion defining the (leading) electrolyte phase ahead of the boundary and T refers to the terminating ion defining the (terminating) electrolyte phase behind it.

$E(A)$ = Electric field gradient in the electrolyte phase defined by the ion A, in units of Volts/cm i.e. $\mathrm{V\,cm^{-1}}$.

$[I,A]$ = Concentration of ion constituent I in the electrolyte phase defined by ion constituent A. The symbol refers to the total concentration of the constituent, i.e. to the summed concentration of its dissociated and non-dissociated form. In the text, the word 'constituent' is used to underline this inclusion of the non-dissociated form. In contexts where no particular electrolyte phase is referred to, the variable A is omitted. Units are moles/millilitre.

$m(i,j)$ = electrophoretic mobility of constituent i at integer charge level j. This mobility is the electrophoretic migration velocity of constituent i carrying j integer net charges in a field gradient of $1\ \mathrm{V\,cm^{-1}}$. In equations that only involve monovalent ions, the variable is omitted. Unless otherwise specified, all mobilities and net mobilities are assigned with positive values irrespective of the charge sign of the ion constituent involved. The unit is $\mathrm{cm^2\,V^{-1}\,s^{-1}}$.

$nm(I,A)$ = electrophoretic net mobility of constituent I at the running conditions in the electrolyte phase defined by the ion constituent A. The net mobility refers to the constituent as a whole rather than to any single charged state. Ideally, calculation of net mobility should involve allowance for any difference between the conditions at which the calculation constants (pK and mobility values) are obtained, and those prevailing in the experimental set-up. These differences may be due to changes in the viscosity, the ionic composition or the temperature of the medium. However, while changes in viscosity may be allowed for through the introduction of frictional coefficients, the effect of changes in temperature and ionic composition remains less predictable, and beyond narrow limits no general correction can be deduced from present day ionic association theory.

The net mobility of an ion constituent is dependent on the net charge carried by it at the pH value of the medium. Quantitative accounts must therefore involve calculation of the charges carried by the constituents. In an aqueous solution containing an ion constituent with n chargeable groups, the constituent can be considered to be present in 2^n charged forms corresponding to the idea that each of the groups are present in either their charged or their uncharged form. Each of the 2^n forms corresponds to one of $(n+1)$ integer charge levels (including the zero charge level). If the electrophoretic mobility corresponding to each charge level is known, the net mobility of the constituent can be calculated from:

$$nm(i) = \sum_{j=1}^{n} m(i, j) y(i, j) \tag{1.1}$$

where $y(i,j)$ is the fraction of the ion constituent present at charge level j. $y(i,j)$ may itself be a sum of fractions since several charge combinations may lead to the same net charge of constituent. Thus, for a 1–1 ampholyte (containing e.g. one amino group and one carboxyl group) the zero net charge level can occur in two ways corresponding to whether both or neither of the groups are charged.

For multicharged molecules, for example proteins, it will often be impossible to obtain precise pK values for each of the chargeable groups. This is due to difficulties in resolution of the smooth titration curves into single pK values, a task that is further complicated by the theoretical dependence of the intrinsic pK values on the net charge of the entire molecule (Linderstrom-Lang and Nielsen 1959). In such cases, the net mobility must be interpolated from experimental data, e.g. as provided by the gel electrophoretic technique of Rosengren *et al* (1977).

While these considerations are crucial for the treatment of multicharged molecules, the calculation of system composition is usually confined to simple buffer components for which adequate mobility and pK data are available from the literature (Jovin *et al* 1970). The degree of ionisation, α, of a chargeable group is calculated from the Henderson–Hasselbalch equation

$$\alpha = 1/\{1 + \exp[s(\mathrm{pH} - \mathrm{pK})]\}, \tag{1.2}$$

where $s = 1$ for basic and $s = -1$ for acidic groups. Having made this calculation for each chargeable group on the constituent in question, it remains in principle a simple job to calculate the fraction of the constituent present on each of the possible charge levels. However, since 2^n charge combinations have to be calculated for any n-valent constituent, the process may be time consuming for other than small values of n. A more direct way is to calculate $v(i, j)$ directly using the expression

$$y(i, j) = \frac{10^{j\mathrm{pH} - \Sigma_{k=1}^{j} \mathrm{pK}(i,k)}}{1 + \sum_{k=1}^{n} 10^{k\,\mathrm{pH} - \Sigma_{l=1}^{k} \mathrm{pK}(i,l)}} \tag{1.3}$$

which is deduced from the assumption that the pK values are associated with the transition of the constituent from one charge level to the next. This assumption is not in agreement with the idea that each chargeable group dissociates in a manner independent of the remaining chargeable groups in the constituent. However, equation (1.3) is simpler to computerise, and the deviation between results obtained with equation (1.3) and the combination model are minor if the pK values are more than one unit apart.

1.6 Conductivity

According to Ohm's law the conductivity, $\varkappa$, is defined through the relation

$$\varkappa = I/E,$$

where I is the electric current and E the electric field gradient. The conductivity of an electrolyte thus equals the numerical value of the current density carried by its ion constituents at unit field strength (CGS units = $\mathrm{ohm}^{-1}\,\mathrm{cm}^{-1}$ or mho cm^{-1}). At unit field strength, the current density carried by a single charged species with n charge levels is

$$I(i) = F\,[I] \sum_{j=1}^{n} j\, m(i, j)\, y(i, j)$$

where F is the Faraday constant and $[I]$ is the constituent concentration (in moles/millilitre). The molar conductivity, Λ, of constituent i is defined as

$$\Lambda(i) = F \sum_{j=1}^{n} j\, m(i, j)\, y(i, j)$$

and the total conductivity, $\varkappa$, for an electrolyte containing m ion species is accordingly

$$\varkappa = \sum_{j=1}^{m} \Lambda(i)\ [I]. \qquad (1.4)$$

1.7 Equations for Calculation of System Composition

The composition of steady-state electrophoresis systems is defined by one relationship describing the steady state, plus three fundamental relationships applicable to all electrophoresis systems. The relations are stated below together with their mathematical equivalents in a form that is readily adapted for numerical processing.

1.7.1 The Velocity Relation

This relation defines a characteristic feature of steady-state electrophoresis systems, namely that the relative migration velocities of the constituents on either side of a steady-state electrolyte boundary are constant. For the ion constituents T and L that define the terminating and the leading electrolyte on each side of a boundary, the relation may be written as follows:

$$v(b) = k(\mathrm{L})nm(\mathrm{L})E(\mathrm{L}) = k(\mathrm{T})nm(\mathrm{T})E(\mathrm{T}), \qquad (1.5)$$

where $v(b)$ is the migration velocity of the boundary. The product of net mobility and field gradient equals the migration velocity of the ion constituents, and the k values are accordingly dimensionless velocity constants. For ion constituents that migrate with the same velocity as the boundary (isotachophoresis), the velocity constant equals one.

1.7.2 The Principle of Electroneutrality

This principle states that the number of positive equals the number of negative charges in a given volume of an electrolyte. For an electrolyte with m ion constituents $I(1), \ldots, I(m)$ each occurring at $n(i)$ charge levels, the principle may be stated as follows

$$\sum_{i=1}^{m} [I(i)] \sum_{j=1}^{n(i)} s\ j\ y(i, j) = 0$$

where s is assigned the value $+1$ or -1 according to the sign of charge of the group j. Calculation of pH in complex systems is readily carried out by iteration routines ('root finders') in which deviations from zero on the right hand side of the equation after insertion of a false pH value is used for the generation of a more correct value for the next iteration cycle.

1.7.3 Ohm's Law

A one dimensional electrophoresis system is characterised by equal current density in all electrolyte phases of the system. In such a system, Ohm's equation for two electrolyte phases defined by the ion constituents L and T may be combined to give:

$$\frac{E(\mathrm{L})}{E(\mathrm{T})}=\frac{\varkappa(\mathrm{T})}{\varkappa(\mathrm{L})},$$

which combined with equations (1.4) and (1.5) gives

$$\frac{k(\mathrm{T})nm(\mathrm{T},\mathrm{T})}{k(\mathrm{L})nm(\mathrm{L},\mathrm{L})}=\frac{\varkappa(\mathrm{T})}{\varkappa(\mathrm{L})}=\frac{\sum_{i=1}^{m}\Lambda(i,\mathrm{T})[i,\mathrm{T}]}{\sum_{i=1}^{n}\Lambda(i,\mathrm{L})[i,\mathrm{L}]} \tag{1.6}$$

where m and n are the number of ion constituents in the respective phases. By rearrangement of the expression, the concentration of one of the constituents can be expressed as a function of the remaining variables. The expression thus obtained is the only available relation between the concentration of a given ion constituent in one electrolyte phase and the constituents in another phase not containing the constituent in question. Thus in pure isotachophoresis systems, the expression is the only relation between the concentrations of leading and terminating ion constituent.

1.7.4 Conservation of Mass

This principle may be stated in one of a variety of forms depending on the system considered. A general way is to express the mass conservation in a differential form

$$\frac{\mathrm{d}J}{\mathrm{d}C}=-\frac{\mathrm{d}x}{\mathrm{d}t}$$

which in the present context can be taken to state that for each constituent present on both sides of an electrolyte boundary, the change in flux divided by the change in concentration of the constituent across the boundary equals the velocity of the boundary.

For electrophoresis systems in which the same ion species is present on both sides of a steady-state boundary, the relation between the concentration of the ion in the terminating zone behind the boundary and in the leading zone in front of the boundary is described by one general equation

$$[I,\mathrm{T}]=[I,\mathrm{L}]\,\frac{k(\mathrm{T})nm(\mathrm{T},\mathrm{T})\{k(\mathrm{L})nm(\mathrm{L},\mathrm{L})\pm nm(I,\mathrm{L})\}}{k(\mathrm{L})nm(\mathrm{L},\mathrm{L})\{k(\mathrm{T})nm(\mathrm{T},\mathrm{T})\pm nm(I,\mathrm{T})\}} \tag{1.7}$$

where $[I,\mathrm{T}]$ is the concentration of the ion in the terminating electrolyte and $[I,\mathrm{L}]$ is the concentration in the leading electrolyte. The equation only has meaning for ion constituents that are present in both electrolyte phases. The minus sign in the nominator and denominator of the equation should be used for ion constituents that migrate in the same direction as the boundary, whereas the plus sign is used for the counter ions.

A given system containing *n* ion species that are present on both sides of a boundary is characterised by *n* equations of this type. The concentration of these constituents in the two electrolytes represents $2n$ variables. The pH values in the two phases on each side of the boundary represent two additional variables (to be calculated by the equations describing electroneutrality in the two phases), and the velocity constants constitute two variables described by the velocity relation (one equation). The system is thus described by $n + 3$ equations containing $2n + 4$ variables, from which it is evident that $n + 1$ variables must be defined by the experimenter. This, of course, may be done in many ways, but in the experience of the author a natural choice is to define the composition of one of the zones (pH or one of the *k* values, anion concentration and ratio between the concentration of cations).

1.8 Isotachophoresis

The name isotachophoresis was introduced as a reference to electrophoresis systems in which the steady-state boundaries migrate with the same velocity as the leading and the terminating ion constituents in the electrolyte phases defining the boundary. The general characteristic of the composition of these systems is that each electrolyte phase is defined by an ion constituent that is present in that phase only. In addition to this characteristic ion constituent, the electrolyte phase may contain any number of additional ion constituents that are present in balanced amounts in all electrolyte phases of the system.

Figure 1.3*a* is a schematic representation of the simplest possible isotachophoretic set-up in which only three ion constituents are present. The two anions L (leading ion) and T (terminating ion) each define an electrolyte phase with the common counter ion C. In the steady state, both anions, and therefore the boundary between them, migrate with equal velocity towards the anode. For the attainment of this steady state, the only requirement is that ion constituent L possess a higher electrophoretic net mobility than the ion constituent T in both phases of the system. The electrolyte phases containing the ion constituents L and T are referred to as the leading and the terminating electrolytes respectively.

As illustrated in figure 1.3*b* an isotachophoresis system may contain several electrolyte phases arranged according to different net mobility of the ion constituents characterising the phases. In this figure, the ion I^- symbolises one or more ion constituents with the same charge as the ion constituents defining the electrolyte phases. Ion constituents of the I type have been assigned the name 'interfering ions'. This term is justified by the fact that such ions, when present in uncontrolled amounts, may prevent the attainment of an isotachophoretic steady state. The explanation of this breakdown is linked with an answer to the question of why ions with different net mobilities can be brought to migrate with equal migration velocities in adjacent electrolyte phases.

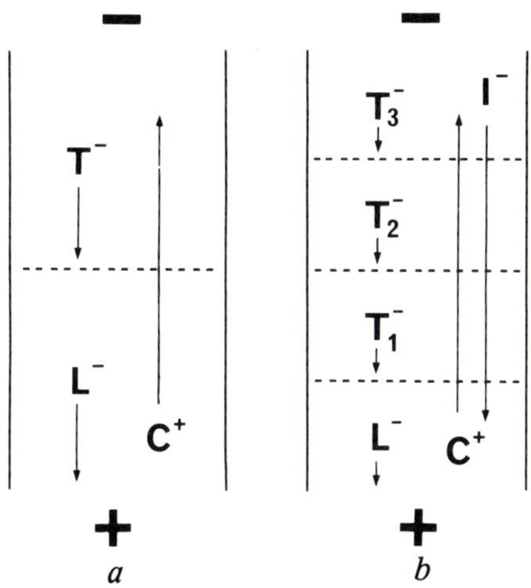

Figure 1.3 Schematic representation of isotachophoresis systems:
(*a*) simplest possible system. The ion constituent L^- (leading ion) possesses a higher electrophoretic net mobility than the ion constituent T^- (terminating ion). C^+ represents the common counterion(s). (*b*) composite system with three terminating ions (T_1–T_3) arranged in order of decreasing net mobility in the electrolyte phases following the leading electrolyte. I^- represents an interfering ion.

Consider again the simple isotachophoresis set-up in figure 1.3*a*. Here, the L anions have a higher electrophoretic net mobility than the T anions, and therefore they might be expected to migrate away from the front of T anions. This would, however, leave an intermediate zone containing only the C cations, which in turn would be a violation of the electroneutrality principle. As a counter-measure to this situation, the electric field gradient behind the escaping boundary of L anions increases just sufficiently so that the T anions attain a velocity equal to that of the L anions. The increase of the electric field behind the L anion boundary may be seen as the combined result of two effects:

(i) the conductivity behind the boundary drops due to the absence of L anions and

(ii) free C cations left behind the boundary induce an increase in electric potential with a polarity that results in attraction of the T anions.

In actual runs, thermal diffusion results in an overlap of the L and T anions in the boundary. The 'width' of the boundary is therefore determined as a compromise between diffusion and the steepness of the field gradient across the boundary.

In the argument above, it was assumed that only the T anions were present to fill in the anion gap left by the L anions. However, most experiments are carried out in aqueous solutions. Thus, if the three-component system in figure 1.3*a* were made by addition of the ion constituents to water, it would in reality be a five-component system due to the presence of hydrogen and hydroxyl ions. Now, it is easy to imagine that the hydroxyl ions could compete with the T ions during the filling of the conductivity gap behind the L ions. Whether this happens or not depends on the relative conductivity contribution of the T ions and the hydroxyl ions in the terminating electrolyte. This in turn is determined by the concentration and net mobility of the T ions and by the pH in the terminating electrolyte. At a critical relative concentration of hydroxyl ions in the terminating electrolyte, the hydroxyl ions will overtake the T ions and fill the gap behind the L ions in the leading electrolyte. In this situation the T ions attain a steady-state migration velocity that is lower than that of the L ions, and the isotachophoretic steady state is broken. The general conclusion is that the presence of ions carrying the same charge sign as the characteristic ions (i.e. interfering ions as defined above) may prevent the attainment of an isotachophoretic steady state. The argument cannot be employed for the counter ions in the system and, except for their ability to induce hydroxyl ions, they may be added to the system in any number and amount without interference with the isotachophoresis *per se*.

The breakdown of isotachophoresis due to the presence of interfering ions may be regarded as a transition of the isotachophoretic steady state into a state of zone electrophoresis in which the leading and the terminating ion migrate independently of each other in a background electrolyte. The relative concentration of interfering ions necessary to induce the breakdown depends on the net mobilities and concentrations of all ion constituents of the system. However, as will be illustrated by a numerical example, the point of breakdown can easily be calculated from the steady-state equations given in the previous section.

As an example, consider a simple isotachophoresis system consisting of only the ions T^-, L^-, I^- and C^+ where I^- is the interfering ion, and in which all the ions are monovalent and fully charged. In the example, the steady-state composition of the terminating or the leading electrolyte is calculated from a defined composition of the other (reference electrolyte). The choice of monovalent fully charged ion constituents is not essential, but simplifies the calculations since charge numbers and pH dependent changes in net mobility need not be considered. The conductivity contribution of a monovalent fully charged ion constituent, I, is

$$\varkappa(I) = Fm(I)[I].$$

In the present example, the steady-state concentration of the terminating ion constituent T can be isolated from equation (1.6) to give

$$[\mathrm{T,T}] = (m(\mathrm{L})[\mathrm{L,L}] + m(\mathrm{C})[\mathrm{C,L}] + m(\mathrm{I})[\mathrm{I,L}])/m(\mathrm{L}) - (m(\mathrm{C})[\mathrm{C,T}] + m(\mathrm{I})[\mathrm{I,T}])/m(\mathrm{T}) \quad (1.8)$$

where [C, T] and [C, L] are the concentrations of the counter ion in the terminating and the leading electrolyte respectively, with a similar nomenclature used for the interfering ion, I.

By substitution of net mobilities with mobilities and setting the velocity constants $k(\mathrm{L})$ and $k(\mathrm{T}) = 1$ (isotachophoresis), equation (1.7) is modified to

$$[\mathrm{C,T}] = \frac{m(\mathrm{T})(m(\mathrm{L}) + m(\mathrm{C}))}{m(\mathrm{L})(m(\mathrm{T}) + m(\mathrm{C}))}[\mathrm{C,L}] \quad (1.9)$$

and

$$[\mathrm{I,T}] = \frac{m(\mathrm{T})(m(\mathrm{L}) - m(\mathrm{I}))}{m(\mathrm{L})(m(\mathrm{T}) - m(\mathrm{I}))}[\mathrm{I,L}]. \quad (1.10)$$

From equation (1.10) it is apparent that the mobility of the interfering ion must be either higher than that of the leading ion or lower than that of the terminating ion. Outside this interval the equation gives negative values for the concentration of interfering ion. The physical interpretation is that interfering ions possessing intermediate mobilities would form a zone of their own between the leading and the terminating ion and thus prevent them from isotachophoretic migration. While this is one way of breaking the steady state, it can be demonstrated that a breakdown can also be induced by a sufficient concentration of interfering ion even though its mobility is outside the interval between the mobilities of the leading and the terminating ions.

Assume that the mobility of the leading ion is 40 arbitrary units and that the mobility of the counter ion and the terminating ion is 20 arbitrary units. The actual units are unimportant in this example, since the common factors cancel out in the equations. As the next step, define the mobility of the interfering ion together with the composition of the reference electrolyte including its concentration of interfering ion. From these data the steady-state concentrations of all the ions in the other electrolyte can be calculated by means of equations (1.8)–(1.10).

The diagram in figure 1.4*a* shows the steady-state concentration of the leading ion as a function of the concentration of interfering ion in the terminating (reference) electrolyte. Figure 1.4*a* is the corresponding diagram for the terminating ion, when the leading electrolyte serves as the reference. In both diagrams calculations have been performed corresponding to different mobilities of the interfering ion, and a breakdown of the isotachophoretic steady state is illustrated by a steady-state concentration of zero of the calculated ion at a critical concentration of interfering ion in the reference electrolyte.

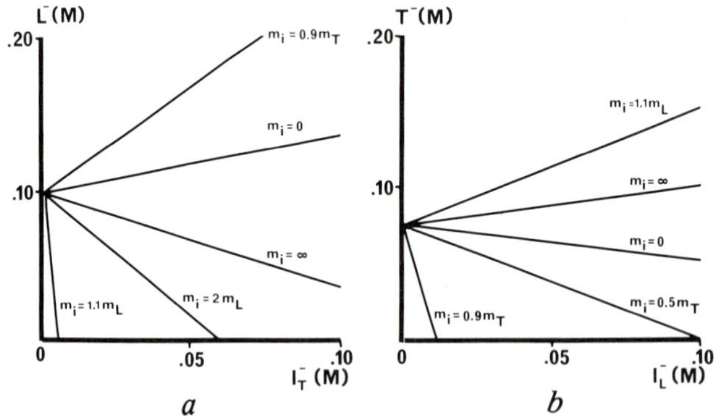

Figure 1.4 Influence of interfering ions on an isotachophoresis system.
(*a*) steady-state concentration of leading ion (L^-) as a function of the concentration of interfacing ion (I_T) in the terminating electrolyte. The calculation was performed assuming a constant concentration (= 0.075 M) of terminating ion (mobility = 20 arbitrary units). Mobility of leading ion = 40 arbitrary units, mobility of counter-ion = 20 arbitrary units and mobility of interfering ion (m_i) as indicated.
(*b*) steady-state concentration of terminating ion (T^-) as a function of the concentration of interfering ion (I_L) in the leading electrolyte. The calculation was performed assuming a constant concentration (= 0.10 M) of leading ion. Mobility values as stated for (*a*).

From figure 1.4*a* it appears that the breakdown can be induced by interfering ions whose net mobilities are less than that of the terminating ion. Similarly, from figure 1.4*b* it appears that in this system the breakdown can be induced by interfering ions whose net mobilities are greater than that of the leading ion. Since the breakdown (as stated above) is inevitable if the interfering ion possesses a net mobility intermediate to those of the leading and the terminating ion, it appears that any interfering ion can induce a breakdown if it is added above a certain uniform concentration to the initial electrolyte phases of an isotachophoresis system. Note, however, from figure 1.4 that it is perfectly possible to design isotachophoresis systems with a balanced amount of interfering ions.

The equations used for the preceding example are only applicable to isotachophoresis systems in which all ion constituents are monovalent and fully charged, and in which the dissociation of water can be neglected. An equally simple set of equations that in addition includes weak monovalent ion constituents can be derived, albeit still under the assumption that the concentration of hydroxyl and hydrogen ions is negligible compared to that of the constituent ions (Schafer-Nielsen *et al* 1980). These equations, given below, are useful in the majority of cases where all the constituents carry a maximum of one net charge per molecule. Note that, since it is the net charge rather than the number of charged groups that is limited to a maximum of one, the equations are also applicable to most amino acids.

Relation between [T, T] ***and*** [L, L]

$$[\mathrm{T},\mathrm{T}] = [\mathrm{L},\mathrm{L}]\,\frac{m(\mathrm{T})(m(\mathrm{C})+m(\mathrm{L}))}{m(\mathrm{L})(m(\mathrm{C})+m(\mathrm{T}))}. \tag{1.11}$$

This is a representation of the well known Kohlrausch regulating function for isotachophoresis systems involving strong monovalent electrolytes. As shown by Ornstein (1964), the expression is also applicable for weak monovalent electrolytes in the near neutral pH range. A useful feature of the equation is that only mobilities are involved, i.e. the expression is independent of the degree of dissociation of the ion constituents and thus of the pH values in the system. The concentrations refer to the total concentrations of the constituents rather than to the concentrations of their ionised forms.

Relation between [C, T] ***and*** [C, L]

$$[\mathrm{C},\mathrm{T}] = [\mathrm{C},\mathrm{L}] + [\mathrm{L},\mathrm{L}]\,\frac{m(\mathrm{C})}{m(\mathrm{L})}\left(1-\frac{m(\mathrm{L})+m(\mathrm{C})}{m(\mathrm{C})+m(\mathrm{T})}\right) \tag{1.12}$$

Note, that the form of equation (1.12) is not identical to that of equation (1.9). The difference is due to inclusion of weak monovalent electrolytes in the calculations (Schafer-Nielsen *et al* 1980).

Relation between [L, L] ***and*** [C,L]

The following equation can be derived from the Henderson–Hasselbach equation and the electroneutrality principle (again under the assumption that the dissociation of water can be neglected in the equation describing electroneutrality)

$$[\mathrm{C, L}] = [\mathrm{L, L}]\,\frac{1 + \exp\{s(\mathrm{pH(L)} - \mathrm{pK(C)})\}}{1 + \exp\{s(\mathrm{pK(L)} - \mathrm{pH(L)})\}}, \tag{1.13}$$

where $s = 1$ if L is an anion and $s = -1$ if L is a cation. For reasons of symmetry, all Ls and Ts can be interchanged in equations (1.11)–(1.13). Furthermore, it can be shown that in isotachophoresis systems involving multiple electrolyte phases, L and T may refer to the isotachophoretically migrating constituent ions from any two of these phases.

In equation (1.13) the concentration of the counter ion is found directly by inserting the defined values of [L, L] and pH(L). Alternatively, one may choose to define [C, L] and [L, L], in which case pH(L) must be found by an iteration procedure, e.g. by solving the electroneutrality equation by a computerised *regula falsi* technique. The procedure is mandatory, since pH can be defined only in one of the electrolyte phases, after which pH, and hence the net mobilities of constituents in the remaining phases, is dictated by the constituent concentrations obtained with equations (1.11) and (1.12).

The net mobility of monovalent constituents is calculated from

$$nm(I) = \frac{m(I)}{1 + \exp\{s(\mathrm{pK}(I) - \mathrm{pH})\}} \tag{1.14}$$

in which it is assumed that the net mobility varies in linear proportion to the net charge of the constituent and where $s = 1$ for anions and $s = -1$ for cations.

From equations (1.11) and (1.12) it appears that the steady-state concentration of the terminating ion constituent is defined solely by the concentration of the leading ion constituent. Also, at a given concentration of the latter, the concentration of counter ions in the terminating electrolyte varies linearly with its concentration in the leading electrolyte. In anodically migrating systems, i.e. systems where the leading ion constituent is an acid, an increase in concentration of counter ion in the leading electrolyte will increase the pH value of both the leading and the terminating electrolyte. The relative increase depends on the relative values of the mobility and pK values of all the constituents and thus must be calculated for each system. One way of depicting the results is shown in the diagrams in figures (1.5)(*a*)–(*d*). These diagrams show the key parameters characterising five electrolyte zones of an isotachophoresis system. The system was defined by a leading electrolyte containing MES (morpholinoethane sulphonic acid) as the leading ion and Tris (Tris hydroxymethyl aminomethane) as the common counter ion of the system. The four terminating ions are Asp (asparagine), Gly (glycine), b-Ala (beta-alanine) and EACA (epsilon amino caproic acid). The system parameters are depicted as a function of the concentration of Tris in the leading electrolyte at a constant concentration

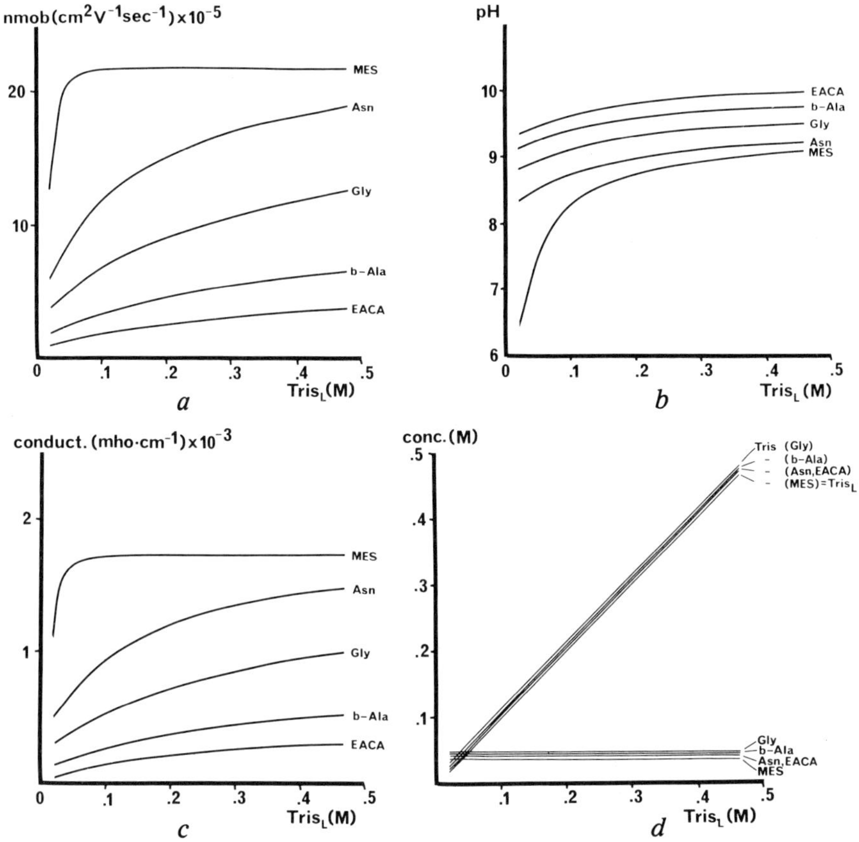

Figure 1.5 Changes in the key parameters characterising the electrolyte phases of a five-spacer isotachophoresis system as a function of the concentration of counter ion in the leading electrolyte. Spacer ions: MES = Morpholino ethane sulphonic acid; Asn = asparagine; Gly = glycine; b-Ala = beta-alanine; EACA = epsilon amino caproic acid. Counterion: Tris (tris hydroxymethyl aminomethane). Constituent data: MES, pK = 6.1, $m = 22$; Asn, pK = 8.8, $m = 26$; Gly, pK = 9.7, $m = 35$; b-Ala, pK = 10.4, $m = 32$; EACA, pK = 10.8, $m = 25$; Tris, pK = 8.1, m = 24. Mobilities (m) stated in units of $cm^2 V^{-1} s^{-1}$. During calculations the amino acids were treated as monovalent acids with pK values as indicated.

(0.04 M) of MES. The diagrams therefore represent a continuous series of isotachophoresis systems that are each distinguished solely by the ratio of leading ion constituent and counter ion constituent in the leading electrolyte. The diagrams were constructed by means of equations (1.11)–(1.14) using a standard iteration procedure for the calculation of pH. With exception of the counter ion, all the constituents are amino acids possessing different pK and mobility values. In the pH ranges involved in the example, all the acidic groups can be considered fully dissociated, and the net mobility of the constituents as a function of the pH is therefore mainly controlled by the degree of protonisation of their amino groups. The constituents have therefore entered the equations as monovalent acids possessing pK values corresponding to those of the amino groups.

Any two or more of the electrolyte phases in figure 1.5 can be combined to form an isotachophoresis system with the same composition of electrolyte phases as shown in the figure. As an example, one may chose to run a system involving all five components or to use only the glycine electrolyte as the leading electrolyte for the EACA electrolyte.

Two general rules governing steady-state electrophoresis systems can be extracted from the preceding outline of isotachophoresis systems. The rules are not without exceptions, but they are useful during the initial phases of system design, where the task is to select the ion constituents that may be combined to give a system with the desired properties

(i) In the steady state, the electric field gradient increases at the boundaries between the electrolyte phases in the direction from leading to terminating electrolyte (cf the conductivity profiles in figure 1.5*c*). The increase is proportional to the ratio between the net mobility of the leading and the terminating ion constituent (compare figure 1.5*a* and *c*).
(ii) In systems defined by weak electrolytes, the increase in field gradient at the electrolyte boundaries is accompanied by a change in the pH value (figure 1.5*b*). In systems defined by weak acids (anodically migrating systems) the pH-value increases, and in systems defined by weak bases (cathodically migrating systems) the pH-value decreases in the direction from leading to terminating electrolyte. The change in field gradient and pH at the boundaries is most pronounced at low relative concentration of counterion.

A number of quantitative aspects of these rules can be recognised from figure 1.5. However, the figure is a static representation in the sense that it gives little information about the development of the electrolyte phases in the time period from the onset of the experiment until the steady state is achieved. While calculations involving the time course of the development of non-steady-state electrophoresis systems are complex and outside the scope of the present outline, it may nevertheless be useful to study a graphical representation of such calculations performed on an isotachophoresis system (figure 1.6). This figure illustrates the time course of a five-component isotachophoresis system with the same constituents as in figure 1.5, from the onset of the electrophoresis until the steady state is established. The calculations were performed on a computer with simulation software designed by the author. The program calculates time dependent changes in the composition of freely defined particle gradients as a result of particle diffusion, solvent flow and electrophoresis and will be described elsewhere.

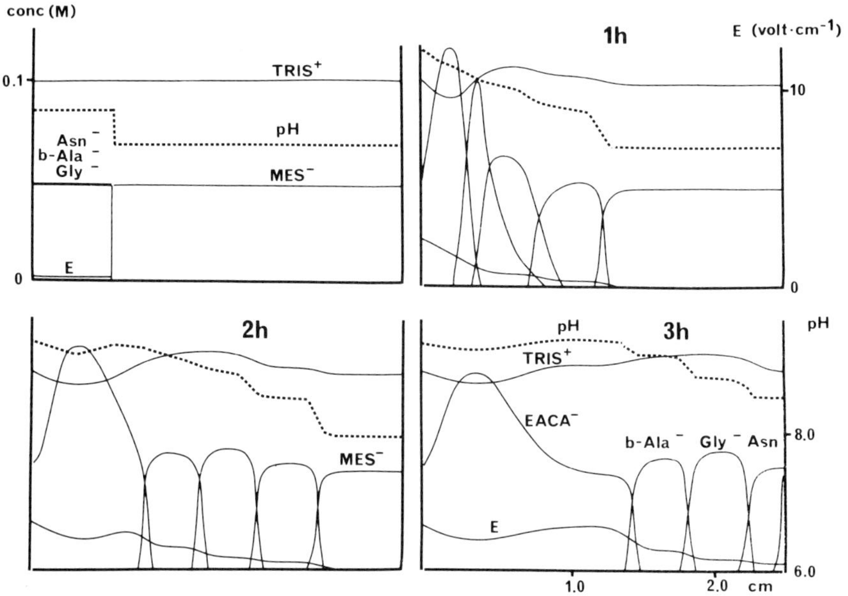

Figure 1.6 Calculated time development of an isotachophoresis system run at a current density of 2 mA cm^{-2} in an open vessel. Leading electrolyte: MES/Tris with initial concentrations as indicated (upper left diagram). Terminating constituents (b-Ala, Gly, Asn) initially placed in left-most 5 mm of separation vessel. Catholyte, EACA/Tris; anolyte, MES/Tris. Anode positioned to the right in the diagrams. Constituent data as in figure 1.5. Each x-axis represents the distance from the cathode on the same scale as the bottom right figure.

In figure 1.6 the distribution of the constituents at three different times after the onset of electrophoresis is shown with the length of the electrophoresis vessel depicted on the abscissae and the system parameters on the ordinates. The initial distribution of the constituents is shown in the upper left diagram. In this diagram, a mixture containing 0.05 M each of asparagine, beta-alanine and glycine with 0.10 M Tris as a counter ion is placed in the leftmost 5 mm of the electrophoresis vessel adjacent to a leading electrolyte containing 0.05 M MES, 0.10 M Tris in the remaining part of the vessel. The calculations were performed under the assumption that the vessel is connected to a catholyte reservoir with a constant concentration of 0.05 M EACA, 0.10 M Tris and an anolyte reservoir of constant composition equal to that of the leading electrolyte. This set-up closely resembles that of a real experiment, where the electrophoresis vessel is initially filled with the leading electrolyte embedded in the anticonvection medium on top of which the sample constituents are layered together with the spacer ions under the catholyte. The systems in figure 1.6 were 'run' with a constant current density of 2 $mA\,cm^{-2}$. After one hour the terminating ion constituents are clearly seen to separate into zones that are each dominated by one of the constituents. Comparison with the development after two and three hours shows that at one hour none of the constituents has achieved its steady-state concentration. With the exception of the terminating ion in the catholyte (EACA), all the terminating ion constituents migrate at their steady-state positions after two hours, and at three hours the leading electrolyte has migrated out of the vessel soon to be followed by the next constituent in the sequence (Asn).

The simulation reveals several general features of isotachophoresis systems. First, that part of the electrophoresis time (and of the separation vessel) is consumed by rearrangement of the sample constituents before the steady state is achieved. Second, that the sample constituents arrange themselves in an ordered sequence with equal migration velocity. Third, that the field gradient and (for weak electrolytes) the pH value change stepwise at the boundaries between the electrolyte phases. The slight overlap of the concentration profiles at the electrolyte boundaries is due to diffusion and will be less pronounced at higher field strengths. Note that the amounts of terminating ion constituents used in the simulation in figure 1.6 are just adequate for the terminating ions to reach the steady-state concentrations predicted by the equations in this section. Addition of larger amounts would have resulted in an increased length of the individual zones (and therefore in a longer 'equilibration' phase), and addition of significantly smaller amounts would have resulted in zones so narrow that the ascending and the descending limb at the electrolyte boundaries would have merged at a maximum below the steady-state concentration calculated by the equations. In actual runs this is often the case for sample components that are introduced in small amounts compared to those of the sample ions. This effect partly explains why steady-state electrophoresis of proteins in the analytical scale often gives much higher resolution than that obtainable in preparative runs, where the relatively large amount of protein results in higher steady-state concentrations of the proteins between the spacer ions, often to an extent where the limit of solubility is exceeded.

The practical use of isotachophoresis systems is to a large extent limited to analytical devices. Nevertheless, preparative equipment is commercially available that with proper selection of the materials fit for this kind of separation has a resolving power that compares well with that of the best chromatographic systems. However, the equipment is still rather expensive and less straightforward in its operation than the chromatographic systems. It therefore remains a subject for future research to develop devices that bring preparative isotachophoresis within reach of the non-expert.

1.9 Moving Boundary Electrophoresis

Moving boundary electrophoresis is strictly speaking a common term for steady-state electrophoresis in general. The isotachophoresis systems described in the previous section thus constitute a subset of the moving boundary systems. However, in the present context the term is used as a reference to steady-state electrophoresis in which the constituent defining an electrolyte phase is present on both sides of at least one of the boundaries that delimits the electrolyte phase.

A simple moving boundary system is shown in figure 1.7*a*. The only difference between this set-up and the isotachophoresis system in figure 1.3*a* is that the terminating ion constituent T is present in both the leading and the terminating electrolyte. During electrophoresis the faster L^- ions will leave some of the T^- ions behind the boundary. This boundary has a migration velocity which is defined by the migration velocity of the L^- ions. Since the terminating electrolyte is supplied by T^- ions from the leading electrolyte in this way, as well as by migration of T^- ions down through the terminating electrolyte, it follows that the terminating ion constituent in the terminating electrolyte must migrate with a velocity that is lower than

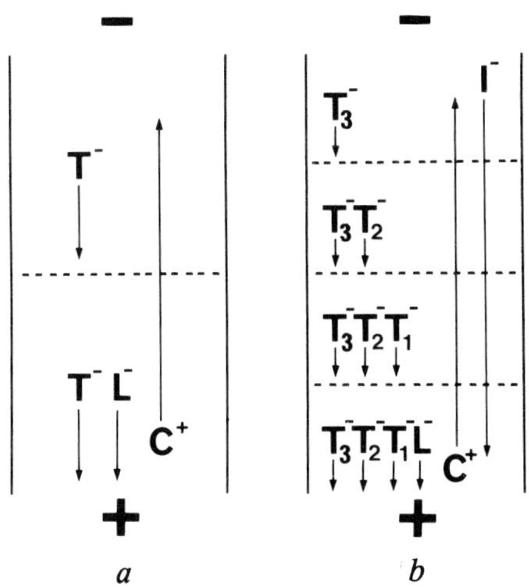

Figure 1.7 Schematic representation of moving boundary systems: (*a*) simplest possible moving boundary system (TEP type); (*b*) composite system with three terminating ions (T_1–T_3) defining electrolyte phases according to decreasing net mobility. I^- represents an interfering ion and C^+ represents the common counter ion(s).

that of the boundary. This situation contrasts with that found in isotachophoresis systems, where by definition all constituents migrate with the same velocity. As in the isotachophoresis systems, a moving boundary system may involve many electrolyte boundaries that are each characterised by a terminating ion constituent, but in some of the zones the picture is complicated by the simultaneous presence of ions that define boundaries of their own elsewhere in the system (figure 1.7*b*).

A closer analysis of the dynamics of the moving boundary systems reveals that the steady-state migration velocity of the terminating ion constituent relative to that of the boundary depends on the relative concentration and net mobility of terminating and leading ion constituents in the leading electrolyte. Thus, at one extreme the relative concentration is close to zero, and the system therefore resembles an isotachophoresis system. At the other extreme the leading electrolyte is dominated by the terminating ion constituent with only traces of leading ion constituent. In this case the relative migration velocity of the terminating and the leading ion constituent approaches the minimum defined by the ratio between their net mobilities at the given pH. Since the two electrolyte phases in the latter case are of nearly identical composition, it follows that the change in field gradient at the boundary is small compared to that obtained in the isotachophoresis set-up. The complete absence of changes in the field gradient is characteristic for zone electrophoresis systems, and in fact the moving boundary systems may be regarded as the general steady-state electrophoresis systems with isotachophoresis and zone electrophoresis as borderline cases (Schafer-Nielsen and Svendsen 1981).

Two extremes exist in the initial set-up of moving boundary systems. In one extreme (shown in figure 1.7*a* and *b*) the initial set-up consists of a leading electrolyte containing all the spacer ions and a terminating electrolyte containing only the slowest terminating ion. The other extreme is represented by an initial set-up in which the leading electrolyte only contains the fastest spacer ion while the remaining spacer ions are present in the terminating electrolyte. In the first case, the boundaries develop behind the initial mixture of spacer ions, whereas in the second case the boundaries are developed in front of the spacer train. These processes resemble the mechanism of boundary formation in the descending and ascending arm respectively of the Tiselius apparatus. As a means for quick reference to the first kind of systems, I have previously coined the term 'telescope electrophoresis' (TEP) systems, whereas the second kind of systems are referred to as 'snow electrophoresis' (SNEP) systems. However, the systems may be combined in a given set-up and both are adequately described by the general equations above. For reasons of simplicity, only the TEP systems will be dealt with.

The possibility for controlled regulation of the relative boundary velocity adds a degree of freedom to the definition of TEP systems as compared to isotachophoresis systems, but at the same time the calculation of system composition is made slightly more complicated by the introduction of velocity constants in the equations. A major advantage of the TEP systems is that the steady-state migration of the spacer ions is achieved almost immediately after the onset of electrophoresis. In the Tiselius apparatus this means that relatively short electrophoresis vessels can be used, the lower limit being defined by diffusion processes and by the resolving power of the detection system. In this apparatus it is customary to introduce the sample ions throughout the major portion of the separation vessel and measure the changes in light absorption or refractive index at the zone boundaries. However, none of the sample ions are completely separated from the rest, and accordingly the set-up has limited value, for example, in the separation of proteins in gels. For TEP systems in gels, the rational approach is therefore to use an initial set-up, where only the spacer ions are introduced throughout the separation gel and the sample molecules are placed in a discrete zone on top of the gel. In this way, the sample ions will be physically separated by the spacers much the way they do in isotachophoresis systems, while only the spacer ion constituents migrate in the fashion of a moving boundary system.

Figure 1.7*b* shows schematically a TEP system defined by four constituents (L and T_1–T_3). The initial set-up in this system is a vessel containing the four spacer ions plus the counter ion(s) and possible interfering ion(s) suspended between a catholyte containing T_3 as the only anion (in addition to possible interfering ions) and an anolyte of the same composition as the initial electrolyte in the vessel. As soon as the electric field is applied, the spacer ions start migrating towards the anode at the bottom of the vessel thereby forming a series of boundaries between electrolyte phases that are each characterised by the 'loss' of one spacer ion at each boundary. While it is obvious that the boundaries migrate with unequal migration velocity in contrast to the situation in isotachophoresis systems, it is perhaps less evident that their relative migration velocities can be controlled at will by proper choice of the relative initial concentration of constituents in the separation vessel.

Calculation of the initial concentrations necessary to obtain a series of boundaries with predetermined relative migration velocities requires definition of the numerical values of the velocity constants in equation (1.5). The general strategy for such calculations is to start by defining a sufficient number of variables describing the composition of the simplest electrolyte phase (i.e. the catholyte defined by T_3 in figure 1.7*b*) including the velocity constants k(T) and k(L) connected with the slowest boundary. From these data one can calculate the composition of the next electrolyte phase in the sequence. This process is repeated down through the sequence of electrolytes until the composition of the most complex electrolyte phase is known. The relative migration velocities of the boundaries are defined by the appropriate setting of velocity constants for each boundary.

Figure 1.8 shows the time development of a TEP system involving the

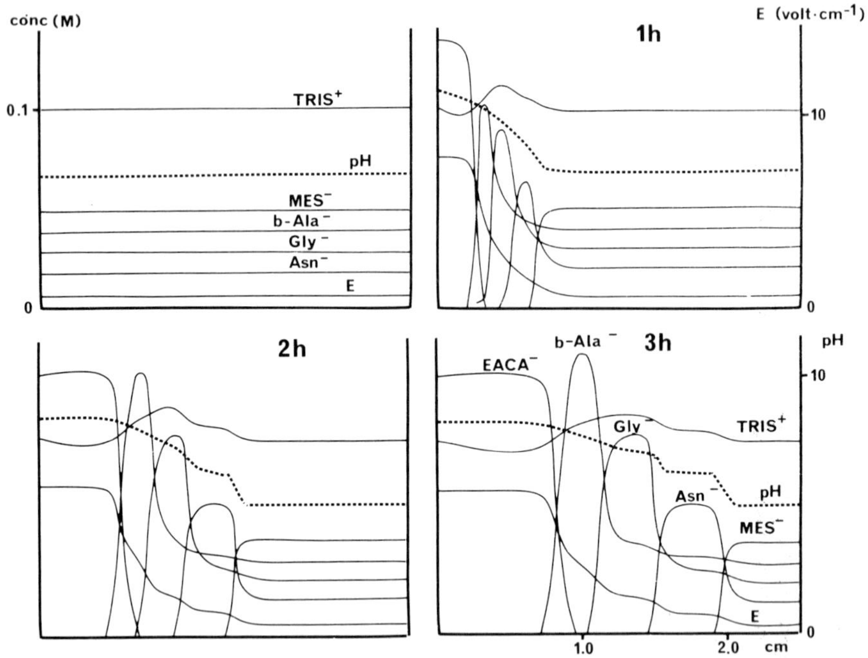

Figure 1.8 Calculated time development of a moving boundary system (TEP type) run at a current density of 2 mA cm^{-2} in an open vessel. Initial distribution of constituents is indicated in the upper left diagram. Catholyte, EACA/Tris; anolyte, MES/Tris. Anode positioned to the right in the diagrams. Constituent data as in figure 1.5. Each *x*-axis represents the distance from the cathode on the same scale as the bottom right figure.

same components as used in the simulation of the isotachophoresis system in figure 1.6. As in figure 1.7 the system in figure 1.8 is run at a current density of 2 mA cm^{-2} over a distance of 25 mm with the anode positioned to the far right in the diagrams. In the initial set-up the separation vessel is filled with a mixture of four spacer constituents (0.05 M MES, 0.02 M Asn, 0.03 M Gly and 0.04 M b-Ala) together with a common counter ion (0.10 M Tris). The anolyte is of the same composition as the spacer mixture and the catholyte contains the spacer constituent with the lowest net mobility (0.13 M EACA) together with the counter ion. As seen in the diagram representing a running time of one hour, a set of electrolyte zones have been induced that is each dominated by one spacer constituent. At this point none of the spacer constituents that were initially present in the separation vessel has reached its final steady-state concentration, but both the pH gradient and the field gradient across the spacer sequence have been established in spite of the fact that the sequence at this point occupies less than 10 mm of the separation vessel. The steady-state concentration of all the constituents has been reached at two hours, and from this point on the system does not undergo any quantitative changes except for the continuous increase in the length of the spacer zones. Note from figure 1.8 that EACA and beta-alanine migrate isotachophoretically since, apart from the overlap induced by diffusion processes, no EACA is present in the zone defined by beta-alanine.

The moving boundary systems conform entirely to the general rules outlined in connection with the isotachophoresis systems. Thus, both the field gradient and the pH gradient are seen to increase at the boundaries in the direction from the leading to the terminating electrolytes in the same way as in the isotachophoresis system. Apart from the difference in migration velocity of the boundaries, a notable difference between the isotachophoresis systems and the TEP systems is that the steady-state concentration of the spacers in the TEP system differs significantly in the zones they each define. Although the summed concentration of spacers in the electrolyte phases is nearly constant, the concentration of a given spacer in the zone as defined by it depends in a rather complex way on the total concentration and the net mobility distribution of the spacer constituents in the initial electrolyte. In analogy with the case for isotachophoresis systems, the presence of interfering ions in TEP systems reduces the field gaps between the electrolyte zones. A consequence of this fact is that in moving boundary systems, where the concentration of interfering ion(s) is high relative to the concentration of spacer ions, the change in concentration of the individual spacer ions at the zone boundaries is smaller than in systems where no interfering ions are present. A practical aspect of this is found in the Tiselius set-up, where a background electrolyte (containing the interfering ion(s)) is necessary if one wishes the relative position of the protein (spacer) boundaries to be reasonably independent of the relative concentration of proteins in the sample.

The calculation of electrolyte composition of moving boundary systems is more complicated than in the case of the simpler isotachophoresis systems. A set of empirical equations that require mobility—rather than net mobility—data has been developed for simple TEP systems with two spacer ions (Schafer-Nielsen and Svendsen 1981). However, for more complicated systems the only rational approach to system design involves the use of a computer program based on the general equations. Such a program has been designed by the author in collaboration with Dr P J Svendsen and is available on request.

1.10 Isoelectric Focusing Systems

Until 1982, the term isoelectric focusing referred solely to the separation of proteins in pH gradients generated during electrolysis of polyampholyte mixtures. Basically, only one rule concerning aqueous solutions of ampholytes is exploited during formation of these pH gradients. The rule is that when a single ampholyte species is dissolved in water the pH value is changed towards the isoionic point of the ampholyte. The isoionic point is defined as the pH value at which the net charge of the ampholyte is zero. For reasons to be discussed below, it should be kept in mind that this value does not necessarily coincide with the isoelectric point of the ampholyte, which is defined as the pH value at which the ampholyte does not migrate in an electric field.

A consequence of the basic rule is, that when two or more ampholytes with different isoionic points are dissolved in water, the pH value obtained will fall within the range defined by the isoionic points of the ampholytes. The actual pH value obtained in an aqueous solution containing a single ampholyte species depends both on the concentration of the ampholyte and on its isoionic point. Complete coincidence between the pH value and the isoionic point can only be obtained for ampholytes with an isoionic point of 7.0. The reason for this is easily visualised by an example in which we consider an aqueous solution of an ampholyte with an isoionic point of 12.0. At this pH value the concentration of hydroxyl ions is 10 mmoles per litre and in comparison the concentration of hydrogen ions is virtually zero. Electroneutrality of a solution of the ampholyte at its isoionic point would in this case require a net concentration of 10 mmoles per litre of positive charges supplied by the ampholyte, i.e. even at a concentration of 1.0 moles of ampholyte per litre approximately 1% of the ampholyte molecule will carry one positive net charge. The ampholyte would therefore not be isoionic, and the pH value of the solution would be slightly lower than the isoionic point of the ampholyte. However, for an ampholyte with an isoionic point of 10, the same percentage will be obtained at an ampholyte concentration of only 10 mmoles per litre. The general conclusion is that only ampholytes possessing isoionic points in the near neutral pH range can generate pH values close to their isoionic points at modest concentrations. In pure aqueous solution the ampholyte will carry a net positive charge if its isoionic point is above the neutral point of the solvent, and a net negative charge if its isoionic point is below this value.

A qualitative insight into the isoelectric focusing process can be gained by studying a graphical representation of the time course of an electrolysis performed on a mixture of ampholytes (figure 1.9). This figure represents data from a computer simulation of the isoelectric focusing of ten different ampholytes over a distance of 1 cm in a closed vessel. Each ampholyte was defined as having one acidic and one basic group with pK values 4 units apart (i.e. each group has a pK value that lies 2 units from the isoionic point of the ampholyte). The isoionic points of the ampholytes were defined to cover the range from 8.0 to 8.9 in steps of 0.1 pH units and the total voltage across the system was defined to be 100 volts. At time $t = 0$ the ampholytes were distributed in uniform concentration throughout the separation vessel (0.01 M for ampholyte 8.2–8.8 and 0.005 M for ampholyte 8.0 and 8.9). The ends of the separation vessel were defined to be reflective barriers (represented by the ordinate axes) with the anode positioned to the right.

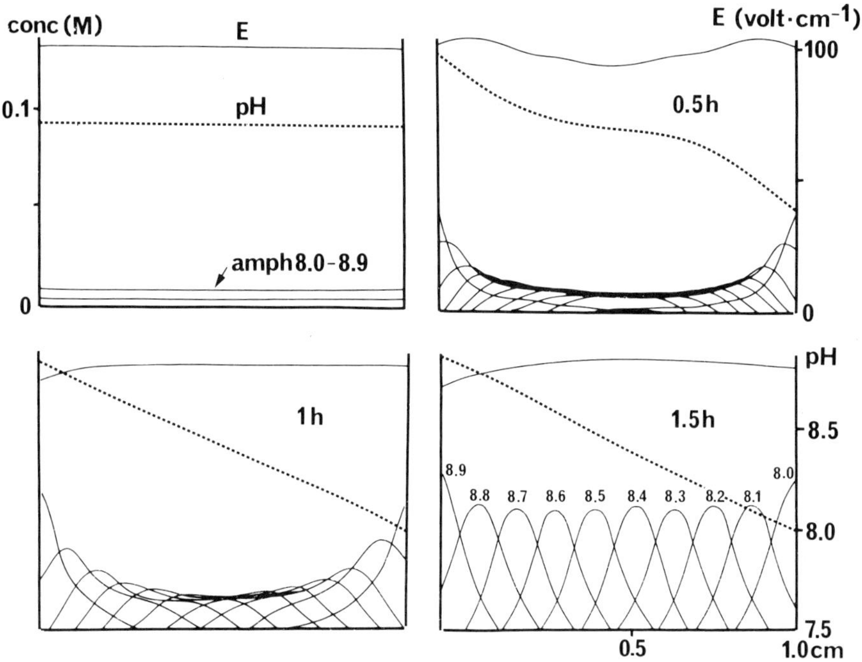

Figure 1.9 Calculated time development of an isoelectric focusing process involving 10 ampholytes in a closed vessel. The isoionic points of the ampholytes are evenly distributed in the pH range from 8.0–8.9. Initial distribution of the ampholytes is indicated in the upper left diagram. The calculation was performed assuming a constant voltage (100 V) across the system. The anode is positioned to the right in the diagrams. Each *x*-axis represents the distance from the cathode on the same scale as the bottom right figure.

In the initial set-up the pH value throughout the vessel is approximately 8.45, being equal to the average value of the isoionic points of the ampholytes. At this pH value the ampholytes with isoionic points from 8.5 to 8.9 carry a net positive charge, whereas the ampholytes with isoionic points from 8.0 to 8.4 carry a net negative charge. The two groups therefore start migrating towards the cathode and the anode respectively and, since the net charge of an ampholyte increases the further it is removed from its isoionic point, the highest migration velocity is attained by the ampholytes whose isoionic points are furthest removed from the initial pH value of the ampholyte mixture.

After half an hour the ampholyte with an isoionic point of 8.0 is present only in the anodic half of the separation vessel, whereas the ampholyte with an isoionic point of 8.9 is present only in the cathodic half (the first crossover point (Svendsen and Schafer-Nielsen 1980)). The remaining ampholytes still occupy larger parts of the separation vessel, but a partial separation of the ampholytes is evident at both ends. As a result of the selective accumulation of ampholytes with extreme isoionic points at the ends of the vessel, the pH value changes locally towards higher values at the cathode and lower values at the anode. In turn some of the ampholyte fronts slow down as they migrate into regions where the pH value approaches the isoelectric points of the respective ampholytes. At a later stage the ampholyte fronts reverse their direction of migration, so that all the molecules of a given ampholyte population migrate towards the point on the pH gradient that coincides with the isoelectric point of the ampholyte.

After one hour the ampholytes in the system of figure 1.9 have separated further, and at this point a linear pH gradient has been established that spans the pH range defined by the isoelectric points of the ampholytes.

After one and a half hours the ampholytes have separated into symmetrical zones. The overlap between the zones defined by single ampholytes results from the two directly opposed forms of transport, i.e. a compromise dictated by diffusion and electrophoresis. At this point the system has achieved a steady state maintained by the electric field, and no further changes in the distribution of ampholytes will be observed as long as the field gradient is maintained.

The computer simulation in figure 1.9 reveals the essential dynamics in the generation of a pH gradient during electrolysis of polyampholyte mixtures. The migration of ampholytes in the early part of the run is seen to be quite similar to that observed in moving boundary systems of the TEP type. In the isoelectric focusing system, however, two TEP systems are initiated on a collision course from each end of the vessel, and the steady state of both systems is broken when the first boundaries collide near the centre of the vessel. At this point the system enters a readjustment phase characterised by reversal of migration of some of the ampholyte boundaries and, although even this process can be quantitatively accounted for by the steady-state equations (Svendsen and Schafer-Nielsen 1980), the focusing process in the readjustment phase must be considered as somewhat chaotic. The final outcome of the process is entirely dependent on the physical set-up. Thus, if the system is confined to a closed vessel, the system will stabilise as shown in the computer simulation. In practical runs this closely resembles the situation found in horizontal slab gels, where the electrodes are placed more or less directly on the ends of the gels. However, in many practical set-ups the system is open, i.e. the separation vessel containing the ampholyte mixture is suspended between comparatively large reservoirs containing a free base in the catholyte and a free acid in the anolyte. In such systems the ampholytes may not obtain a steady state characterised by a complete standstill of the separated ampholytes. This is a simple consequence of the fact that only ampholytes possessing an isoionic point of about 7.0 can induce a pH value that equals their isoionic points in water.

For the closed system in figure 1.9 the immobility of the ampholytes in the steady state is conditioned by the presence of impermeable walls at the ends of the vessel. These walls act as reflective barriers that make it possible to obtain a balance between transport by electrophoresis and diffusion. In the open systems the constituents in the electrode reservoirs migrate electrophoretically towards the electrodes, i.e. away from the separation vessel, and a standstill of the ampholytes can only be obtained if the concentration of the constituents in the electrode reservoirs is so high that their electrophoretic migration out of the vessel is exactly balanced by their influx due to passive diffusion. The concentration necessary for this balance depends on the current density and the net mobility of the constituents in the reservoirs. In practical runs the balance is difficult to obtain, and one must therefore expect a certain drift of the pH gradient towards one or both of the electrode reservoirs during prolonged runs. The drift will be especially pronounced in systems involving ampholytes with extreme isoionic points, and in practice it is difficult to obtain useful gradients that extend beyond the pH range from 3 to 10. The drift may be further aggravated by migration of ionic contaminants from the electrode reservoirs into the separation vessel (e.g. bicarbonate ions introduced by the uptake of carbon dioxide from the air) or by the presence of fixed net charges in the gel matrix (e.g. from acrylic acid residues contaminating the acrylamide preparation).

The problem of gradient drift together with the ever present problem of making polyampholyte mixtures with an even and reproducible distribution of single ampholyte species was effectively eliminated in 1982, when LKB introduced the Immobilines (Righetti *et al* 1982). These reagents comprise a set of monovalent weak organic acids and bases containing a vinyl group that makes them able to copolymerize with acrylamide. By making a polyacrylamide gel in the presence of a pH gradient formed by an acid and a base from the Immobiline series, the gradient becomes immobilised through covalent coupling of the reagents into the polyacrylamide fibres. In the near neutral pH range the net charge of such gels will be low since the immobilised acids and bases outbalance each other, and therefore the usual problems associated with fixed charges, i.e. ion exchange effects and electroendosmosis, are to a large extent avoided. In fact, one may consider these gels as giant isoelectric ampholyte molecules which surround the solvent and whose isoionic points change along their length.

The Immobiline principle not only solves the problem of gradient drift and of discontinuities in the pH gradient often seen in connection with the traditional polyampholytes. It also makes it possible to create very narrow pH gradients in which an increased separation can be obtained of sample ampholytes with closely spaced isoelectric points.

The enormous resolving power of the technique has already been convincingly demonstrated (Bjellqvist *et al* 1982), and apart from certain problems associated with measurements of pH in the gels and in the generation of linear gradients covering more than a few pH units, the technique seems to hold every promise for the future. The general mechanisms involved in the focusing process in Immobiline gels are quite simple. Figure 1.10 shows a computer simulation in which a pH gradient has been defined over a distance of 25 mm by a constant concentration of an immobile acid with a pK value of 4.7 (A_{im}) and a linear gradient of an immobile weak base with a pK value of 8.0 (B_{im}). The pH gradient created by the two constituents spans the range from about 4.2 to 5.0. In the initial set-up a mixture containing 10 mmoles each of three 1,1-ampholytes (isoionic points 4.5, 4.7 and 4.9, all with pK values 1.5 units from the isoionic point) was placed in the cathodic (left) end of the separation vessel. The total voltage across the system was defined as 250 volts. Under these conditions the steady state of the system was achieved after about two hours running time. At this point the ampholytes are all focused at stable positions in the pH gradient defined by the immobile constituents. The simulation was performed under the

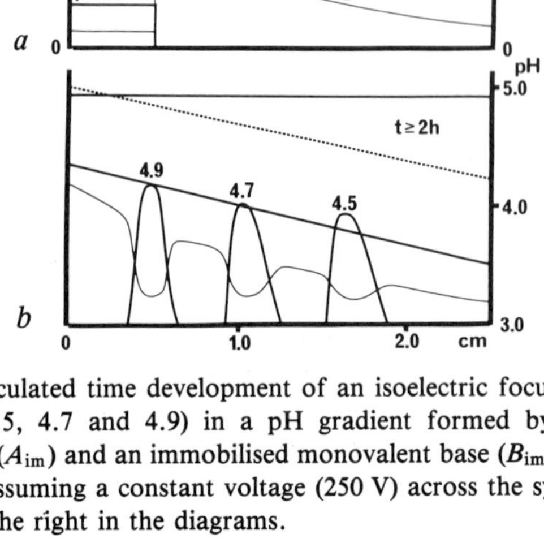

Figure 1.10 Calculated time development of an isoelectric focusing process (isoelectric points 4.5, 4.7 and 4.9) in a pH gradient formed by an immobilised monavalent acid (A_{im}) and an immobilised monovalent base (B_{im}). The calculation was performed assuming a constant voltage (250 V) across the system. The anode is positioned to the right in the diagrams.

assumption that the immobilised constituents did not contribute to the conductivity of the system. The shape of the field gradient in figure 1.10 is therefore determined solely by the conductivity conferred on the system by the ampholytes and the hydrogen ions generated by protolysis of water. Since the concentration of the latter increases about tenfold in direction from the anode to the cathode in this particular system, a corresponding decrease is seen in the local field gradient. The asymmetric shape of the concentration profiles of the focused ampholytes is a result of this decrease in the field gradient, and in systems running closer to the neutral point the concentration profiles will be more symmetric due to the less drastic change in concentration of hydrogen (and hydroxyl) ions.

From a theoretical point of view, the mechanism involved in the focusing process in immobilised pH gradients seems to be quite simple. While this impression may well be correct as far as the major mechanisms are concerned, it should not be forgotten that our experience with the special ionic conditions prevailing in such gels is limited. Does it for instance have a meaning to use concepts like ionic strength and, indeed, pH in a medium where the charges are restricted from free movement? Or is it possible to make improved steady-state electrophoresis systems with continuously migrating sample ions in gels, where the major part of the buffer capacity is supplied by immobile buffer constituents? The answers to these and probably more interesting questions as well are undoubtedly on their way as contributions to our understanding and eventual mastery of the electrophoresis principle.

Acknowledgements

This work was supported by grants from the Harboe Foundation and the Danish Medical Research Council.

References

Alberty R A 1950 *J. Am. Chem. Soc.* **72** 2361–7

Bjellqvist B, Ek K, Righetti P G, Gianazza E, Gorg A and Westermeier R 1982 *J. Biochem. Biophys. Meth.* **6** 317–39

Bode H J 1980 *Electrophoresis '79* ed. B J Radola (Berlin: de Gruyter) pp 39–52

Davis B J 1964 *Ann. NY Acad. Sci.* **121** 404–27

Dole V P 1945 *J. Am. Chem. Soc.* **67** 1119–26

Durrum E L 1950 *J. Am. Chem. Soc.* **72** 2943

Everaerts F M, Beckers J L and Verheggen T P 1976 *Journal of Chromatography Library* vol. 6 (Amsterdam: Elsevier)

Gasparic V, Bjellqvist B and Rosengren A 1975 *Swedish Patent Specification* 140 49–1.

Jovin T M 1973 *Ann. NY Acad. Sci.* **209** 477–96

Jovin T M, Dante M L and Chrambach A 1970. *Multiphasic Buffer Systems Catalogue* publ. no 196-090 (Springfield: NTIS)
Kohlrausch F 1897 *Ann. Phys., Lpz* **62** 209–16
Kolin A 1954 *J. Chem. Phys.* **22** 1628–9
—— 1955 *Proc. Natl. Acad. Sci. USA* **41** 101–10
Laemmli U K 1970 *Nature* **227** 680–5
Linderstrom-Lang K and Nielsen S O 1959 *Electrophoresis, Theory, Methods and Applications* ed. M Bier (New York: Academic)
Longsworth L G 1945 *J. Am. Chem. Soc.* **67** 1109–19
Maizel J V 1969 *Fundamental Techniques in Virology* (New York: Academic) p 334
O'Farrell P H 1975 *J. Biol. Chem.* **250** 4007–21
Ornstein L 1964 *Ann. NY Acad. Sci.* **121** 321–49
Picton H and Linder S E 1892 *J. Chem. Soc.* **61** 148
Raymond S and Weintraub L S 1959 *Science* **130** 711
Righetti P G and Gianazza E 1980 *Electrophoresis '79* ed. B J Radola (Berlin: de Gruyter) pp 23–8
Righetti P G, Gianazza E, Bjellqvist B, Ek K, Görg A and Westermeier R 1982 *Electrophoresis '82* ed. D Stathakos (Berlin: de Gruyter)
Rosengren A, Bjellqvist B and Gasparic V 1977 *Electrofocusing and Isotachophoresis* (Berlin: de Gruyter) pp 165–71
Routs R J 1971 *Electrolyte Systems in Isotachophoresis and their Application to some Protein Separations* Diss. Technische Hogeschool, Eindhoven
Schafer-Nielsen C and Svendsen P J 1981 *Anal. Biochem.* **114** 244–62
Schafer-Nielsen C, Svendsen P J and Rose C 1980 *J. Biochem. Biophys. Meth.* **3** 97–128
Smithies O 1955 *Biochem. J.* **61** 629
Svendsen P J and Schafer-Nielsen C 1980 *Electrophoresis '79* ed. B J Radola (Berlin: de Gruyter) pp 265–274
Svensson H 1948 *Acta Chem. Scand.* **2** 841–55
—— 1961 *Acta Chem. Scand.* **15** 325–41
—— 1962 *Acta Chem. Scand.* **15** 456–66
—— 1962 *Arch. Biochem. Biophys.* **suppl. 1**
Tiselius A 1930 *Nova Acta Reg. Soc. Sci. Upsal.* **IV7** 1–105
Vanderhoff J W and Micale F J 1979 *Electrokinetic Separation Methods* ed P G Righetti, C J Van Oss and J W Vanderhoff (Amsterdam: Elsevier) pp 81–95
Vesterberg O 1969 *Acta Chem. Scand.* **23** 2653–66
Vesterberg O and Svensson H 1966 *Acta Chem. Scand.* **20** 820–34
Williams R A and Waterman R E 1930 *Proc. Soc. Exp. Biol. Med.* **27** 56–9

Chapter 2

One-dimensional PAA-gel Electrophoretic Techniques to Separate Functional and Denatured Proteins

by
G M Rothe and W D Maurer

2.1 Introduction

The high resolution of biologically active or denatured proteins achievable with modern electrophoretic techniques is inseparably bound to the introduction of polyacrylamide (PAA) as a supporting medium (Raymond and Weintraub 1959).

PAA-gel electrophoresis first became a routine method in protein laboratories when Ornstein and Davis (1962) suggested separating human serum proteins by the use of three different buffer systems, two of which are present in two different concentration PAA gels. Their principle of concentrating proteins before separating them according to size and net charge is called 'disc-electrophoresis' and has been modified many times (e.g. Clarke 1964, Laemmli 1970, Panyim and Chalkley 1969, Reisfeld *et al* 1962, Rodbard and Chrambach 1971, Thomas and Hodes 1981).

However, PAA has also been used for many other electrophoretic techniques of which the most important one-dimensional ones are:

(i) the separation of catalytically active enzymes (Markert and Møller 1959);
(ii) the differentiation of size and charge isomeric proteins (Hedrick and Smith 1968);
(iii) the determination of the size of denatured proteins and sub-units, named SDS-electrophoresis (Shapiro *et al* 1967, Weber and Osborne 1969);

(iv) the separation and size estimation of non-denatured and denatured proteins by using PAA gels with increasing polymer concentrations from top to bottom—also called 'pore-gradient gel electrophoresis' (Lambin and Fine 1979, Lasky 1978, Margolis and Kenrick 1968);
(v) the determination of specific interactions between biologically active proteins and an immobilised affinity partner, known as 'affinity electrophoresis' (Bøg-Hansen 1973, Hořejší and Kocourek 1974, Nakamura *et al* 1960);
(vi) the study of membrane proteins (Hjelmeland *et al* 1983a b, King 1963, McCarthy and Racker 1966, Tzagaloff and Penefsky 1971).

PAA-gel electrophoresis is usually performed with gel columns approximately the size of a pencil (Davis 1962, 1964), but flat gels (Raymond and Weintraub 1959) are also used, as are 5 μl glass capillaries in specialised micro-methods (Bispink and Neuhoff 1977, Gainer 1973, Grossbach 1965, Hyden *et al* 1966, Neuhoff 1968) and 20 ml gels in sophisticated macro methods (Hrkal 1979).

The aim of this chapter is to survey the separation of biologically active and denatured proteins by use of one of the above named one-dimensional methods at an ordinary analytical scale.

2.2 Structure and Physico-chemical Properties of Polyacrylamide Gels

2.2.1 Concentrations of the Monomers

Polyacrylamide (PAA) is the polymerisation product of acrylamide and *N,N′*-methylene bisacrylamide (Bis)(cf figure 2.1) (Davis 1964, Raymond and Weintraub 1959). The function of Bis is to link the growing polymer chains at the sites where it (Bis) has been incorporated into the reactive sites of other polymer chains (Maurer 1968, Vollmert 1962). Consistency and transparency as well as the separation properties of polyacrylamide (PAA) gels depend:

(i) on the total polymer concentration T(T: g acrylamide + g Bis per 100 ml of solution);
(ii) the ratio of acrylamide to Bis;
(iii) the degree of polymerisation of both (Maurer 1968, Ornstein 1964).

Flexible and transparent gels are produced from acrylamide and Bis at a ratio of approximately 30 : 1 with a Bis concentration exceeding 3%.

A PAA gel of low concentration can probably be thought of as being an open-meshed three-dimensional lattice of long PAA chains tied at distant intervals by 'knots' of crosslinkers. When T is increased and a low concentration of Bis maintained, the frequency of chains increases and protein mobility is reduced due to frictional retardation. This has been shown for total polymer concentrations up to 40 and even 50%, at 5% Bis (Blattler *et al* 1972, Rodbard and Chrambach 1970).

A second mechanism for increasing mechanical obstruction in the gel is to increase the frequency of crosslinks between gel strands. Presumably this has the effect of making the chains less flexible and thus more difficult for the migrating molecules to force apart. If the degree of crosslinking is greatly increased without also increasing the number of chains to be crosslinked, however, there comes a point beyond which more links cannot be incorporated between chains and it also further reduces pore size. The extra crosslinkages in fact produce 'thick rigid fibres' (Fawcett and Morris 1966). The resulting opaque large pore gels have been studied and used for fractionating large macromolecules or for minimal sieving applications (Rodbard *et al* 1972). Many more crosslinks can be formed before producing opaque large pore gels provided the number of chains, i.e. the concentration of acrylamide, is raised concomittantly. Campbell *et al* (1983) have shown that gels of a total polymer concentration of 40% containing 12% Bis as crosslinker are as transparent as 3% gels that contain only 4% Bis.

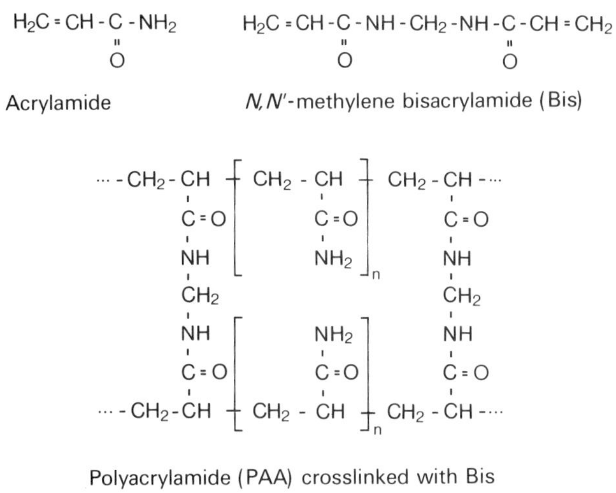

Figure 2.1

2.2.2 Catalyst Systems

The polymerisation of acrylamide is commonly initiated by one of the two following catalyst–redox systems (cf Maurer 1968):

(i) Ammonium persulphate (0.07%)–*N,N,N′,N′*-tetramethylethylenediamine (0.03%), or
(ii) Riboflavine (1 mg/100 ml)–*N,N,N′,N′*-tetramethylethylenediamine (0.03%) (Hansl, 1964, Ornstein 1964, Raymond and Weintraub 1959).

The indicated concentrations are used to polymerise PAA gels of a total polymer concentration of 7.5%.

System (i) is the most popular one and is used to polymerise gels in the concentration range of 3–40%. The polymerisation of low concentration gels in the range of 2.5–4% is often initiated by system (ii) where ultraviolet light is needed to initiate free radical formation at the riboflavine molecule. The catalyst systems (i) and (ii) also serve to polymerise the monomers when soluted in glycerol, ethyleneglycol, 8 M urea, 20% sucrose or 0.2% of a detergent such as sodium dodecyl sulfate or Triton X-100. An increase in viscosity accelerates polymerisation so that for gels of lower concentration the catalyst concentrations should be increased, while for highly concentrated gels they should be decreased. Ammonium persulphate and 3-dimethylaminopropionitrile have also been recommended for the polymerisation of PAA gels (cf Maurer 1968) but they are not as effective as the two other systems. Additionally, ammonium persulphate releases practically no molecular oxygen, which causes the polymerisation to be strongly inhibited (Maurer 1968).

2.2.3 Sieving Properties for Gels of Different Concentrations

Polyacrylamide was first employed as a support medium for the separation of serum proteins by Raymond and Weintraub (1959) and Ornstein and Davis (1962). They recognised that low concentration gels have practically no molecular sieving effect on proteins with molecular masses $< 2 \times 10^6$. Low concentration gels, however, can be used to separate very large particles such as viruses. For example tomato bushy stunt virus (a spherical particle with a molecular mass of 9.3×10^6 and a diameter of 30 nm, Martelli *et al* 1971) could be separated from southern bean mosaic virus (a spherical particle of 6.6×10^6 molecular mass and 25 nm diameter) using low concentration gels (Shepherd 1971). *Tipula* iridescent virus (a spherical particle of ~130 nm diameter and a molecular mass of $\sim 1000 \times 10^6$, Bahr *et al* 1976) and latex spheres up to 0.5 μm, however, are excluded from low concentration gels. Low concentration gels ($T = 3\%$; acrylamide : Bis = 4 : 1) are used in disc gel electrophoresis to concentrate diluted sample proteins into a narrow zone before entering the separation gel. Gels of 7.5% are recommended to separate human plasma proteins by disc electrophoresis (Ornstein and Davis 1962, Maurer 1968) while 5% gels have been used in the continuous method (Raymond and Weintraub 1959, Clarke 1964). PAA gels with concentrations continuously increasing from 3–30% separate proteins exclusively according to their molecular size (Margolis and Kenrick 1968) and surpass homogeneous gels severalfold in band sharpening and resolution. This type of gel, also called gradient gel, provides—together with isoelectric focusing—the optimum medium in protein separation presently available.

2.2.4 *Gel Structure*

Results on the ultrastructure of PAA gels are few and somewhat controversial (Clavert 1975). Rüchel *et al* (1978) examined the structures of PAA gels by transmission electron microscopy of freeze-etched specimens and found the structural details correlated with empirical findings concerning the molecular sieving effects of PAA gels. All PAA gels visualised so far by electron microscopy have been revealed as cellular structures in the submicron range. This is in contrast to the generally held assumption of a random meshwork of crosslinked individual PAA rods. The presence of comparatively large structures, especially in the low concentration gels as discovered by Rüchel *et al* (1978), does not seem to be an artefact since complementary PAA-gel surfaces have equivalent structures. The formation of relatively large caverns in low concentration gels can be explained by the observations of Flory (1953) and Rüchel *et al* (1978). According to them, polymerisation of synthetic chain molecules, including PAA, first produces high molecular aggregates that are in the sol state and therefore are not interconnected. The formation of very large aggregates before crosslinkage to a three-dimensional gel occurs is also indicated by an abrupt start of gelation. Rüchel *et al* (1978) assume that the formation of sponge-like structures is caused by the stepwise process of polymerisation. The rapid reduction in the concentration of sol particles means that the concentration of monomers increases rapidly in some areas while it becomes necessarily depleted in others.

It is difficult to extract separated proteins quantitatively out of PAA gels because the proteins are partially attached to the gel matrix. Rüchel *et al* (1978) think that the migration of protein molecules through the sponge-like structure of a PAA gel consists of an unhindered passage through the large gel caverns and a restricted passage through membranous areas. They believe that the latter is responsible for the sieving properties of the matrix.

2.2.5 *Other Crosslinkers*

Two principally different methods have been used to extract proteins from PAA gels:

(i) physical methods
(ii) chemical methods.

Physical methods use thorough mincing or homogenisation of the gels following electrophoresis or extended soaking to make proteins migrate out of the matrix (Bishop *et al* 1967, Content and Duesberg 1971, Cottolico and Jones 1975, Öfverstedt *et al* 1983, Otto and Šnejdárková 1981, Schuerch *et al* 1975, Shuster 1971, Strålfords and Belfrage 1983, Young and Young 1974). Although recoveries as high as 60–80% have been reported (Young and Young 1974), these techniques are not recommended for routine work because the recoveries vary considerably depending on the nature and size of the protein. Solubilisation of the matrix has been suggested as an alternative, but the proteins from solubilised gels can only be used to estimate their ^{14}C and ^{3}H contents. On the contrary, ribonucleic acids can be isolated in a biologically active state (Hansen 1976).

2.2.5.1 *Ethylene diacrylate and N,N′-diallyltartardiamide*

Choules and Zimm (1965) used ethylene diacrylate (EDA) (cf figure 2.2) in place of Bis as a crosslinking agent for PAA gels. The total polymer concentration was 7.5% and the ratio of acrylamide to EDA was 15:1. Using a modified buffer system of Davis (1964) they separated rabbit anti-m-dinitrophenol light chains at pH 6.5 and trypsin and chymotrypsin at pH 2.5 with a buffer system containing 1 M propionic acid.

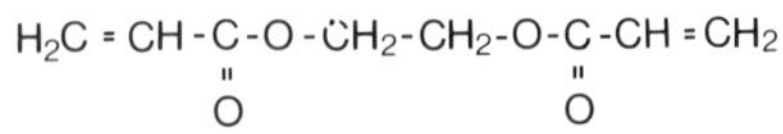

Ethylene diacrylate (EDA)

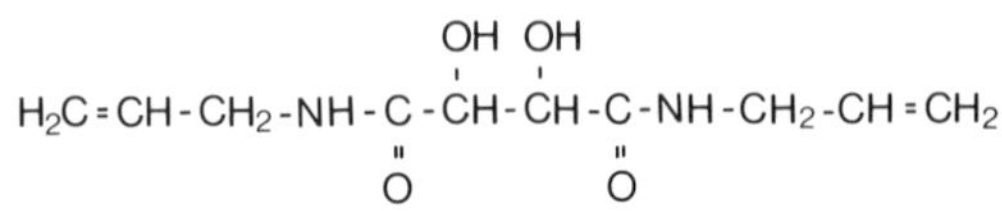

N,N′-diallytartardiamide (DATD)

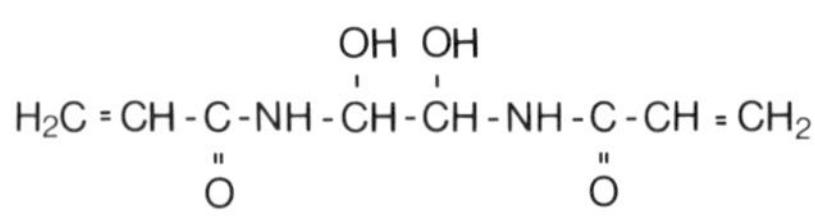

N,N′-(1,2-dihydroxyethylene) bisacrylamide (DHEBA)

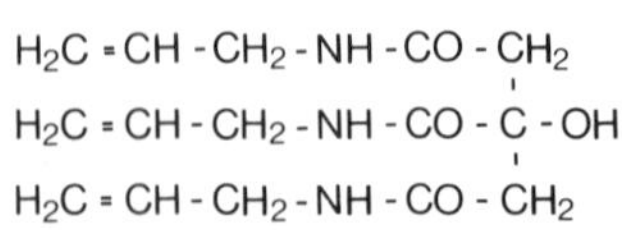

N,N′,N′-triallyl citric triamide (TACT)

Figure 2.2

Following electrophoresis, gels were stained with the dye aniline blue-black and duplicate columns were frozen for 5 min in dry ice and cut into 1 mm slices. Each slice was placed in a 20 ml counting vial, and the digestion was started by adding 0.5 ml of a solution that contained 9 vol 1 M piperidine and 1 vol alcoholic thiamine solution. Depending on the extent of crosslinking the gels dissolved within 1 to 4 h. After the digestion,

10 ml of Kinard's scintillation fluid (Kinard 1957) was added to each vial and mixed briefly. The clear homogeneous solution was used for scintillation counting, and reasonably efficient and accurate tritium counting was obtained.

The crosslinker EDA cannot be used in alkaline buffer systems because of its ester bonds which undergo a base-catalysed hydrolytic cleavage (Alpers and Glickman 1970, Paus 1971). Several different authors used *N,N'*-diallyltartardiamide (DATD) as a crosslinker (cf figure 2.2)(Anker 1970, Baumann and Chrambach 1976, Späth and Koblet 1979, Tas *et al* 1979, Young *et al* 1980, Ziegler and Hengartner 1977). Gels co-polymerised with this substance can be solubilised with periodic acid which selectively oxidises the 1,2-diol structure of the DATD molecule.

2.2.5.2 N,N'-(1,2-dihydroxyethylene)bisacrylamide

O'Connell and Brady (1976) found that both of the aforementioned crosslinkers have poor physical and molecular sieving qualities. They recommended using *N,N'*-(1,2-dihydroxyethylene)bisacrylamide (DHEBA) instead. Gels polymerised with this crosslinker are soluble in dilute periodic acid or dilute aqueous solutions of bases. DHEBA together with acrylamide can be used to generate PAA gradient gels. Flat gels ($140 \times 120 \times 3$ mm^3) of that type were cast in glass cassettes and generated from both light and dense solutions, which were used to produce the linear gel gradient with the aid of a two-chamber gradient mixer and the gel casting apparatus of Margolis and Kenrick (1968). Catalyst concentration was adjusted so that setting began at the top of the gradient, and the reaction was allowed to proceed for 2 h at 22°C (for details see table 2.1).

The gradient gels were used to separate protein mixtures from banana fruit tissues labelled with carbon-14 or tritiated amino acids. After electrophoresis gels were sectioned according to the pattern of stained zones. Each slice was placed in a glass scintillation counting vial fitted with a 0.005-inch teflon-lined plastic cap, and 1.0 ml of 0.025 M periodic acid was added. The vials were sealed and incubated for 48 h at 50°C. 10 ml of Mix I scintillation fluid (Patterson and Greene 1965) was added, and the vials were cooled overnight before counting in a liquid scintillation counter (O'Connell and Brady 1976).

PAA gels produced with the crosslinker *N,N'*-(1,2-dihydroxyethylene)-bisacrylamide are more stable to alkaline pH than gels co-polymerised with EDA (Cain and Pitney 1968) and may be used with the common alkaline buffer systems (see table 2.1). Borate buffers should not be used, however, because of their ability to form negatively charged complexes with the *cis*-1,2-diol structure of the crosslinker DHEBA. During periodate oxidation the 1,2-diol structure is cleaved while the amidomethylol bonds are labile to strong bases (O'Connell and Brady 1976 see figure 2.3).

Table 2.1 Light and dense solutions to create *N,N'*-(1,2-dihydroxyethylene) bisacrylamide acrylamide gels (O'Connell and Brady 1976)

Reagent	*Light solution*	*Dense solution*
1 Acrylamide	40.00 g	105.00 g
2 *N,N'*-(1,2-dihydroxyethylene) bisacrylamide	1.94 g	6.16 g
3 Sodium azide	0.026 g	0.020 g
4 Buffer	40 ml	30 ml
5 Distilled water to	398 ml	298.5 ml
6 TEMED	0.010 ml	0.0025 ml
7 Ammonium persulphate (50% w/v)	2 ml	1.5 ml

Distilled water may be used to replace buffer in the above formulation. Gelatination was initiated 26 min after addition of the persulphate to the light solution. Components 1 to 5 were dissolved in distilled water, the solutions filtered, degassed under vacuum, and the TEMED and ammonium persulphate solution added immediately before filling the gradient mixer. The overlaying solution was 5% (v/v) in aqueous ethanol and the underlaying solution was 25% (w/v) in sucrose. All solutions were kept at 22°C. Buffer: 18.75 g glycine, 7.5 g Tris, 0.79 g calcium acetate, 0.163 g sodium azide and distilled water to 250 ml. Before sample loading, gels were prerun for 3 h at 300 V (30 mA/gel) at 4°C. The electrophoresis buffer was composed of: 25 mmol Tris, 100 mmol glycine and 2 mmol calcium acetate; pH 8.5. Electrophoresis was performed at 300 V, and 4°C for 18–22 h.

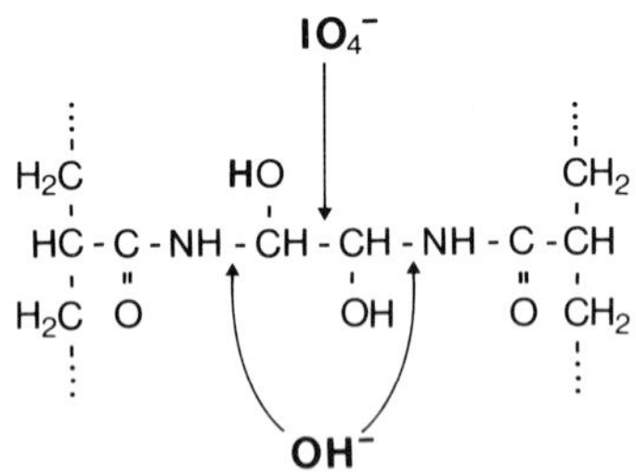

Figure 2.3 Structures of PAA gels crosslinked with *N,N'*-(1,2-dihydroxyethylene) bisacrylamide as proposed by O'Connell and Brady (1976). Gels of this type have regions with a 1,2-diol structure which is cleavable by periodate (IO_4^-) as well as amidomethyl bonds which are labile to bases (OH^-)(cf O'Connell and Brady 1976).

2.2.5.3 *N,N',N'-triallylcitric triamide*

Crude plant and animal extracts contain many small proteins which migrate out of the dense end of 3–30% PAA gradient gels prepared from acrylamide and Bis, when electrophoresis at high voltage gradients is conducted for extended periods (Margolis and Wrigley 1975). To improve the retention of PAA gradient gels for low molecular mass proteins O'Connell and Brady (1976) suggested preparing the gels from acrylamide, Bis and *N,N′,N′*-triallylcitric triamide (TACT). Gels were produced essentially as described for PAA gels crosslinked with DHEBA but using the light and dense solutions indicated in table 2.2.

The small pore structure in the dense part of such gels eliminated the possibility of the loss of low molecular mass proteins and allowed the use of high voltage gradients, which in turn improved the resolution of protein zones.

Table 2.2 Light and dense solutions to create acrylamide, Bis, triallylcitric triamide gels (O'Connell and Brady 1976)

Reagent	*Light solution*	*Dense solution*
1 Acrylamide	20 g	120 g
2 *N,N′*-methylenebisacrylamide	1 g	6 g
3 *N,N′,N′*-triallylcitric triamide	—	21 g
4 Sodium azide	0.025 g	0.020 g
5 Buffer	40 ml	30 ml
6 Distilled water to	398 ml	298.5 ml
7 TEMED	0.025 ml	0.0125 ml
8 Ammonium persulphate (50% w/v)	2 ml	1.5 ml

2.2.6 *The Pore Radius of PAA Gels, its Calculation and its Influence on the Separation of Proteins*

White (1960) first attempted to relate the average pore size of homogeneous PAA gels to the total polymer concentration. The results he obtained are fundamentally in accordance with theoretical models proposed by Ornstein and Davis (1962), Raymond and Nakamichi (1962) and Tombs (1965), although White himself stated that the pore diameters he calculated by diffusion measurements might be too small. Based on the so-called Ogston model (Ogston 1958) which assumes dextran gels to be assembled by arbitrarily arranged gel rods, Raymond and Nakamichi (1962) suggested that the average pore radius (r_{av}) of PAA gels is inversely related to the total PAA concentration (T)

$$r_{av} = \frac{1}{2} \times \frac{Kd}{(T/100\,\rho)} = 4.108\,T^{-1/2} \qquad (2.1)$$

with r_{av}, average pore radius (nm); d, diameter of a PAA gel rod: 0.5 nm; ρ, density of a gel rod: 1.2 g cm^{-3}; K, factor resulting from the angle in which the gel rods are linked together: 1.5.

Ornstein and Davis (1962) used an even simpler model to roughly estimate the average pore size of PAA gels (cf figure 2.4). If we assume—in accordance with them—that the PAA matrix is to be built by lattices forming cubes of equal size and the monomers polymerising quantitatively, then we can compute:

(i) the total length per unit volume (L in $\mathrm{m\,l^{-1}}$) of a fibre that could be built by the linkage of all monomers and
(ii) the average pore diameter ($\bar{p}$) if this fibre is used to build lattices forming cubes of equal size (cf figure 2.4).

The theoretically possible length (L) of the fibre contained in the unit volume of one litre is given by

$$L = \frac{10\,T}{\mathrm{mol}\,M_{\mathrm{ac}}} N_{\mathrm{A}} l' = 18.63 \times 10^{12}\,T \tag{2.2}$$

with T, the total polymer concentration: acrylamide plus Bis per 100 ml (g); mol M_{ac}, molecular mass of acrylamide: 71.1 ($\mathrm{g\,mol^{-1}}$); the thickness of the fibres is not taken into account; N_{A}, Avogadro's number: 6.022×10^{23} ($\mathrm{mol^{-1}}$); l', length of an acrylamide molecule: 0.22×10^{-9} (m).

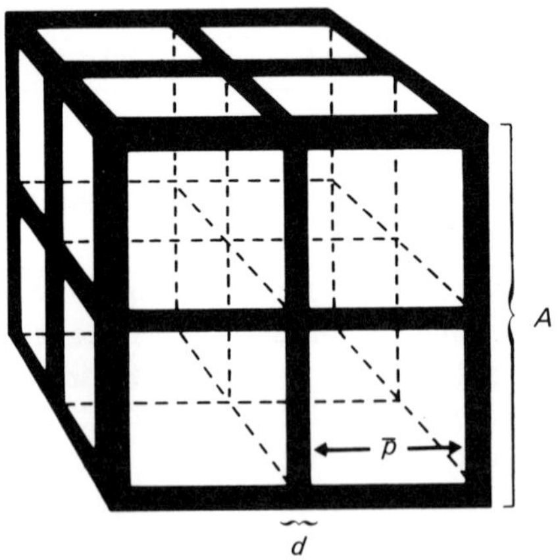

Figure 2.4 Model of the PAA matrix as used by Ornstein and Davis (1962) to calculate average pore diameters ($\bar{p}$) when a total polymer concentration of T is given. A, edge length of a cube having the unit volume of 1 litre; d, diameter of a PAA-gel fibre.

It is evident from figure 2.4 that for the construction of the edges of an internal cube, only three lengths of an edge are needed because it shares the rest with the adjoining cubes. Therefore the total chain length required for the construction of cubes with a certain edge length is the product resulting from the number of cubes contained in the unit volume and three times the length of the edge

$$L = (A/\bar{p})^3 \times 3\,\bar{p} \qquad (2.3)$$

with A, length of one edge of the unit volume: 0.1 m; $\bar{p}$, average pore diameter in nm.

After rearranging the terms the following equation is obtained

$$\bar{p} = (3\,A^3/L)^{-1/2}. \qquad (2.4)$$

Inserting $18.63 \times 10^{12} T$ for L and $A = 0.1$ m we obtain $\bar{p}$ in nm

$$\bar{p} = 12.67\sqrt{\mathrm{T}}. \qquad (2.5)$$

In table 2.3. the values calculated by this model are compared with those estimated by Raymond and Nakamichi (1962). If it is true that PAA matrices have a sponge-like structure (Rüchel *et al* 1978) then the pore diameters calculated by assuming a random meshwork of gel fibres are too small. When performing PAGGE (polyacrylamide gradient gel electrophoresis) one can use the Stokes' radius of marker proteins to mark the largest pores available in a PAA gel of a known concentration of T. Figure 2.5 demonstrates that the largest gel pores of a certain PAA concentration are:

Table 2.3 Estimation of the average pore diameter of PAA gels using a simple lattice model

%T	$L(10^{12}$ m)†	$\bar{p}$(nm)‡	$\bar{p}$(nm)§
5	93.2	5.68	3.67
10	186.3	4.01	2.60
15	279.5	3.28	2.12
20	372.7	2.84	1.84
25	465.8	2.54	1.64
30	559.0	2.32	1.50
35	652.2	2.15	1.39
40	745.3	2.01	1.30

† Quantitative polymerisation to gel rods only is assumed; L, total length of all gel rods: 18.63×10^{12} T (m); T (total % polymer concentration).

‡ Same assumption as in †. The average pore diameter $\bar{p}$ is calculated by $\bar{p}$ (nm) $= 12.69/\sqrt{t}$, with T being the total polymer concentration.

§ Same assumption as in †. The average pore diameter $\bar{p}$ is calculated according to Raymond and Nakamichi (1962): $\bar{p}$ (nm) $= 2 \times 4.108/\sqrt{t}$, with T being the total polymer concentration.

(i) much larger than the average pores calculated and
(ii) the largest pore radii deviate increasingly from the average pore radii with decreasing gel concentration.

This confirms that the pore radii of PAA gels are to be treated as a statistical quantity (Tombs 1965). The standard deviations, however, have yet to be found out. It is also unclear whether the pore radii are distributed according to a Gaussian function, or whether they show a logarithmic distribution. Future gel models must consider these points as well as the probability that the uniformity of the pore radii increases with increasing PAA concentrations.

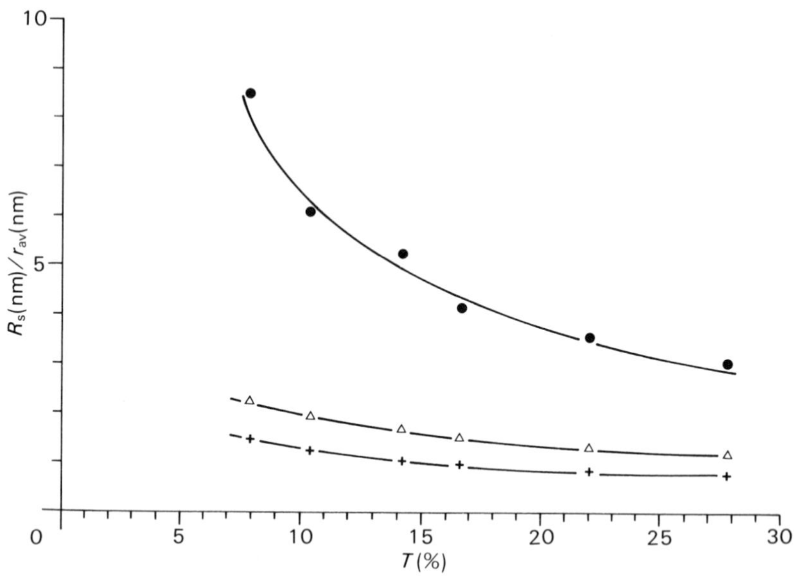

Figure 2.5 Plot of average pore radius (r_{av}(nm)) against PAA-gel concentration (T(%)). Triangles, pore radii calculated using the formula $r_{av} = \frac{1}{2}\, 12.67\ (T^{-1/2}(\%))$; crosses: pore radii calculated using Raymond and Nakamichi's formula: $r_{av} = 4.108\ (T^{-1/2}(\%))$; full circles = Stacking gel concentration of marker proteins: thyroglobulin, ferritin, catalase, lactate dehydrogenase, bovine serum albumin, and ovalbumin.

2.3 Analytical Polyacrylamide Gel Electrophoresis

2.3.1 The Separation of Biologically Active Proteins

Non-denatured proteins may separated by homogeneous and heterogeneous PAA gel and buffer systems. Homogeneous buffer systems have been used for example by Raymond (1962) who separated serum proteins on 5% homogeneous PAA gels (acrylamide : Bis = 19 : 1) using a 0.1 M Tris-HCl buffer of pH 9.0 as gel and electrode buffer. It must be pointed out, however, that almost all buffer systems suggested to separate catalytically active enzymes on a PAA matrix are discontinuous with respect to gel and electrode buffer. Numerous discontinuous buffer systems were suggested (cf e.g. Maurer 1968, McLellan, 1982, Rodbard and Chrambach 1971a, Thomas and Hodes 1981), but in practice only a few are used now to separate biologically active proteins. These are:

(i) the original system of Davis (1964),
(ii) the Davis system without sample gel (Clarke 1964),
(iii) the buffer system of Laemmli without SDS (King and Laemmli 1971, Laemmli 1970),
(iv) the systems of Reisfeld (Reisfeld *et al* 1962; Williams and Reisfeld 1964) and
(v) the system of Neuhoff (1968).

Systems (i)–(iii) and (v) separate proteins at an alkaline pH while system (iv) is used to separate proteins under acidic conditions.

As a rule, protein samples are loaded on top of PAA gels no matter whether continuous or discontinuous buffer systems are used. Therefore buffer conditions must be chosen which force all proteins to migrate in one direction (mostly to the anode). Under these conditions it is difficult to adapt the buffer systems to the requirements of biologically active proteins, especially enzymes. A possible solution would be to separate proteins horizontally by using PAA flat gels inserted in an apparatus similar to the one suggested by Brewer and Sing (1970). This device would permit the combination of the better separation properties of PAA gels with the carefully adapted continuous buffer systems developed in starch gel electrophoresis (e.g. Harris and Hopkinson 1978).

2.3.1.1 *Disc Electrophoresis*

2.3.1.1.1 GENERAL ASPECTS AND BUFFER SYSTEMS

Usually disc electrophoresis is performed in round gel columns, but flat gels have the advantage that several protein samples can be run on the same gel and then separated under more similar conditions. Round gels prepared from the same solutions have slight differences in gel structure leading to slightly different protein mobilities. Disc electrophoresis according to Ornstein and Davis (1962) is the most popular method—besides starch gel electrophoresis—to separate biologically active proteins. It uses 2–3 different gels and 4 different buffer systems:

(i) a sample gel and buffer system containing the diluted protein mixture,

(ii) a concentration gel and buffer system (also called 'large pore gel') in which all proteins are concentrated in one single narrow zone,
(iii) a separation gel and buffer system in which the proteins are separated according to net negative charge and size and
(iv) an electrode buffer system which is different from the three aforementioned systems.

The three gel systems are polymerised in glass tubes of a length of 100 mm and an internal diameter of 6 mm. Nowadays the sample gel is omitted and protein samples containing ~20% sucrose or 10 to 20% glycerol together with 0.003% of a tracking dye (Bromophenol blue or Bromophenol red) are applied directly on top of the concentration gel. Usually the total polymer concentration of the concentration gel is 3.125 (g (100 ml)$^{-1}$; acrylamide: Bis = 4:1), while the total polymer concentration of the separation gel is 7.5% (w/v) and the ratio of acrylamide to Bis is ~30:1. The length of the concentration gel is approximately 5–10% of the length of the separation gel.

In the large pore gel concentration of proteins into a narrow zone is achieved by a different distribution of fast migrating buffer ions (also called 'leading ions') and slowly migrating buffer ions (also called 'trailing ions'). At the beginning of electrophoresis the leading ions (Cl^-) are in the total gel column while the trailing ions (glycine) are exclusively in the electrode buffer. Under the influence of the electric field both types of ions migrate to the anode. However, the leading ions migrate faster than the proteins and the trailing ions, leaving a zone of reduced conductivity behind. Since the specific conductivity is inversely proportional to the electric field strength, the migration of the negatively charged proteins and the trailing ions is accelerated until all the ions 'migrate' at the same velocity thereby concentrating the proteins into a very thin zone (cf figure 2.6). Consequently a moving boundary is formed between leading and trailing ions which traverses the concentration gel in the direction of the spacer gel.

When the migrating protein zone enters the separating gel it strikes discontinuities in gel concentration, buffer system and pH value. The pH value of the separating gel is chosen in a way that the dissociation of the trailing ions increases manyfold. An increase in dissociation means an increase in electrophoretic mobility, and as a consequence the trailing ions overtake the proteins and start to migrate closely behind the leading ions (cf figure 2.6). The proteins now migrate in a homogeneous electric field and at a constant pH value. In 7.5% PAA gels they are separated according to their net negative charge and their Stokes' radius. One must be aware, however, that if the mobility achieved by the trailing ion in the unstacking step remains lower than that of some sample constituents, those constituents will remain stacked and migrate with the (new) ionic boundary, and thus these will not be subjected to separation (Anderson *et al* 1983).

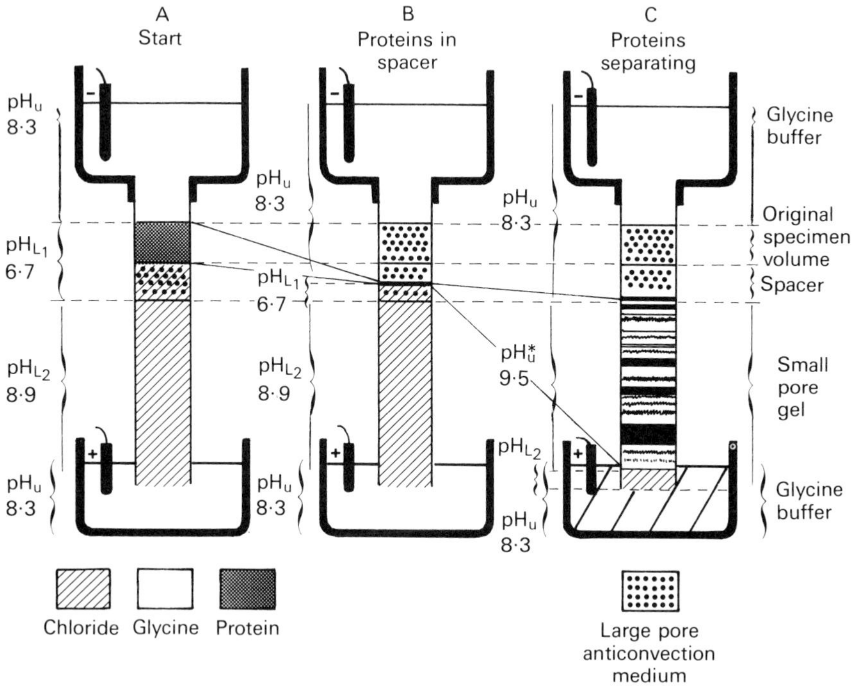

Figure 2.6 Separation of proteins in disc-electrophoresis when using a neutral stacking gel buffer (pH_{L_1} 6.7), an alkaline separation buffer (pH_{L_2} 8.9) and an alkaline electrode buffer (pH_U 8.3)(Ornstein 1964).

To sum up, both concentration and molecular sieving cause the high resolution of disc electrophoresis. The most popular disc electrophoretic systems to separate non-denatured proteins are compiled in table 2.4. A number of catalytically active enzymes which have been separated by disc electrophoresis are summarised in table 2.5.

2.3.1.1.2 PREPARATION OF DISC GELS

Rod gels are mainly used in disc electrophoresis while flat gels are preferred for SDS and pore-gradient gel electrophoresis. Methods to prepare a number of flat gels simultaneously are described in §2.2.1.4.1. Gel columns are preferably prepared in glass tubes 70 to 150 mm long with an inner diameter of 5 to 6 mm. All tubes should be made from the same length of glass tubing, they should have plain and smooth ends and their outer diameter must be of a size that will fit tightly into the rubber gaskets of the upper electrode vessel. They must all be of the same length and inner diameter. According to Davis (1964) the length of the separating gel ($T = 7.5\%$) should be 75 mm when separating serum proteins. Before use

Table 2.4 The most popular buffer systems used to separate biologically active proteins in disc electrophoresis

Sample gel	*Stacking gel (acrylamide: bis)*	*Separating gel (acrylamide: bis)*	*Electrode buffer*	*Sample buffer*	*Electrophoretic conditions* (V)	(A/gel)	(h)	*Remarks*	*Reference*
Same as stacking gel	0.06 N HCl 0.0062 M Tris pH 6.7 *T*: 3.125% (10 : 2.5)	0.06 N HCl 0.038 M Tris pH 8.9 *T*: 7.7% (30 : 0.8)	0.005 M Tris 0.0384 M glycine pH 8.3	Not specified. 0.003 ml of serum are mixed with 0.15 ml stacking gel solution	300	0.001	2	(i)	Ornstein and Davis 1962
None used	Replaced by 2 to 5% urea or sucrose in water	0.0062 M Tris 0.0483 M glycine pH 8.15 (7.5% acrylamide 0.125% Bis)	0.387 M glycine 0.05 M Tris 0.005 M HCl pH 8.1	Not specified. Foreign ions must be of low concentration	400	0.004	0.5	(ii)	Clarke 1964
Same as stacking gel	0.06 N KOH 0.063 M acetic acid, pH 6.8 *T*: 3.125% (10 : 2.5)	0.06 N KOH 0.376 M acetic acid, pH 4.3 *T*: 7.7% (30 : 0.8)	0.14 M β-alanine 0.35 M acetic acid, pH 4.5	Not specified 5–10 ml sample are mixed with 0.15 ml stacking gel solution	300	0.006	0.33	(iii)	Reisfeld *et al* 1962

0.059 M Tris, 0.032 M H_3PO_4, pH 6.7 without gel	0.031 M Tris, 0.0164 M H_3PO_4, 0.2% Triton X-100, pH 6.7 *T*: 5% (20 : 0.2)	0.089 M Tris, 0.02 N H_2SO_4 pH 8.8 0.05 M hydantoin, 1% Triton X-100 *T*: 20.2% (20 : 0.2)	0.05 M Tris, 0.384 M glycine, pH 8.4	0.059 M Tris 0.032 M H_3PO_4 pH 6.7	60	0.07	0.3 $\times 10^{-3}$	(iv)	Neuhoff 1973
None used	0.125 M Tris–HCl, pH 6.8 *T*: 3% (30 : 0.8)	0.375 M Tris–HCl, pH 8.8 *T*: 8 or 10% (30 : 0.8)	0.025 M Tris 0.192 M glycine pH 8.3	0.0625 M Tris–HCl, pH 6.8, 10% glycerol, 0.001% bromphenol blue		0.03	7	(v)	Laemmli 1970

(i) Photopolymerisation may be replaced by chemical polymerisation using TEMED and ammonium persulphate. A sample gel is not necessary. The albumin front is marked by the addition of 1 ml of 0.001% bromophenol blue to the upper buffer reservoir (~1 litre). Electrophoresis should be started within 1 h after the spacer gel has been set. Gels are prepared in 9 cm long glass tubes with an internal diameter of 6 mm. The separating gel is prepared from 1 vol *A*, 2 vol *C*, 1 vol H_2O, and 4 vol *F* with: *A* = 48 ml 1 N HCl, 36.6 g Tris, 0.46 ml TEMED and H_2O to 100 ml, pH 8.9; C: 30.0 g acrylamide, 0.8 g Bis, 15 mg $K_3[Fe(CN)_6]$ and H_2O to 100 ml; *F*: 140 mg ammonium persulphate in 100 ml H_2O (fresh prepared). The stacking gel is prepared from 1 vol *B*, 2 vol *D*, 1 vol *E* and 4 vol H_2O with *B*: 48 ml 1 N HCl, 5.98 g Tris and H_2O to 100 ml, pH 6.7; *D*: 10 g acrylamide, 2.5 g Bis and H_2O to 100 ml; *E*: 4 mg riboflavin in 100 ml H_2O. To separate biologically active enzymes it is recommended that electrophoresis is performed in the cold (4–6°C) and to reduce the current to 1 mA.

(ii) Separation gels were prepared by mixing 2 vol *A*, 1 vol *B*, 4 vol *C* and 1 vol *D* with *A*: 30.0 g acrylamide, 1.0 g Bis and 123 ml H_2O; *B*: 0.28% (v/v) TEMED; *C*: 0.14% (w/v) ammonium persulphate; *D*: 29 g glycine, 6 g Tris and 980 ml H_2O. Stacking and sample gels were replaced by 0.2 ml of a 5% (better 10%) sucrose solution containing 200–400 μg of total serum protein (5 to 50 μg of a given protein). Dialysis prior to electrophoresis is recommended only if foreign electrolyte concentrations are high, as in urine. The electrode buffer was prepared from a stock of 29 g glycine, 6.0 g Tris, 5 ml N HCl and 975 ml H_2O (final pH 8.1) and diluted 1 : 10 prior to use. Gels were prepared in pyrex tubes 6 mm in diameter and 75 mm long. Electrophoresis was performed for 30 min at 4 mA per tube.

Table 2.4 continued

(iii) This system is used to separate basic proteins and peptides at acid pH. 50 μg protein can be separated per gel rod. The separation gel has a volume of 0.85 ml and is prepared from a mixture of 1 vol *A*, 2 vol *C*, 1 vol H_2O and 4 vol *F*, with *A*: 48 ml N KOH, 17.2 ml acetic acid (glacial), 4 ml TEMED and water to make 100 ml; *C*: 30 g acrylamide, 0.8 g Bis and water to make 100 ml; *F*: 0.28 g ammonium persulphate per 100 ml of water. The stacking gel has a volume of 0.15 ml and is prepared by mixing: 1 vol *B*, 2 vol *D*, 1 vol *E*, and 4 vol H_2O, with *B*: 48 ml N KOH, 2.87 ml acetic acid (glacial), 0.46 ml TEMED and water to make 100 ml; *D*: 10 g acrylamide, 2.5 g Bis and water to make 100 ml; *E*: 4 mg riboflavin dissolved in 100 ml H_2O. Polymerisation of the stacking (and sample) gel is initiated by a 15 W fluorescent lamp placed about 8 cm behind the tubes. The remaining part in the tubes is filled with electrode buffer (31.2 g β-alanine, 8.0 ml acetic acid (glacial) and water to make 1 litre, pH 4.5). Gels were prepared in glass tubes 5 mm in diameter and 70 mm long. Electrophoresis is performed for 20 min at a current of 6–8 mA per tube.

(iv) Gels are prepared in glass capillaries (Drummond) of a volume of 2, 5 or 10 μl and a length of 54.5, 32.5, or 41.5 mm. They must be cleaned before use with chromic acid and Javelle's solution. The concentration of the separating gel must be increased when using capillaries smaller than 5 μl. For 5 μl capillaries a total polymer concentration of 20% is recommended while it is 25% for 2 μl and 30% for 1 μl capillaries. The capillaries are filled by capillary attraction to 75% of their total volume by dipping them into the separating gel mixture. Then they are placed upright, against the inner wall of a beaker, the bottom of which is covered with a 50% (w/v) aqueous sucrose solution containing 7 mg ammonium persulphate per 100 ml. Separation gels are prepared by mixing 0.5 ml *A* with 1.5 ml *C* and taking a 1 ml aliquot and mixing it with 1 ml *D*, where *A*: 860 mg Tris, 8 ml H_2O, 0063 ml TEMED, 3.6 N H_2SO_4 to pH 8.8 ($\sim$0.45 ml) and H_2O to 10 ml; *C*: 20 g acrylamide, 200 mg Bis, 3.75 mg $K_4[Fe(CN)_6]$ and H_2O to 37.5 ml; *D*: 70 mg ammonium persulphate, 25 ml 4% Triton X-100 in H_2O, and 25 ml H_2O. Triton X-100 is incorporated into the gels to facilitate their removal from the capillaries. The addition of 100 mg hydantoin to 2 ml of separation gel solution improves the band sharpening. Separation gels should stand for at least 10 h at room temperature. Stacking gels are prepared from a mixture of 0.25 ml *E*, 0.75 ml *C*, 2.6 ml H_2O, and 0.40 ml *F*, with *E*: 5.98 g Tris, 50 ml H_2O, 1 M H_3PO_4 to pH 6.7, 0.46 ml TEMED and H_2O to 100 ml; *C*: 20 g acrylamide, 200 mg Bis, 3.75 mg K_4 [$Fe(CN)_6$)], and H_2O to 37.5 ml; *F*: 200 mg ammonium persulphate, 5 ml 4% Triton X-100 (in water) and 5 ml H_2O. The volume of stacking gel is 0.15–0.30 μl with 5 μl capillaries. The sample volume is 0.1–1 μl. The remaining capillary volume is filled up with solution *B*, diluted 1 : 8 with water, with *B* being composed of 2.85 g Tris, 1 M H_3PO_4 to pH 6.7 ($\sim$12.5 ml) and H_2O to 50 ml. The electrode buffer is composed of 3.0 g Tris, 14.4 g glycine and H_2O to 500 ml, pH 8.4. Separation is performed for 20 to 30 min at 60 V ($\sim$7 μA).

(v) The system was developed to separate denatured proteins in the presence of SDS, but omitting the detergent it can also be used for the separation of biologically active proteins. Stacking and separating gels were prepared from a stock solution containing 30% acrylamide and 0.8% Bis. The separation gel contained 0.1% SDS. The gels were polymerised chemically by the addition of 0.025% ammonium persulphate. 10 cm gels were prepared in glass tubes with a length of 15 cm and an internal diameter of 6 mm. The stacking gels also contained 0.1% SDS and were polymerised chemically in the same way as the separating gels. 0.1% SDS was also included in the electrode buffer. Samples (0.2–0.3 ml) contained (final concentration) 0.0625 M Tris–HCl (pH 6.8), 2% SDS, 10% glycerol, 5% 2-mercaptoethanol and 0.001% bromophenol blue. Proteins were completely dissociated by immersing the samples for 1.5 min in boiling water. Electrophoresis was performed with a current of 3 mA per gel until the marker dye had reached the bottom of the gel.

Table 2.5 Native enzymes which have been separated by polyacrylamide gel electrophoresis and finally identified by enzyme visualisation techniques

Enzyme	*EC numbers*	*Reference*
Acid phosphatase	3.1.3.2	Uehave *et al* 1974
Alkaline phosphatase	3.1.3.1	Nimmo and Nimmo 1982
5-aminolaevulinate synthase	2.3.1.37	Davies and Neuberger 1973, 1979
Aminopeptidase (cytosol)	3.4.11.1	Strongin *et al* 1976
Anthranilate phosphoribosyltransferase	2.4.2.18	Grove and Levy 1975 Henderson *et al* 1970
Anthranilate synthase	4.1.3.27	Henderson *et al* 1970 Grove and Levy 1975
ATP pyrophosphatase	3.6.1.8	Risi *et al* 1977 Schäfer *et al* 1978
Catechol oxidase	1.10.3.1	Honold and Stahmann 1968 Honold *et al* 1966, 1967
Catalase	1.11.1.6	Gregory and Fridovich 1974 Graft and Gaafer 1974 Honold and Stahmann 1968 Honold *et al* 1966, 1967
Cellulase	3.2.1.4	Goren and Huberman 1976
Chorismate synthase	4.6.1.4	Nimmo and Nimmo 1982
Chymotrypsin	3.4.21.4	Andary and Dabich 1974
3′,5′-Cyclic-nucleotide phosphodiesterase	3.1.4.17	Solomon *et al* 1977 Tsou *et al* 1974
Cystathione β-synthase	4.2.1.22	Willhardt and Wiederanders 1975 De Issaly 1968
Deoxyribonuclease I	3.1.21.1	Porter 1981
DAHP synthase	4.1.2.15	Nimmo and Nimmo 1982
Dihydrouracil dehydrogenase	1.3.1.2	Hallock and Yamada 1973 Smith and Yamada 1971
Dipeptidase	3.4.13.11	Sugiura *et al* 1977
DNA nucleotidyltransferase	2.7.7.7	Zöllner *et al* 1973 Neuhoff and Lezius 1968
3-Enol pyruvoylshikimate-5-phosphate synthase	2.5.1.19	Nimmo and Nimmo, 1982
Fructose-bisphosphatase	3.1.3.11	Colombo and Marcus 1973 Hubert and Marcus 1974
D-fructose 5-dehydrogenase	1.1.99.11	Ameyama *et al* 1981
β-D-fructofuranosidase	3.2.1.26	Faye 1981
1,4-α-D-glucan branching enzyme	2.4.1.18	Satoh *et al* 1978 Yonezawa and Hori 1975
Glutamate dehydrogenase (NAD(P))	1.4.1.3	Kimura *et al* 1977

(*Continued*)

Table 2.5 Continued

Enzyme	*EC numbers*	*Reference*
Glutamine synthetase	6.3.1.2	Miller *et al* 1974
		Meyer and Stadtman 1981
Guanine deaminase	3.5.4.3	Bieber 1974
Hypoxanthine phosphoribo-syltransferase	2.4.2.8	Bieber 1974
Lactose synthase	2.4.1.22	Pierce *et al* 1980
Malate dehydrogenase (decarboxylating)	1.1.1.39	Zink and Katz 1973
Mannitol dehydrogenase	1.1.1.67	Ueng *et al* 1976
NADPH-cytochrome reductase	1.6.2.4	Fan and Masters 1974
		Ichihara *et al* 1973
Nitrite reductase	1.7.99.3	Hucklesby and Hageman 1973
Nucleotide pyrophosphatase	3.6.1.9	Balakrishnan *et al* 1974
Oestradiol 17 β-dehydrogenase	1.1.1.62	Kavavolas *et al* 1970
Peroxidase	1.11.1.7	Honold and Stahmann 1968
		Honold *et al* 1966 1967
Proteinases (unspecified)		Andary and Dabich 1974
Pyruvate decarboxylase	4.1.1.1	Nimmo and Nimmo 1982
RNA polymerase	2.7.7.6	Nimmo and Nimmo 1982
Superoxide dismutase	1.15.1.1	De Rosa *et al* 1979
Tetrahydrofolate dehydrogenase	1.5.1.3	Blakley 1960
		Nixon and Blakley 1968
Tripeptide aminopeptidase	3.4.11.4	Sugiura *et al* 1977
Trypsin	3.4.21.4	Andary and Dabich 1974
Tyrosine aminotransferase	2.6.1.5	Shuler and Tryfiates 1977
UDP glucose pyrophosphorylase	2.7.7.9	Nimmo and Nimmo 1982
Urease	3.5.1.5	Shaik-M *et al* 1980

the tubes are cleaned by treating them with 10% (w/v) dichromate sulphuric acid and by rinsing them with distilled water and acetone. Then one end is capped with a plastic cap surrounding the tube. The capped tubes are mounted in a rack.

Before pouring the gels the various stock solutions needed (cf table 2.4) are warmed to room temperature while the electrode buffer solution is cooled down to between 4 and 8°C. The ammonium persulphate solution has to be prepared freshly. All other solutions can be prepared as stock solutions and kept at low temperatures for several weeks. The separating gel solution must be used immediately after the two catalysts (TEMED, ammonium persulphate) have been mixed together. During stirring the mixing of air into the gel solution must be avoided. Using a pasteur pipette the tubes are filled three-quarters full with separating gel solution. The gel solution is then carefully overlayered with 0.2 to 0.4 ml of distilled water. After the polymerisation is over, approximately one hour, the water is removed without damaging the surface. The surface is rinsed twice with sample gel solution and then the sample gel solution is layered to a height of approximately 5 mm. Finally it is overlayered with 0.3 ml of distilled water and photo- or chemically polymerised (cf also table 2.4). Disc gels should be used the same day they are prepared.

Before using the gels the cap is carefully removed without letting air penetrate between the gel and the glass wall. The outer walls are wetted and the tubes are inserted, with the stacking gel uppermost, into the gaskets of

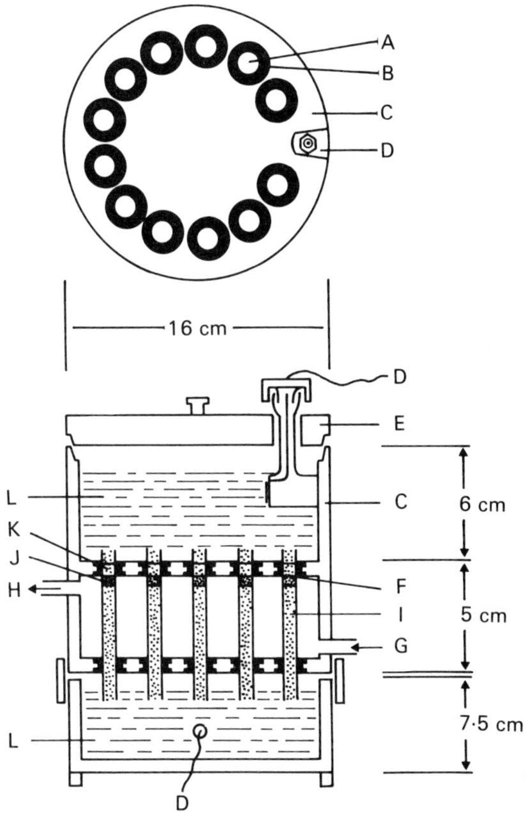

Figure 2.7 Round disc gel apparatus to separate functional proteins.
A, glass tubes (total length, 100 mm; internal diameter, 6 mm); B, F, silicon gaskets; C, upper electrode vessel; D, electrode (platinum wire); E, cover; G, cooling water (← inlet); H, cooling water (← exit); I, separation gel; J, stacking gel; K, sample solution; L, electrode buffer.

the disc gel apparatus. They are left there for approximately 30 min to cool the gels down to the temperature of the circulating water. The water layer is taken off by aid of a syringe carrying a piece of tygon tube, and finally the sucrose enriched sample solution (in a diluted phosphate buffer) is applied. The sample volume should not be larger than the volume of the stacking gel. There should be no air bubbles between the lower end of the gels and the electrode buffer. Finally the sample solutions are overlayered with electrode buffer and then the buffers are filled into the upper and lower buffer reservoirs. A round disc gel apparatus suitable to separate functional proteins is shown in figure 2.7.

Following electrophoresis, the glass tubes are taken out of the apparatus and the gels removed using a syringe filled with water and fitted with a long hypodermic needle. The needle is inserted a few millimetres into the bottom end of the tube between the separation gel and the glass wall while rotating the tube and squirting water out of the needle. The gel will finally slip out of the tube and can be placed in the staining solution.

2.3.1.2 Determination of the Size of Non-denatured Proteins by use of Discontinuous Gel and Buffer Systems

So far two different methods have been suggested to estimate the size of native proteins in PAA-gel electrophoresis. The older methods use homogeneous gels of different concentrations while the more recent methods use PAA gels with increasing concentrations from top to bottom, so-called gradient gels. When using homogeneous gels of different concentrations, two-step mathematical procedures serve the estimation of the molecular mass of proteins: first the retardation coefficient (K_R) is determined by a so-called Ferguson plot (Ferguson 1964): protein mobility is studied in homogeneous separation gels of 6 to 12% total polymer concentration. Then the log of the protein mobility relative to the dye front (Bromphenol blue)(log R_f) is plotted against the total polymer concentration (T). The slope of the resulting straight line is defined as K_R. Secondly, the K_R values are related to the molecular mass of the proteins investigated (cf for details table 2.6). Hedrick and Smith (1968) suggested to correlate K_R linearly to the molecular mass of native proteins in the range of 50×10^3 to 500×10^3. The buffer systems they used did not influence the linearity of the K_R–molecular mass relationship and the accuracy of the method was found to be $\pm 4\%$ with a range of 0–11%. The unknown molecular mass of a protein is calculated by inserting its K_R value in the equation found for the regression line in a K_R–molar mass plot established with proteins of known sizes. It is one of the advantages of this method that impure proteins can also be used to estimate the molecular mass of a certain protein, provided a specific chemical or biological test for its identification exists.

A linear extrapolation of Hedrick and Smith's relationship to molecular masses smaller than 50×10^3, however, would lead to serious interpretation errors. A linear extrapolation of Hedrick and Smith's data to zero molecular mass results in $K_R \simeq 11 \times 10^{-2}\ \text{cm}^3\ \text{g}^{-1}$ (Gonenne and Lebowitz 1975). This would imply that an infinitely small molecule would behave in PAA-gel electrophoresis as if it were many orders of magnitude larger, i.e. equivalent to a $\sim 30 \times 10^3$ molecular mass protein. But this is not consistent with experimental results on proteins having molecular masses between 10×10^3 and 50×10^3. Gonenne and Lebowitz related the K_R values of proteins with molecular masses less than 50×10^3, with the corresponding log mol M. Moreover, using 18 different proteins and the experimental data of Hedrick and Smith they confirmed a nonlinear relationship between K_R and molecular mass. They tried to correlate their data with those of Hedrick and Smith using a modified equation of Rodbard and Chrambach (1971a, b)

$$K_R = \pi l' (\beta \text{ mol } M^{1/3} + r)^2 \tag{2.6}$$

where, for a spherical protein, $\beta = (3\bar{v}/4\pi N_A)^{1/3}$, N_A is Avogadro's number, and $\bar{v}$ is the average partial specific volume of a protein ($0.74\ \text{cm}^3\ \text{g}^{-1}$). Although the regression lines they calculated for proteins in the range of 10×10^3–50×10^3 are satisfying, those which were calculated for mol $M > 50 \times 10^3$ are not convincing. Based on Ogston's model (Ogston and Phelps 1961) Rodbard and Chrambach (1970, 1971a, b) developed a theoretical model on the electrophoretic separation of native proteins in PAA gels. The related K_R and the radius of spherical unhydrated proteins ($\bar{R}$) in the following equation

$$K_R = \pi l' (\bar{R} + r)^2 \tag{2.7}$$

where $l' \simeq 0.45 \times 10^{12}\ \text{cm/cm}^3$, and r(nm) is the radius of a PAA-gel fibre. As a result they assumed a linear correlation between $\sqrt{K_R}$ and $\bar{R}$. It must be pointed out, however, that they did not include more than three proteins of mol $M < 50 \times 10^3$ in any given set of electrophoretic conditions. Gonenne and Lebowitz (1975) were already aware of the two main drawbacks of this model. These are:

(i) most probably the Ogston model developed for dextran gels cannot be applied to PAA gels and
(ii) as in chromatography, the separation behaviour of proteins has to be correlated with the Stokes' radius (R_s) and not with the radius of an unhydrated weight equivalent sphere ($\bar{R}$)(Fawcett and Morris 1966, Laurent and Killander 1964, Siegel and Monty 1966).

The postulate that Stokes' radius is the critical parameter in gel electrophoresis (Ornstein 1964, Zwaan 1967) was first substantiated by Felgenhauer who demonstrated a linear correlation between the Einstein Stokes' radius and the exclusion limit of PAA gels (Felgenhauer 1974).

Table 2.6 Determination of the size of non-denatured proteins by use of discontinuous gel and buffer system

Buffer in large pore gel (% acrylamide + Bis, acrylamide : Bis)	*Buffer in small pore gel (% acrylamide + Bis, acrylamide : Bis)*	*Sample buffer (electrode buffer)*	*First derivation, second derivation*	*Molecular mass range ($\times 10^3$) (± degree of accuracy)*	*Remarks*	*References*
0.06 M imidazole HCl pH 5.7 (3.125; 4 : 1)	0.12 M Tris–HCl, pH 7.9 (3–12; 30 : 1)	0.06 M imidazole HCl in 50% glycerol, pH 5.7 (0.034 M asparagine, titrated to pH 7.3 with Tris)	log R_f against % T; K_R against mol M	50–500 (± 5%)	(i)	Hedrick and Smith 1968
0.125 M H_3PO_4, 4.71 M Tris, pH 6.9 (3.125; 4 : 1)	0.378 M Tris, 0.125 N HCl, pH 8.9 (3–10; $C = 3.2\%$)	0.05 M Tris, 50% (w/v) sucrose, pH 7.5 (4.96 mM Tris, 38.4 mM glycine, pH 8.3)	log R_f against %T; K_R against log mol M	11.4–82.5 (± 5%)	(ii)	Gonenne and Lebowitz 1975
Several different discontinuous buffer systems at pH 10.2, 8.88, 7.5 and 5.5 (3–12; C = 5.0%)			log R_f against % T; $\sqrt{K_R}$ against R	40–450 (not determined)	(iii)	Rodbard and Chrambach 1971a
0.125 M H_3PO_4, 4.71 mM Tris, pH 6.9 (3.125; 4 : 1)	0.378 M Tris, 0.125 N HCl, pH 8.9 (3.8–6.9; 32.5 : 1)	not specified (4.96 mM Tris, 38.4 mM glycine, pH 8.3)	log % T against R_f; log K_R against log mol M	43–900 (± 5%)	(iv)	Bryan 1977
0.125 M H_3PO_4, 4.71 mM Tris, pH 6.9 (3.125; 4 : 1)	0.378 M Tris, 0.125 N HCl, pH 8.9 (3–35; $C = 2.6\%$)	not specified (4.96 mM Tris, 38.4 mM glycine, pH 8.3)	log % T against R_f; R_s against $1/T_{Lim}$	36–2300 (not determined)	(v)	Felgenhauer 1974

(i) K_R: slope of a regression line obtained by plotting the logarithm of protein migration relative to Bromphenol blue (log R_f) against the total acrylamide plus Bis concentration (% T), mol$M$: mol mass. A change in the acrylamide : Bis ratio resulted in different separation patterns. Size isomeric proteins give a family of nonparallel lines in Ferguson plots (log R_f against % T) while charge isomeric proteins give a parallel family of lines. Proteins differing in both charge and size give non-parallel lines. The polymerisation of the large pore gel is initiated by 0.5 mg riboflavin/100 ml and UV light. The polymerisation of the small pore gel is initiated by 0.115 ml TEMED and 0.035 g ammonium persulphate 100 ml. It is supposed that some protein molecules will behave anomalously, especially when joined to lipids or carbohydrates.

(ii) Buffer systems according to Davis (1964), with the exception that the spacer gel was omitted. The percent cross-linking concentration C is defined as $100 \times$ Bis/(Bis + acrylamide).

(iii) The polymerisation is performed at 0°C. Experimentally the authors were not able to differentiate between the equation $K_R = a + b$ (mol M) (Hedrick and Smith 1968) and the equation based on the Ogston model $\sqrt{K_R} = \pi l' (\bar{R} + r)$; l' = length of a PAA-gel rod per unit volume and unit concentration of T; $\bar{R}$ = radius of a weight equivalent sphere having a density of 1.3 g cm^{-3}; r = radius of a gel rod = 0.5 nm.

(iv) Buffer system as described by Davis (1964) except that 1 mM L-threonine was included in both the large and small pore gels as well as in the upper reservoir buffer. The separation behaviour of multimeric enzymes was studied. These were: bovine serum albumin (1–3; 69 000), anthranilate synthetase-phosphoribosyltranspherase (2, 4, 8; 140 000), homoserine dehydrogenase (1, 4; 82 500), secretory Ig A (1, 2; 390 000), and apoferritin (1, 2; 450 000); the numbers in brackets indicate the polymer forms and molecular mass of the smallest form. The average error resulting from the use of either K_R against R or log K_R against log mol M is about $\pm 5\%$, whereas the average error using K_R against mol M was found to be $> 18\%$.

(v) R_s = Stokes radius; % T = total polymer concentration; R_f = protein migration relative to the dye front; log % T is plotted against R_f and the points of intersection of the regression lines with the log T axis = T_{Lim} are determined. In a second plot the reciprocal exclusion limits ($10^2/T_{Lim}$) are plotted against the corresponding Einstein Stokes radii (R_s). Proteins with frictional ratios of about 1.48 are situated above the regression line of an R_s against $1/T_{Lim}$ plot while proteins with frictional ratios of about 1.22 are situated below that line. Buffer systems according to Davis (1964).

Using gel gradients from 3 to 35% and discontinuous gel electrophoresis, Felgenhauer found that most of the 25 proteins he used came to zero mobility on prolonged times of electrophoresis, while smaller proteins continued migrating. Zero mobilities, however, already postulated by Slater (1968) and also noticed by Margolis and Kenrick (1968) and others are contradictory to Ferguson's equation

$$\log m_{\mathrm{T}} = \log m_0 - K_{\mathrm{R}} T \tag{2.8}$$

with T = sum of monomer and comonomer concentration (Hjertén 1962), m_{T} = mobility at T% gel, and m_0 = free mobility at $T = 0\%$.

The Ferguson equation predicts:

(i) the free mobility can be determined as the intercept of a straight line with the log m axis, and
(ii) electrophoretic mobility does not cease even at very high polymer concentrations.

However, progressive deviations from linearity are observed if the data obtained with polymer concentrations exceeding 15% T are plotted according to Ferguson (Felgenhauer 1969, 1974). Instead, linearity is found over a broader pore size range if relative mobilities referred to Bromphenol red are plotted against log PAA concentration provided a constant monomer–comonomer proportion is used. As a result Felgenhauer presented the following equation

$$\log T = \log T_{\mathrm{EL}} - K m_{\mathrm{T}} \tag{2.9}$$

with: T, PAA concentration: g/100 ml; T_{EL}, monomer concentration at zero velocity (exclusion limit); m_{T}, relative mobility at T %; Bromphenol red = 100; K, slope of retardation line: constant.

In contrast to the Ferguson equation Felgenhauer's equation predicts

(i) that the free mobility cannot be extrapolated from the retardation line, and
(ii) there will be zero mobility above a certain PAA concentration ('limiting monomer concentration')(cf figure 2.8).

Extrapolation of the regression lines of log T against R_{f} plots to zero R_{f} values (the point of intersection with the log T axis) leads to %T values which were defined by Felgenhauer as exclusion limits (T_{Lim}). A plot of the reciprocal of the exclusion limits of proteins against the corresponding Stokes' radii results in a number of points which can be considered as scattered around a straight line (see figure 2.9). The same coefficient of correlation ($r = 0.98$) was established with the mean geometric radii (R). Felgenhauer also treated the mobilities of his test proteins according to Ferguson's equation and calculated the retardation coefficients from those parts of the retardation curves which are acceptably straight (generally below 15% *T*). He then correlated the square roots of the retardation coefficients and the corresponding Stokes radii or the radii of weight equivalent spheres (Rodbard and Chrambach 1971a, b), and again found identical coefficients of correlation ($r = 0.97$). He explains this phenomenon by electrophoretic alignment.

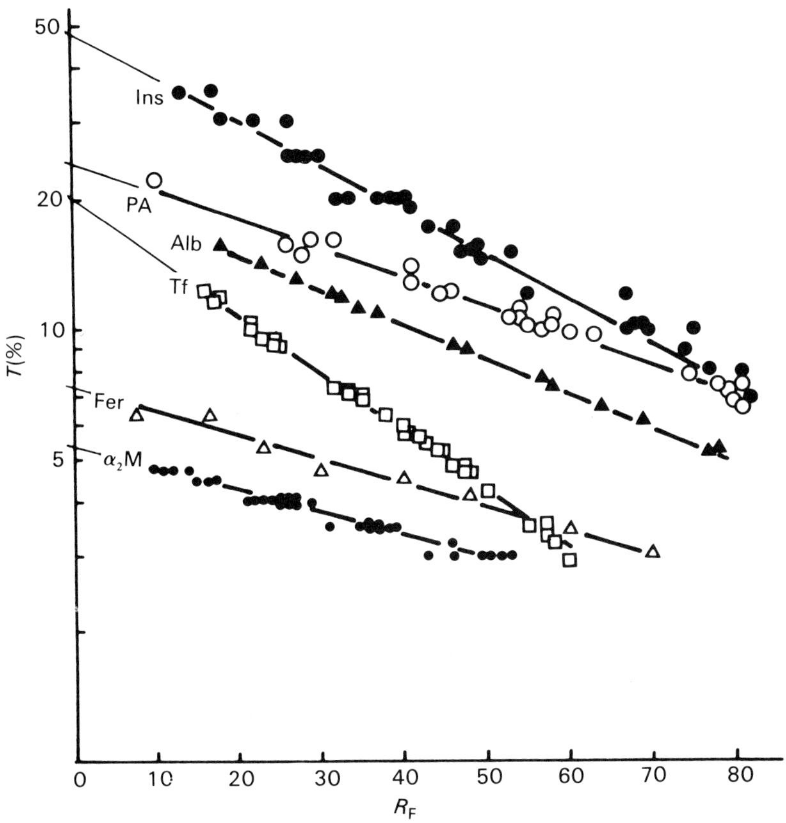

Figure 2.8 Plot of log PAA concentration, *T* (%) against the relative migration (R_f) of six different proteins. The stacking concentration (*T* (%)) of each protein results from the intersection of the regression line with the *T* axis ($R_f = 0$). Ins, insulin; PA, prealbumin; Alb, albumin; Tf, transferrin; Fer, ferritin; α_2H, α_2-macroglobulin. Generally the proteins are stacked around $R_f = 85$ below their retardation limits (taken from Felgenhauer 1974 with permission of the author and the publisher).

An interesting subsidiary effect of Felgenhauer's data treatment is the following: proteins with larger frictional ratios ($f/f_0 \simeq 1.48$) are situated above the regression line of an R_s against $1/T_{Lim}$ plot while proteins with smaller friction ratios ($f/f_0 \simeq 1.22$) are situated beneath the line. Again

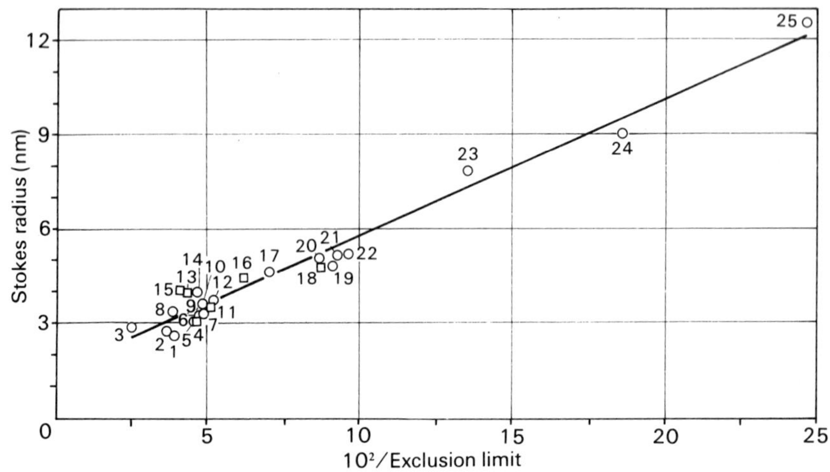

Figure 2.9 Plot of the Stokes radius (nm) of 25 standard proteins against the exclusion limits. Exclusion limits are defined as gel concentrations of zero protein mobility, extrapolated from the retardation lines (log PAA concentration against relative protein mobility). Protein isoelectric points are below 7.0, glycoproteins with carbohydrate contents above 18% are marked by squares (taken from Felgenhauer 1974 with permission of the author and publisher).

Index, to proteins:

1 *Bacterium subtilis* α-Amylase †:
$R = 26.7$ Å, $M_r = 48\,580$,
$f/f_0 = 1.11$

2 Chicken ovalbumin †:
$R = 27.4$ Å, $M_r = 43\,500$,
$f/f_0 = 1.18$

3 Cow β-lactoglobulin †:
$R = 29.0$ Å, $M_r = 35830$,
$f/f_0 = 1.33$

4 Horse radish peroxidase:
$R = 30.8$ Å, $M_r = 41\,280$,
$f/f_0 = 1.34$

5 *Bacterium subtilis* hydroxynitril lyase (gift Dr Aschhoff, Erlangen):
$R = 31.0$ Å, $M_r = 73\,000$,
$f/f_0 = 1.12$

6 Heart mitochondrial malate dehydrogenase ‡:
$R = 33.2$ Å, $M_r = 62\,000$,
$f/f_0 = 1.27$

7 Hog kidney D-amino-acid oxidase ‡:
$R = 33.4$ Å, $M_r = 114\,250$,
$f/f_0 = 1.04$

8 Bovine haemoglobin †:
$R = 33.9$ Å, $M_r = 67\,350$,
$f/f_0 = 1.26$

9 Human albumin †:
$R = 34.5$ Å, $M_r = 69\,000$,
$f/f_0 = 1.27$

10 Bovine albumin §:
$R = 35.5$ Å, $M_r = 66\,200$,
$f/f_0 = 1.34$

11 Swine follicle stimulating hormone [d]:
$R = 35.6$ Å, $M_r = 67\,360$,
$f/f_0 = 1.32$

12 Human transferrin §:
$R = 37.2$ Å, $M_r = 81\,000$,
$f/f_0 = 1.30$

13 Calf fetuin ¶:
$R = 40.0$ Å, $M_r = 48\,470$,
$f/f_0 = 1.66$

14 Chicken conalbumin †:
$R = 40.4$ Å, $M_r = 86\,180$,
$f/f_0 = 1.38$

15 Human acid α_1-glycoprotein §:
$R = 40.6$ Å, $M_r = 44\,100$,
$f/f_0 = 1.74$

16 Human haptoglobin 1-1 §:
$R = 45.1$ Å, $M_r = 80\,000$,
$f/f_0 = 1.59$

17 Human ceruloplasmin §:
$R = 47.2$ Å, $M_r = 152\,200$,
$f/f_0 = 1.34$

18 Human choriogonadotropin
(gift Dr Graesslin, Hamburg):
$R = 48.6$ Å

19 Lens leucine aminopeptidase
(gift Dr Kretschmer, Halle):
$R = 49.0$ Å, $M_r = 326\,000$,
$f/f_0 = 1.08$

20 Muscle phosphorylase ‡:
$R = 49.3$ Å, $M_r = 177\,000$,
$f/f_0 = 1.46$

21 Bovine liver catalase ‡:
$R = 52.2$ Å, $M_r = 241\,170$,
$f/f_0 = 1.27$

22 Pig heart fumarate hydratase ‡:
$R = 52.7$ Å, $M_r = 206\,000$,
$f/f_0 = 1.27$

23 Horse apoferritin †:
$R = 79.0$ Å, $M_r = 473\,450$,
$f/f_0 = 1.53$

24 Human serum α_2-macroglobulin:
$R = 91.0$ Å, $M_r = 797\,750$,
$f/f_0 = 1.49$

25 Human serum β-lipoprotein:
$R = 126.0$ Å, $M_r = 2239\,000$,
$f/f_0 = 1.46$

† Serva, Heidelberg. ‡ Boehringer Mannheim GmbH.
§ Behringwerke, Marburg. ¶ Calbiochem, Luzern.

electrophoretic alignment can be used to explain this phenomenon since, in the absence of electrophoretic alignment, the exclusion limits should scatter irrespective of molecular eccentricity. Felgenhauer also investigated the separation behaviour of three elongated proteins (tropomyosin: $f/f_0 = 3.10$, L-meromyosin: $f/f_0 = 2.90$, fibrinogen: $f/f_0 = 2.33$) and found them strongly deviating from the regression line obtained from the proteins with a more or less globular shape. Overall, globular non-denatured proteins exhibit similar hydrodynamic properties in partition chromatography and gel electrophoresis, and their hydrodynamic radius can be reliably determined with either method. Since elongated proteins are aligned parallel to the electric field in PAA electrophoresis (Ornstein 1964) these proteins give spuriously small sized values (Felgenhauer 1974). As a consequence, two proteins with identical Stokes' radii and isoelectric points may be separable in gel electrophoresis if they differ in their axial ratio.

2.3.1.3 *Differentiation of Size Isomers and Charge Isomers*

Upon electrophoresis with low concentration PAA gels (3 to 12% *T*) both size and charge parameters are responsible for the separation of native proteins (Smithies 1959, Sober *et al* 1965). It has been demonstrated for example that ferritin, a highly charged protein with a molecular mass of $> 10^6$, and ovotransferrin, a slightly charged protein with a molecular mass of 87×10^3, migrate at the same rate in 5.7% PAA gels using disc electrophoresis (Hedrick and Smith 1968). By applying the Ferguson relationship (Ferguson 1964) it can be deduced whether proteins differ in charge or size or both. First the proteins are run in homogeneous gels of 3 to 12% *T* and then the log of protein mobility relative to the dye front is measured and plotted against the corresponding gel concentrations. The slope of the resulting straight line is defined as K_R. Hedrick and Smith (1968) recognised that a size isomer family of proteins (e.g. bovine serum albumin polymers) give—on a Ferguson plot—a family of non-parallel lines extrapolating to a common point in the vicinity of 0% gel concentration while a charge isomer family of proteins (e.g. lactate dehydrogenase isozymes) gives a parallel family of lines. Proteins differing in both charge and size give non-

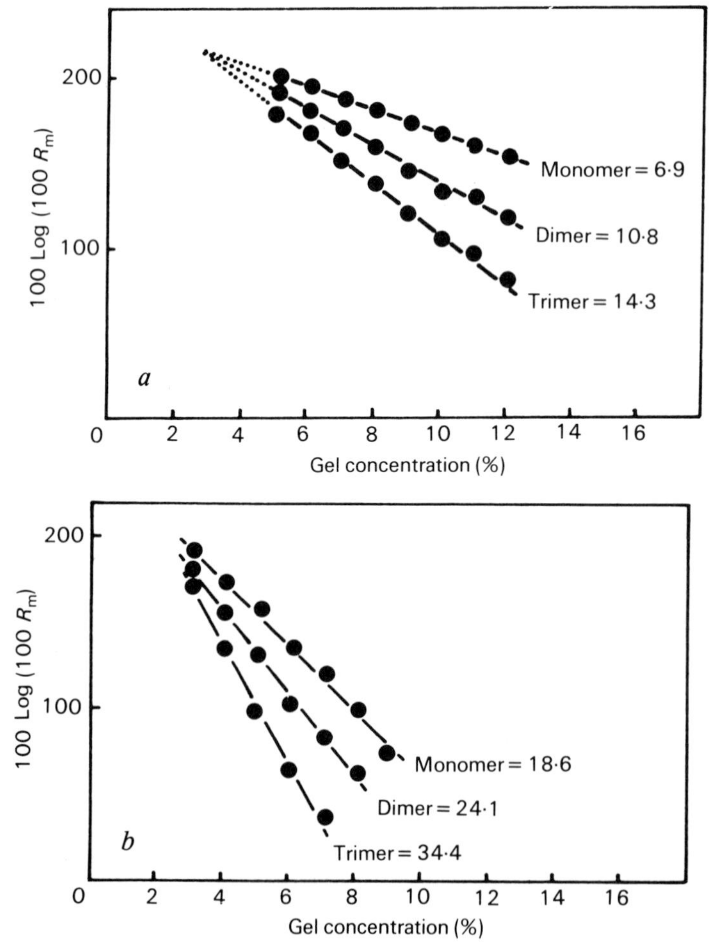

Figure 2.10 The effect of different PAA concentrations (%) on the mobility (R_m) of (*a*) bovine serum albumin polymers; (*b*) ferritin polymers. The negative slopes of the lines are noted on both figures (taken from Hedrick and Smith 1968 with permission of the author and publisher).

parallel lines intersecting at gel concentrations other than 0% gel concentration. Fig. 2.10(*a*) indicates the separation of bovine serum albumin monomer, dimer and trimer at gel concentrations in the range from 5–12%. As can be seen, the value of the negative slope increases with increasing size; the larger the molecule the more it is sieved by increasing gel concentrations. Ferritin, which also exists in monomeric, dimeric and trimeric states, gave similar results yielding non-parallel lines interesecting at ~1% gel concentration (Hedrick and Smith 1968) (fig 2.10(*b*)).

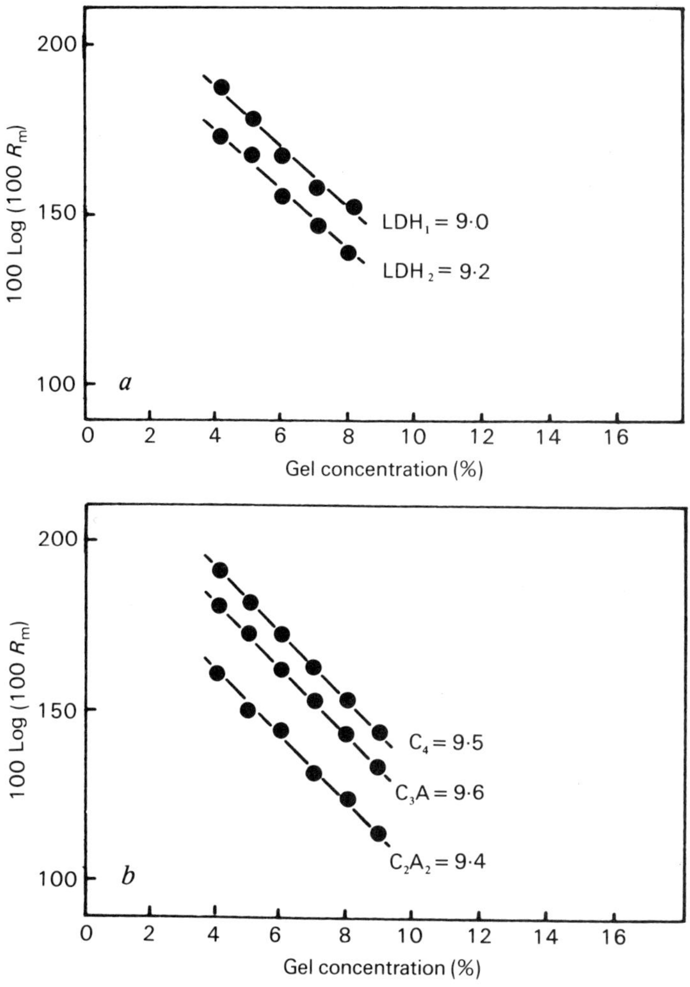

Figure 2.11 The effect of different PAA concentrations (%) on the mobility (R_m) of (*a*) lactate dehydrogenase isoenzymes; (*b*) aldolase isoenzymes. The negative slopes of the lines are noted on both figures (taken from Hedrick and Smith 1968 with permision of the author and publisher).

Proteins with similar sizes but different net charges are also separable by disc gel electrophoresis. Lactate dehydrogenase and aldolase isozymes are examples of this; they yield parallel lines in a Ferguson plot (figure 2.11)(Hedrick and Smith 1968). If two protein molecules differ in both the net charge and the molecular size, a Ferguson plot yields non-parallel lines which do not intersect at the ordinate. In figure 2.12 this is demonstrated for the ferritin monomer and ovotransferrin.

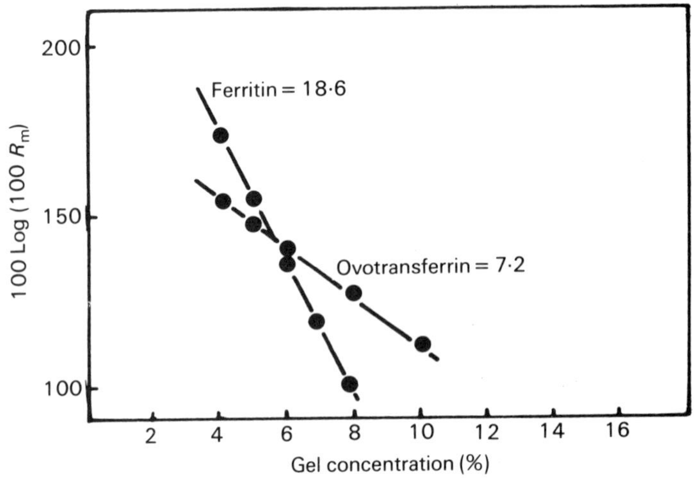

Figure 2.12 The effect of different PAA concentrations (%) on the mobility (R_m) of the ferritin monomer and ovotransferrin. The negative slopes of the lines are noted on the figure (taken from Hedrick and Smith 1968 with permission of the author and publisher).

2.3.1.4 Determination of the Size of Proteins Using Polyacrylamide Gradient Gel Electrophoresis (PAGGE)

The high resolution of homogeneous PAA-gel electrophoresis (Davis 1964, Ornstein 1964) can be improved by using gradient gels (Esposito and Obijeski 1976, Mahadik 1976, Margolis and Kenrick 1967, Rothe and Purkhanbaba 1982 a, b, Wright *et al* 1975). The average pore radius of these gels decreases with increasing gel concentration, i.e. the average pore radius decreases in the direction of the migrating proteins. Band sharpening is much better in gradient gels than in homogeneous gels, and gradient gels separate proteins over a wider range of molecular masses than homogeneous gels. Since the PAA matrix causes no electroendosmosis, gradient gels have also been used to estimate the size of denatured and non-denatured proteins (Anderson *et al* 1972, Kopperschläger *et al* 1969, Lambin 1978, Lambin and Fine 1979, Lasky 1978, Rodbard *et al* 1971,

Slater 1969). Several different devices for the preparation of gradient gels are now commercially available so that the preparation of such gels is no longer restricted to a few highly specialised laboratories. Even ready-to-use pore gradient gels are on the market so that in the near future the highly economical and precise method of PAGGE will probably become a routine method for characterising proteins by their molecular size.

2.3.1.4.1 PREPARATION OF GRADIENT GELS

Gradient gels are prepared from high and low concentration gel solutions usually by using the technique which was first described by Martin and Ames (1961) for the preparation of linear sucrose gradients (cf tables 2.1 and 2.2 also). Gradient gels are cast in glass cassettes of various dimensions. They may be produced as columns or flat gels. Gel rods may be prepared in 10–15 μl glass capillaries having a thickness of 0.2 mm (Neuhoff and Schill 1968) or in glass tubes as cylinders of approximately 10 cm length and 6 mm thickness (Alligood 1983, cf figure 2.13(*b*)). Flat gels usually have dimensions of 82 × 82 (140) mm^2 and thickness of 2.8, 0.8 or 0.2 mm (Margolis and Kenrick 1967, Rothe and Purkhanbaba 1982a). To prepare 13 gradient gels simultaneously, each having the dimensions of 82 × 82 × 0.7 mm^3, we use the following procedure (Rothe 1982). The glass cassettes fitted with slot formers are inserted in a gel casting apparatus (cf figure 2.13 (*a*)) and filled with distilled water whilst avoiding the entrapment of any air bubbles. Linear PAA gradients with a constant ratio of acrylamide: Bis (24:1), ranging from ~5 to ~27% acrylamide plus Bis per 100 ml of buffer, were prepared from a T_{min} solution containing 4.464 g acrylamide and 0.186 g Bis, and a T_{max} solution containing 33.48 g acrylamide and 1.395 g Bis per 100 ml of buffer. The buffer was either a 0.1 M Tris HCl buffer of pH 9.0 or a 90 mM Tris, 80 mM boric acid, 2.5 mM EDTA buffer of pH 8.4. Catalysts are dissolved separately in the buffer: 55 μl TEMED and 105 mg ammonium persulphate in 10 ml, respectively. Prior to use all solutions are degassed and brought to room temperature. The gradients are prepared by using a two-chamber gradient mixer, two reservoirs—one for each catalyst solution—a proportioning pump a 1 ml mixer and the gel casting apparatus (cf figure 2.14). Chamber (A) and the connecting tube to chamber (B) of the gradient mixer are filled with 70 ml of the T_{min} solution, while the tube to chamber (B) is closed. 67 ml of the T_{max} solution are pipetted into chamber (B), reservoir (D) is filled with TEMED and reservoir (E) with ammonium persulphate. The proportioning pump (F) is switched on, all lines are filled, and the first 5 ml of the solution leaving tube (I) are discarded. After the pump has been switched off, this tube (I) is connected to the gel casting apparatus (J), the connections between chamber (A) and (B) of the gradient mixer are opened, and the two stirrers (C and H) and the pump are switched on. Immediately after pumping the PAA gradient into the gel casting apparatus it is

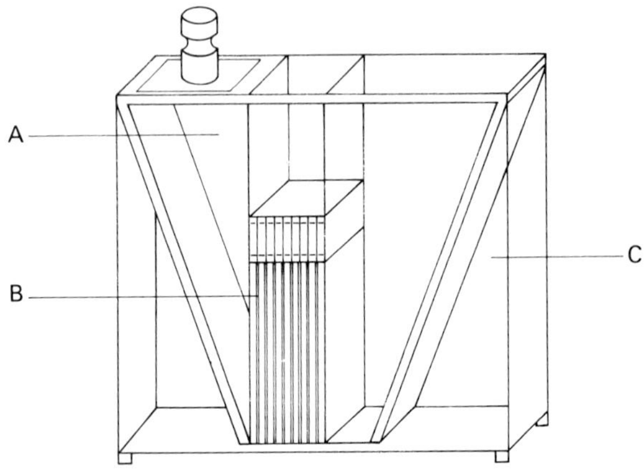

a

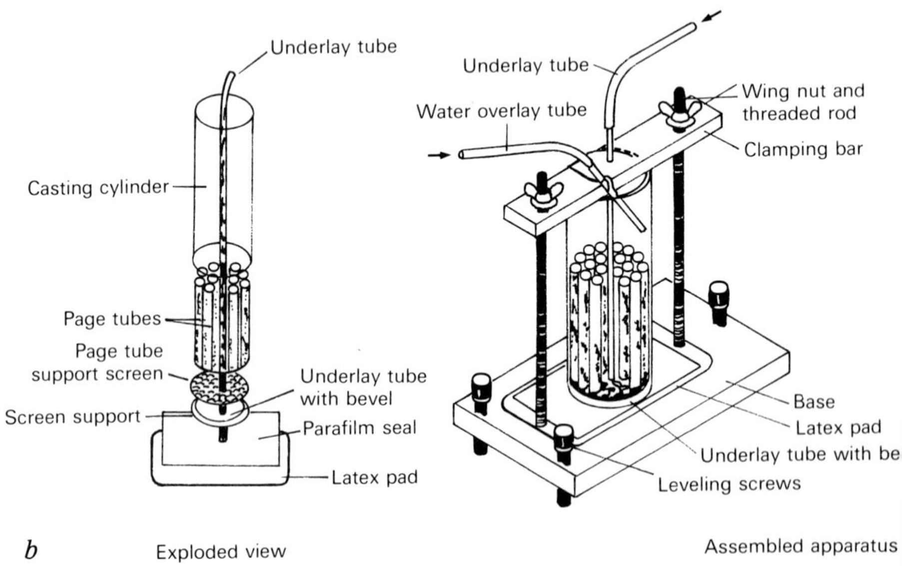

b

Figure 2.13(*a*) Gel casting apparatus to simultaneously produce several flat gradient gels. A, wedge; B, glass cassettes; C, casting apparatus (taken, with permission, from Pharmacia 1980). (*b*) Gel casting apparatus to simultaneously prepare several rod shaped gradient gels (taken from Alligood 1983 with permission of the author and publisher).

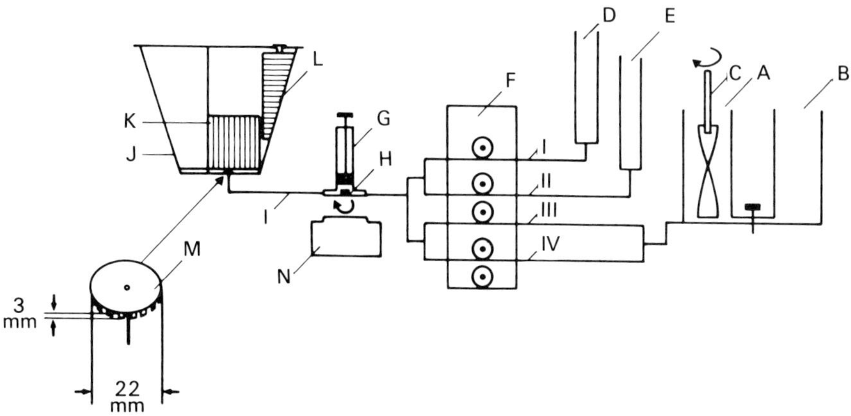

Figure 2.14 Scheme of the apparatus used to simultaneously prepare several gradient gel slabs (taken from Rothe and Purkhanbaba 1982 with permission of the authors and the publisher).
A and B, two-chamber gradient mixer (dimensions of each chamber: inner diameter = 41 mm, height = 100 mm); C, stirrer; D, reservoir for persulphate solution; E, reservoir for TEMED solution; F, proportioning pump; G, modified disposable syringe, used as small chamber mixer; H, magnetic bar; I, tube connecting the stirrer to the gel casting apparatus; J, gel casting apparatus; K, gel cassettes; L, wedge made from plexiglass; M, distributor; N, magnetic stirrer; I, II, Ismatec Tygon tubings, ENE code 08 (0.64 ml min^{-1}); III, IV, ismatec Tygon tubings, ENE code 23 (6.80 ml min^{-1}).

underlayered by a solution containing 40% (w/w) sucrose lifting the whole gradient into the cassettes.

2.3.1.4.2 CONDITIONS OF ELECTROPHORESIS

The glass-cassette cast gel gradients are inserted vertically into an electrophoretic apparatus consisting of an upper and lower electrode vessel. The upper buffer tank has four rubber gaskets which can hold up to four cassettes. After pre-electrophoresis for 30 min at 100 V to remove unpolymerised monomers and catalysts, the electrode buffer in the upper electrode vessel is taken away and the samples are applied. Then the electrode vessel is carefully filled again, and the proteins are allowed to penetrate into the gel slabs for 45 min at a voltage of 200 V without circulating the buffer. Finally, the buffer is circulated from buffer tank to buffer tank at the same voltage. (The voltage per gel plate remains constant no matter whatever number of plates are run because of the construction of the apparatus (cf Rothe and Purkhanbaba 1982a).

2.3.1.4.3 EVALUATION OF THE GEL GRADIENT

To evaluate the actual course as well as the actual range of gel gradients it is recommended that the gel plates are cut into 5 mm wide strips normal to the gel length. Each strip is weighed, dried at 80°C for 24 h, stored for several hours over $CaCl_2$ in an evacuated desiccator, and weighed again. The dry matter of the gel strips is defined as *TM* (g), while the relative dry matter *TM*(%) can be calculated from the wet matter *FM* (g) and the dry matter using the equation:

$$TM\ (\%) = 100\ (TM/FM).$$

The volume of a 5 mm broad gel strip is $82 \times 5 \times 2.8\ \mathrm{mm}^3 = 1.148\ \mathrm{cm}^3$; using a 0.1 M Tris–HCl buffer each strip contains 0.01165 g Tris. Thus the PAA concentration (g Bis + g acrylamide per 100 ml solution = *T*) per gel strip can be calculated using the formula

$$T = 100\ (TM - 0.012\ \mathrm{Tris})\ 1.148^{-1}.$$

As can be expected, the dry matter of unprocessed and processed gels is unaltered. As the relative dry matter (*TM*%) increases linearly only in unprocessed gels, it increases non-linearly in processed gels, indicating that the swelling of PAA gels depends on their total PAA concentration. In processed gels the relative dry matter increases with the square root of the gel length, i.e. $TM\ (\%) = a\sqrt{D} + b$ (with a and b being constants, and D being the gel length, cf Rothe and Purkhanbaba 1982a).

2.3.1.4.4 MATHEMATICAL PROCEDURES TO COMPUTE THE SIZE OF NON-DENATURED PROTEINS

Margolis and Kenrick (1968) introduced 'pore gradient' or polyacrylamide gradient gel electrophoresis (PAGGE) as a means for fractionating non-denatured proteins. Several different laboratories took up their idea, improved the preparation of gradient gel slabs and suggested various different mathematical procedures to estimate the molecular masses of proteins (Anderson *et al* 1972, Kopperschlaeger *et al* 1969, Lambin and Fine 1979, Lasky 1978, Rodbard *et al* 1971, Rothe 1982, Slater 1969). They all demonstrated that PAGGE provides resolution and information unobtainable from gels of a uniform concentration. This is because electrophoresis through a gel gradient from the low to the high gel concentration results in an improved band sharpening and a separation of proteins over a wider range of molecular masses. In PAGGE the velocity of proteins is—

(i) inversely related to time and
(ii) exponentially related to the migration distance (Rodbard *et al* 1971a).

Accordingly, the distance traversed by any one band after a given time is not a physical constant (Rodbard and Chrambach 1971a). In fact, in PAGGE the migration of any protein is governed by—

(i) the percentage *T*-range and the course of the gel gradient,

(ii) the electrical field strength,
(iii) the duration of electrophoresis, and
(iv) the charge density and both the size and the shape of the protein.

In PAGGE—like in gel permeation chromatography—the size of a particular protein is evaluated by correlating its migration properties with the migration properties of proteins of known sizes. For this purpose one-step and two-step mathematical procedures have been suggested (see Tables 2.7 and 2.8). One-step procedures are easier to handle, but two-step procedures are essential for the estimation of the size of non-denatured enzyme proteins. Felgenhauer (1974) provided evidence that the so-called Ferguson relation which relates the logarithm of a protein's migration (log R_f) to the total PAA concentration (T) (Ferguson 1964) is limited to $T < 12\%$, i.e. this relation should not be the base for a two-step procedure when 3–30% PAA gradients are used.

It is consistently implied that molecular mass determinations by PAGGE are restricted to a state where the marker proteins and the proteins under investigation have reached their apparent end-point in migration. However, proteins with molecular masses < 80 000 migrate out of 3–30% T (acrylamide : Bis = 24 : 1) gradients after prolonged times of electrophoresis. Thus the stacking point of these proteins cannot be determined experimentally unless the maximum gel concentration is raised to 40%. In such an extended gradient proteins with molecular masses < 25 000 would not stop migrating. The use of gels with gradients up to 40% T causes some complications: during gelling the gels heat up strongly in regions of high concentration and, if they are subjected to electrophoresis at 5–8°C, large air bubbles are frequently formed between the gel and the supporting glass plates at its lower end. As a result, migrating protein bands diffuse at these locations. Therefore we suggest that linear gel gradients ranging from 3 to 30% T are used. We also suggest that the 'maximum migration distances' of native proteins in the range from 15 000 to 1 000 000 should be estimated using a mathematical extrapolation procedure. In doing so we imply that the time-dependent migration of proteins follows the same laws at gel concentrations > 30% as those which regulate their migration at gel concentrations < 30% T.

It must be emphasised here that the migration of native proteins depends both upon their charge density and their molecular size, especially at low gel concentrations. Therefore the size of a certain protein cannot be determined simply from correlating its migration distance and molecular mass with those of the calibration proteins obtained after an arbitrarily chosen time of electrophoresis. Proteins of equal size may have migrated faster or slower than the calibration proteins according to different charge densities or interactions with the gel matrix.

We suggest that the native molecular size of proteins should be estimated

Table 2.7 One-step procedures used to determine the size of native proteins in PAGGE

Mathematical correlation	*Molecular mass range* *(×10^3)*	*Buffer systems*	*Remarks*	*Authors*
Log mol M was plotted against migration distance (D) and a limited linear correlation was found $\log \text{mol } M = -aD + b$	50–200	Electrode buffer: 0.06 M Tris, 0.4 M glycine, pH 8.3	(i)	Kopperschläger *et al* 1969
Log mol M was plotted against log migration distance and a limited linear correlation was found $\log \text{mol } M = -a' \log D + b'$	100–400	Gel buffer: 0.35 M Tris–HCl, pH 8.9.		
Log mol M was plotted against migration distance (D) and a limited linear correlation was found $\log \text{mol } M = -aD + b$	50–300	Gel and electrode buffer: 0.09 M Tris, 0.003 M EDTA 0.08 M borate, pH 8.3.	(ii)	Anderson *et al* 1972

Log mol M was plotted against electrophoretic mobility (R_f) and a limited linear correlation was found $\log \text{mol } M = aR_f + b$	67–1300	Gel and electrode buffer: 0.09 M Tris, 0.08 M boric acid, 0.93 g($^{-1}$) EDTA-Na_2, pH 8.4.	(iii)	Lasky 1978
Log mol M was plotted against square root of migration distance ($\sqrt{D}$) and a limited linear correlation was found $\log \text{mol } M = -a\sqrt{D} + b$	10–1300	Gel and electrode buffer: 0.1 M Tris–HCl, pH 9.0.	(iv)	Rothe and Purkhanbaba 1982a, b
Log Stokes' radius ($\log R_s$) was plotted against square root of migration distance ($\sqrt{D}$) and a limited linear correlation was found $\log R_s = -a'\sqrt{D} + b'$		Gel and electrode buffer 0.09 M Tris, 0.08 M boric acid, 0.0025 M EDTA–Na_2 pH 8.4.		Rothe 1982, 1983

(i) Electrophoresis was performed at 200–300 V for 4–6 h. In the molecular mass range 1×10^5–5×10^5 a linear correlation was found using the equation mol $M = a \times 1/D^3$, while for the range 4×10^4–2×10^5 a linear correlation was found if log mol M was plotted against migration distance (D): mol $M = a' \times 1/\exp(D)$ T range: linearly increasing from 3–20%.
(ii) The correlation log mol $M = -aD + b$ was found to hold even in gradient gels containing 8 M urea (plus p-chloromercuribenzoate), and proved to be independent of time for an electrophoretic duration of 18–64 h. T range: 4–26%.
(iii) For the calculation of the calibration curve only, proteins with small frictional coefficients (f/f_0) are recommended. Pre-electrophoresis: 20 min at 70 V, electrophoresis at ~12°C, 16 h, 150 V. T range: 4–30%.
(v) Both correlations were independent of time for electrophoretic durations of 0.5–95 h. Pre-electrophoresis: 20 min at 100 V, electrophoresis at ~8°C, 200–300 V. T range: 4–30%. To estimate the monomeric size of dissociating enzymes, such as carbonic anhydrase, a different procedure is needed.

Table 2.8 Two-step methods used in the determination of molecular masses of native proteins in gradient gel electrophoresis.

First derivation	*Second derivation*	*mol M range* ($\times 10^3$)	*Buffer systems*	*Remarks*	*Authors*
Migration distances (cm) were plotted against (V × h) no linear relationship	Log migration distances were plotted against log molecular masses and a linear relation was found: log mol $M = -\alpha \log D + \beta$	30–460	Gel and electrode buffer 1.2 g Tris and 5.76 g glycine per litre, pH 8.3	(i)	Slater 1969
Ferguson plot: log migration distances per time ($\log v$; $v = \text{cm/h}^{-1}$) were plotted against the total acrylamide concentrations the proteins have reached (T) *or* $\log R_s$ values were plotted against T. Determination of the slopes of the regression lines which were defined as K_R	$\sqrt{K_R}$ values were plotted against R. R was defined as the radius of a weight-equivalent, unhydrated sphere of a density of 1.33	40–450	Several different discontinuous buffer systems at pH 3.5, 5.5, 6.5, 7.8, 8.9, 9.4 and 10.2	(ii)	Ferguson 1964, Rodbard *et al* 1971, Rodbard and Chrambach 1970

The square root of the duration of electrophoresis ($\sqrt{t}$) was plotted against the distance migrated by a protein (D), and a linear relation was found $\sqrt{t} = aD + b$	log mol M was plotted against the slope of the regression lines of the first derivation log mol $M = \alpha \log\ a + \beta$	20–950	(*a*) Gel and electrode buffer 10.75 g Tris, 5.04 g boric acid, 0.93 g EDTA-Na_2 per litre, pH 8.2 (*b*) Gel and electrode buffer 0.04 M Na Veronal, 0.04 M Tris, 0.01 M glycine, 0.04 M ethanolamine, 0.001 M EDTA, pH 9.8 (*c*) Gel and electrode buffer 0.01 M sodium phosphate pH 7.2	(iii)	Lambin and Fine 1979

(i) For the construction of the regression line in the log D against log mol M correlation, the asymptotic parts of the first derivation, i.e. the constant Dt^{-1} values, are used; γ-globulin was found to be beyond the calibration curve. T range: 5–30%(Cyanogum 41).
(ii) Using linear gel gradients in the T range of 5–15%, the $\sqrt{K_R}$ against R relation was found to be consistent with the Ogston model developed for fibrous gels. The surface of unhydrated mol mass-equivalent spheres—instead of Stokes' radii—is related to migration distances of proteins. For objections against the use of the Ferguson relationship, cf Felgenhauer (1974).
(iii) A linear correlation was found between $\sqrt{t}$ and D provided that the duration of the electrophoresis exceeded 8 h at 80 V. The carbohydrate content of proteins did not influence the second derivation, but the buffer pH value did.

using a two-step mathematical procedure:

(i) the maximum migration distance of the protein of unknown size and those of the marker proteins are estimated,
(ii) a calibration curve is set up by correlating the maximum migration distances of the marker proteins with their molecular size.

The corresponding equation is used, together with the data for the maximum migration distance of the protein under investigation, to estimate either its Stokes' radius or its molecular mass.

2.3.1.4.4.1 DETERMINATION OF THE MAXIMUM MIGRATION DISTANCE OF PROTEINS IN LINEAR POLYACRYLAMIDE GRADIENT GELS

After prolonged electrophoresis in 3–10% T linear gradient gels, proteins with molecular masses exceeding 80 000 reach a point where their migration

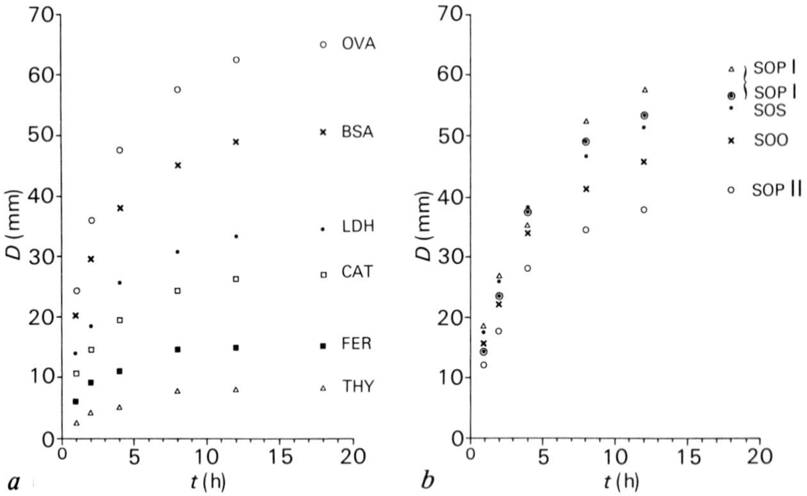

Figure 2.15 (*a*) Time-dependent migration of marker proteins in a linear 5–27% T PAA gel. The concentration of the crosslinker is constant (acrylamide : Bis = 24 : 1). The marker proteins are: ovalbumin (OVA), bovine serum albumin (BSA), L-lactate dehydrogenase (LDH), catalase (CAT), ferritin (FER) and thyroglobulin (THY). Gel and electrode buffer: 0.1 M Tris–HCl, pH 9.0. The total length of the gradient is 73 mm and the voltage applied 200 V. (*b*) Time-dependent migration of shikimate oxidoreductase (EC 1.1.1.25) and its multiple molecular forms occurring in pea leaves (SOP I and II), Scytonema (SOS) and Oscillatoria (SOO). Enzyme extracts were prepared by homogenising 1 g of fresh plant material together with 3 ml of a 0.1 M phosphate buffer, pH 7.6, centrifuging, gel chromatographic desalting against 0.1 M Tris–HCl buffer of pH 9.0, and ultrafiltrating the resulting extracts until activity of 5 microkatals per litre was finally obtained. 10 nanokatals of enzyme activity were applied. Enzyme bands were visualised histochemically (2.4 g NAD, 5.6 g shikimic acid, and 10 mg MTT in 0.1 M Tris–HCl buffer of pH 9.5). The electrophoretic conditions are the same as those in figure 2.15(*a*).

velocity becomes zero. This is obvious from figures 2.15 (*a*) and (*b*) where the migration distances of several different (marker) proteins are plotted against the time of electrophoresis under the influence of a constant electrical field. Since the proteins are large enough to reach a maximum migration distance, their time-dependent migration must obey some sort of a saturation function.

It is evident from figures 2.15 (*a*) and (*b*) that the migration distance of any protein is inversely correlated with time. On the other hand, time-dependent migration does not follow a plain hyperbolic function since the migration distances (D (mm)) are not linearly correlated with the reciprocal of the time of electrophoresis (t(h)). Additionally, both quantities are not interrelated by a simple exponential function. We therefore retained the correlation between D and $(1/t)$ and tried to transform the D and t data

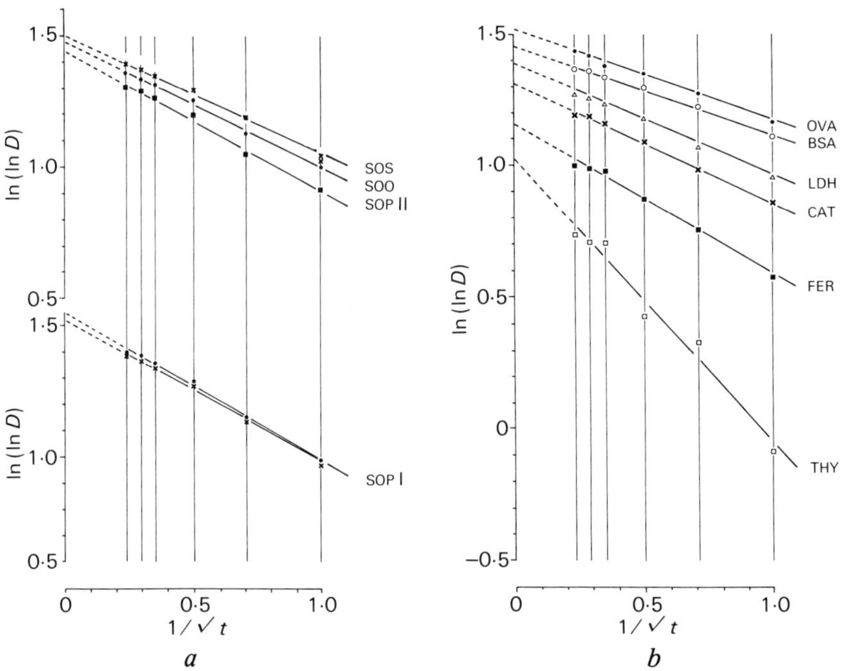

Figure 2.16 (*a*) [ln(ln D)] against $(1/\sqrt{t})$ plot to calculate the maximum migration distances of marker proteins. For electrophoretic conditions and abbreviations see figure 2.15 (*a*). The data for the slopes and the intercepts (and their respective standard errors) are compiled in Table 2.10. (*b*) [ln(ln D)] against $(1/\sqrt{t})$ plot to calculate the maximum migration distance of shikimate oxidoreductase (EC 1.1.1.25) and its multiple molecular forms occurring in pea leaves (SOP I and II), Scytonema (SOS) and Oscillatoria (SOO). Additionally the chromoprotein phycoerythrin (PHY) was investigated. Electrophoretic conditions are the same as those given in figure 2.15(*a*). The data for the slopes and the intercepts (and their respective standard errors) are compiled in Table 2.10.

Table 2.9

Protein		Time t(h) (($1/\sqrt{t}$) given in brackets) 1 (1)	2 (0.70711)	4 (0.5000)	8 (0.35355)	12 (0.28868)	18 (0.23570)
Ovalbumin	D (mm)	24.5	36.0	47.5	57.5	62.5	65.0
	ln(ln D)	1.16274	1.27635	1.35086	1.39916	1.41953	1.42897
Bovine serum	D (mm)	20.5	29.3	38.0	45.0	48.5	50.2
albumin	ln(ln D)	1.10540	1.21716	1.29132	1.33675	1.35624	1.36507
L-lactate	D (mm)	13.8	18.3	25.5	30.7	33.3	34.5
hydrogenase	ln(ln D)	0.96495	1.06709	1.17517	1.23089	1.25435	1.26440
Catalase	D (mm)	10.5	14.5	19.5	24.3	26.2	27.0
	ln(ln D)	0.85500	0.98363	1.08870	1.16017	1.18349	1.19266
Ferritin	D (mm)	6.0	8.3	10.8	14.3	14.5	15.0
	ln(ln D)	0.58320	0.74965	0.86691	0.97842	0.98363	0.99623
Thyroglobulin	D (mm)	2.5	4.0	5.0	7.7	7.7	8.0
	ln(ln D)	−0.09742	0.33663	0.47588	0.71355	0.71355	0.73210
SOR	D (mm)	17.3	26.0	38.0	46.5	51.2	54.5
Scytonema	ln(ln D)	1.04757	1.18114	1.29132	1.34533	1.37010	1.38584
SOR	D (mm)	15.3	22.0	34.0	41.0	45.5	49.0
Oscillatoria I	ln(ln D)	1.00351	1.12851	1.26027	1.31199	1.33965	1.35888
	D (mm)	12.0	17.5	28.0	34.7	37.3	41.0
II	ln(ln D)	0.91024	1.05159	1.20363	1.25606	1.28620	1.30667
SOR	D (mm)	14.5	23.5	37.7	49.0	53.0	56.5
Pea leaves I	ln(ln D)	0.98363	1.14962	1.28914	1.35888	1.37884	1.39482
	D (mm)	14.5	22.5	36.0	44.5	49.3	59.5
II	ln(ln D)	0.98363	1.13575	1.27635	1.33381	1.36044	1.38120

Time-dependent migration distances (kindly provided by Dr H Purkhanbaba) and transformed values of non-denatured marker proteins and shikimate [illegible] plant species. The proteins were run under the conditions given in figures 2.15 a, b.

Table 2.10

Protein	*Slope*	*Standard error of slope*	*Intercept* $[\ln(\ln D)]_{max}$	*Standard error of intercept*	*Linear correlation coefficient*	*Molecular mass* $((mol\ M)^{1/3})$	*Molecular mass (calculated)*	*Percentage deviation*	*Stokes' radius (nm)*	*Stokes' radius (calculated) (nm)*	*Percentage deviation*
Ovalbumin	−0.35255	0.01024	1.52087	0.00593	0.99832	43 000 (35.034)	47 652	+11	3.05	2.87	−6
Bovine serum albumin	−0.34400	0.01129	1.45557	0.00654	0.99785	67 000 (40.616)	76 446	+14	3.55	3.49	−2
L-lactate dehydrogenase	−0.40836	0.01561	1.36944	0.00905	0.99709	140 000 (51.925)	132 972	−5	4.20	4.34	+3
Catalase	−0.45742	0.01315	1.31247	0.00762	0.99835	232 000 (61.446)	185 502	−20	5.25	4.94	−6
Ferritin	−0.56285	0.02683	1.14908	0.01554	0.99549	440 000 (76.059)	435 105	−1	6.10	6.80	+12
Thyroglobulin	−1.08870	0.08116	1.03883	0.04702	0.98907	669 000 (87.460)	728 900	+9	8.50	8.21	−3
SOR Scytonema	−0.45000	0.01415	1.50159	0.00820	0.99803		55 101			3.05	
SOR I Oscillatoria	−0.47663	0.01922	1.47887	0.01113	0.99676		62 976			3.27	
SOR II Pea leaves	−0.53223	0.02504	1.44272	0.01450	0.99560		83 405			3.61	
SOR I	−0.55232	0.02448	1.54314	0.01418	0.99609		40 018			2.67	
Pea leaves II	−0.53039	0.02417	1.52361	0.01401	0.99587		46 657			2.85	

Data comprising the constants of the linear regression lines (slope, intercept, standard errors and correlation coefficient) and the molecular size (molecular mass, Stokes' radius) of the proteins indicated (see figures 2.16 a, b). The molecular sizes calculated according to our method are compared with the literature data and the percentage deviation determined. Additionally the molecular masses and the Stokes' radii of the enzyme shikimate oxidoreductase (EC 1.1.1.25) from three different plant species are presented.

until—with a moderate accuracy—a straight line was obtained in a corresponding plot. As can be seen from figure 2.16(a) and (b) this is the case if ln(ln D) is plotted against $(1/\sqrt{t})$. The graphs result from the data comprised in Table 2.9.

Provided a straight line results when [ln(ln D)] is plotted against $(1/\sqrt{t})$, then D and t are interrelated by the equation:

$$\ln(\ln D) = -a/\sqrt{t} + b \qquad (2.10)$$

where a corresponds to the slope of this line and b is the intercept. This equation predicts that when t reaches very high values (i.e. approximates infinity) then $1/\sqrt{t}$ becomes zero. This means that the maximum migration of any protein (in the mol M range from 15 000 to 1 000 000 migrating in a linear PAA gradient from 4 to 30% T) can be taken from the intercept of the straight line with the ordinate in a [ln(ln D)] against $(1/\sqrt{t})$ plot. Letting t approximate infinity means that equation (2.10) becomes

$$\ln(\ln D) = b \qquad (2.11)$$

or

$$D = \exp(e^b). \qquad (2.12)$$

The values of a and b and their respective standard errors for the straight regression lines of the marker proteins and various shikimate oxidoreductase enzymes (EC 1.1.1.25) are given in Table 2.10.

2.3.1.4.4.2 THE SEPARATION BEHAVIOUR OF MULTIPLE MOLECULAR FORMS UNDER THE CONDITIONS OF PAGGE

Hedrick and Smith (1968) demonstrated that it is possible by electrophoresis in homogeneous PAA gels of various concentrations between 5 and 12% T to distinguish between charge and size isomers of a protein. Charge isomers are proteins having equal molecular masses but differing in their net charges (e.g. L-lactate dehydrogenase (EC 1.1.1.27)) whereas size isomers do not differ in their net charge but in their molecular masses (e.g. the polymers of bovine serum albumin or ferritin (Hedrick and Smith 1968)).

In gradient gels native proteins migrate—as in homogeneous gels—according to their charge and size. Therefore PAGGE can be used to differentiate between size and charge isomers. This can be seen clearly in a [ln(ln D)] against $1/\sqrt{t}$ plot where charge isomers give straight lines of different slopes all intersecting together on the ordinate line point. The molecular mass variants of an enzyme, however, have different slopes and different points of intersection in such a graph.

Where the differences in molecular mass are small, however, the separation of isomeric proteins under prolonged electrophoresis may also result from differences in their shape caused by the phenomenon of 'electrophoretic alignment'. An example for the separation of molecular mass (or

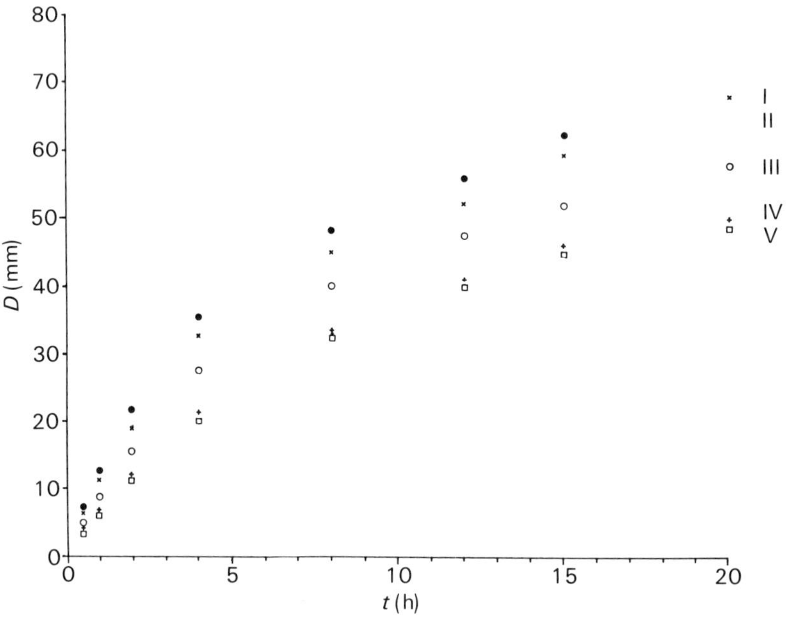

Figure 2.17 Time-dependent migration distances of carbonic anhydrase (EC 4.2.1.1) enzymes from erythrocytes of five different mammalian species. Purified enzyme preparations comprised carbonic anhydrase from bovine (I–III), rabbit (III, IV), human (V) and canine (V) erythrocytes. The proteins migrated in a linear PAA 4–30% *T* gradient (acrylamide : Bis = 24 : 1) under an electrical field strength of 300 V per 73 mm gel length and at a temperature of 5°C. Gel and electrode buffer: 0.09 M Tris, 0.08 M boric acid, 1.25 mM EDTA, disodium salt, pH 8.4. After electrophoretic separation, carbonic anhydrase bands were identified by enzyme visualisation using the following staining mixture: 1 mg 4-methylumbelliferyl acetate dissolved in 0.5 ml dimethylformamide and added to 10 ml of 0.1 M phosphate buffer of pH 6.5. Stained bands were marked under UV light. The enzyme preparations were purchased from Sigma and had the following lot numbers: I and II, C 5024 and C 7500; III, C 5024, C 7500 and C 1266; IV, C 1266; and V, C 4396 and C 3513.

shape) variants has already been given above by the enzyme shikimate oxidoreductase (EC 1.1.1.25) from three different plant species (cf figure 2.16(*b*)). The time-dependent migration of the enzyme carbonic anhydrase (EC 4.2.1.1) from erythrocytes of five mammalian species is demonstrated in figure 2.17(*a*). The corresponding [ln (ln D)] against ($1/\sqrt{t}$) plots (figure 2.18) show that all lines have a point of interesection together on the ordinates. Concludingly they all have the same 'stacking point' which means that they all differ in their charge but not in their molecular mass (see also Tables 2.11 and 2.12).

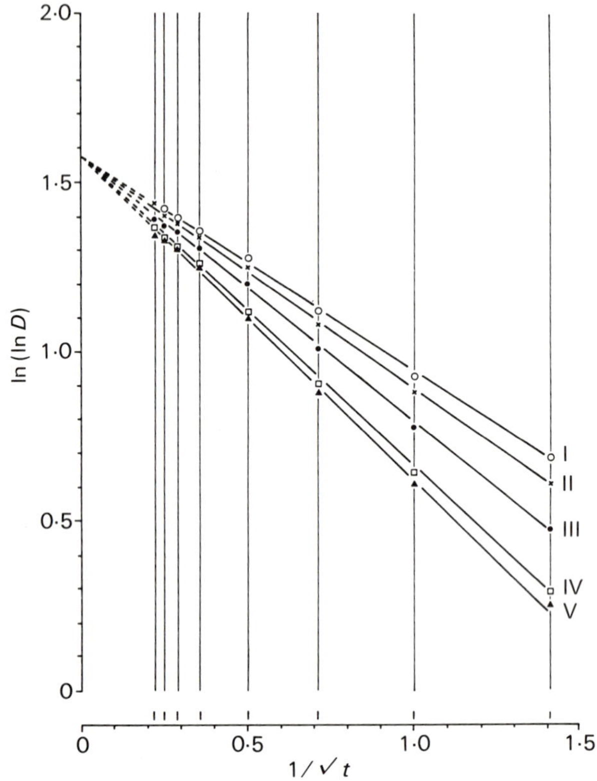

Figure 2.18 [ln(ln D)] against ($1/\sqrt{t}$) plot to calculate the maximum migration distances of carbonic anhydrase (EC 4.2.1.1) from erythrocytes of the following mammalian species: bovine (I–III); rabbit (III and IV); human (V) and canine (V). Electrophoretic conditions were the same as those in figure 2.17. The data for the slopes and the intercepts (and their respective standard errors) are compiled in Table 2.12. Notice the common point of intersection of all lines on the ordinate which is typical for protein charge isomers.

2.3.1.4.4.3 THE CONSTRUCTION OF A CALIBRATION CURVE TO DETERMINE THE MOLECULAR SIZE OF NATIVE PROTEINS

Assuming that the proteins investigated reach (or would reach) zero mobility under appropriate conditions, their maximum migration distance must be correlated with their Stokes' radii. The mean pore radius and the limiting pore radius of PAA gels are both inversely interrelated with the gel concentration. Therefore we tried to find a suitable correlation between ($1/b$) (which is $1/[\ln(\ln D)]_{max}$) and the known Stokes' radii of marker proteins. In conformity with a number of calibration lines such as those shown in figures 2.19 and 2.20 we found the best fit by using the equation

$$\{[\ln(\ln D)]_{max}\}^{-1} = aR_s + b. \qquad (2.13)$$

According to Siegel and Monty (1966) the Stokes' radius of a protein and its molecular mass ('molecular weight') are interrelated by the equation

$$R_s = f/f_0\left(\frac{3\bar{v}MM}{4\pi N_A}\right) \tag{2.14}$$

or

$$R_s = 66 \times 10^{-12}(f/f_0)\ (\text{mol } M)^{1/3}. \tag{2.15}$$

Here, f/f_0: frictional coefficient, equivalent to the ratio of the half-axes of a rotary ellipsoid, a/b = about 1.35 for globular proteins; $\bar{v}$: partial specific volume, $0.725 \times 10^{-6}\ m^3\ g^{-1}$ equivalent to the reciprocal of the specific weight; N_A: Avogadro's number, 6.022×10^{23}; mol M: molecular mass (g/mol); R_s: Stokes' radius (m).

Provided that the proteins under investigation are globular proteins, i.e. that they differ little with respect to their partial specific volume and their molecular asymmetry (f/f_0), their Stokes' radii and molecular masses are then interrelated by the simple equation

$$R_s = \alpha\ (\text{mol } M)^{1/3} \tag{2.16}$$

with α being a constant. Substituting the term α (mol M) for R_s in equation (2.14) we obtain

$$\ln\{[\ln(\ln D)]_{max}\}^{-1} = c\ (\text{mol } M)^{1/3} + d. \tag{2.17}$$

This means that the logarithm of the reciprocal of the maximum migration distance is linearly correlated—at least under the experimental conditions employed and the limitations mentioned above—with the third root of the molecular mass. The respective calibration line obtainable in a [ln $(1/b)$] against (mol $M)^{1/3}$/plot is shown in figures 2.19 and 2.20. The deviations of the calculated molecular masses (and Stokes' radii) using the described procedure from the literature data are compiled in Tables 2.11 and 2.12. The estimated molecular masses and the Stokes' radii deviate approximately ± 10%. The tables additionally give information on the molecular masses of the enzyme shikimate oxidoreductase (EC 1.1.1.25) from several higher plants as well as those for the enzyme carbonic anhydrase (EC 4.2.1.1) from four different mammalian species. The calculated mol M values of between 35 000 to 70 000 for shikimate oxidoreductase and 27 000 for carbonic anhydrase are in the ranges given for these enzymes in the literature.

2.3.2 *The Separation of SDS-denatured Proteins*

2.3.2.1 *The Interaction of Proteins with Sodium Dodecyl Sulphate*

Sodium dodecyl sulphate (SDS), an amphipathic molecule, is known to form complexes with both non-polar side-chains and charged groups of amino acid residues in polypeptides of all possible sizes and shapes without

Table 2.11

		Time t(h)((1/$\sqrt{t}$) given in brackets)							
Protein		*0.5* *(1.41421)*	*1* *(1.00000)*	*2* *(0.70711)*	*4* *(0.50000)*	*8* *(0.35355)*	*12* *(0.28868)*	*16* *(0.25000)*	*20* *(0.22361)*
Ovalbumin	D (mm)	13.05	20.25	31.5	44.0	54.0	61.0	67.5	—
	ln(ln D)	0.9434	1.10132	1.23837	1.33083	1.38354	1.41364	1.43797	—
Bovine serum	D (mm)	11.7	17.8	26.5	36.3	43.5	47.5	50.5	53.2
albumin	ln(ln D)	0.90000	1.05753	1.18698	1.27866	1.32780	1.35086	1.36659	1.37979
L-lactate	D (mm)	7.5	11.9	17.5	24.5	30.2	33.3	35.5	37.5
dehydrogenase	ln(ln D)	0.70057	0.90686	1.05159	1.16273	1.22608	1.25435	1.27243	1.28767
Catalase	D (mm)	5.5	8.8	13.2	18.8	23.5	26.6	28.5	30.0
	ln(ln D)	0.53342	0.77691	0.94787	1.07632	1.14962	1.18812	1.20893	1.22414
Ferritin	D (mm)	3.7	6.5	9.0	12.0	14.3	16.6	17.9	18.9
	ln(ln D)	0.26875	0.62690	0.78720	0.91023	0.97842	1.03297	1.05946	1.07812
Thyroglobulin	D (mm)	1.9	3.5	4.7	6.5	8.0	10.0	10.8	11.6
	ln(ln D)	−0.44339	0.22535	0.43668	0.62902	0.73210	0.83403	0.86691	0.89650

Carbonic anhydrase from erythrocytes									
Bovine I	D (mm)	7.3	12.5	21.5	35.5	48.2	56.0	62.5	—
	ln(ln D)	0.68707	0.92653	1.12104	1.27243	1.35464	1.39261	1.41953	—
Bovine II	D (mm)	6.3	11.0	19.0	32.5	45.0	52.3	58.8	68.0
	ln(ln D)	0.61006	0.87459	1.07992	1.24739	1.33675	1.37549	1.40466	1.43972
Bovine	D (mm)	5.0	8.8	15.5	27.5	40.1	47.5	52.0	58.0
rabbit III	ln(ln D)	0.47588	0.77691	1.00826	1.19821	1.30600	1.35086	1.37400	1.40129
Rabbit IV	D (mm)	3.8	6.7	11.8	21.5	33.6	41.0	45.7	50.2
	ln(ln D)	0.28893	0.64296	0.90345	1.12104	1.25690	1.31199	1.34080	1.36507
Canine	D (mm)	3.5	6.3	11.2	20.0	32.5	39.8	44.5	48.5
human V	ln(ln D)	0.2253	0.61006	0.88208	1.09719	1.24739	1.30396	1.33381	1.35624

Time-dependent migration distances (kindly provided by Dr H Purkhanbaba) and transformed values of marker proteins and carbonic anhydrase (EC 4.2.1.1) variants from erythrocytes of four different mammalian species. Gel length (D (mm)) and gel concentration (% T) are interrelated by the equation $T = \alpha D + \beta$ with $\alpha = 0.3528 \pm 0.0054$ and $\beta = 4.1116 \pm 0.2344$; the correlation coefficient being $r = 0.9985$.

Table 2.12

Protein	*Slope*	*Standard error of slope*	*Intercept* $[\ln(\ln D)]_{\max}$	*Standard error of intercept*	*Linear correlation coefficient*	*Molecular mass* $((\text{mol } M)^{1/3})$	*Molecular mass (calculated)*	*Percentage deviation*	*Stokes' radius (nm)*	*Stokes' radius calculated (nm)*	*Percentage deviation*
Ovalbumin	−0.42521	0.00682	1.53823	0.00517	0.99936	43 000 (35.034)	41 628	−3	3.05	2.70	−11
Bovine serum albumin	−0.40531	0.00519	1.47102	0.00370	0.99951	67 000 (40.616)	78 933	+18	3.55	3.52	0
L-lactate dehydrogenase	−0.49264	0.00404	1.39950	0.00288	0.99980	140 000 (51.925)	141 330	+1	4.20	4.44	+6
Catalase	−0.58024	0.00405	1.35675	0.00289	0.99985	232 000 (61.446)	193 221	−17	5.25	5.01	−5
Ferritin	−0.65059	0.02540	1.22800	0.01812	0.99546	440 000 (76.059)	442 461	0	6.10	6.86	+13
Thyroglobulin	−1.08870	0.08116	1.03883	0.04702	0.98907	669.000 (87.460)	728 900	+9	8.50	8.13	−3

Carbonic anydrase from erythrocytes							
Bovine I	−0.63709	0.00970	1.57848	0.00735	0.99942	26 712	2.22
Bovine II	−0.69533	0.01007	1.58281	0.00718	0.99937	25 380	2.17
Bovine rabbit III	−0.78570	0.01005	1.57668	0.00717	0.99951	27 280	2.24
Rabbit IV	−0.91609	0.01094	1.57135	0.00780	0.99957	29 013	2.25
Human canine V	−0.96001	0.00799	1.57547	0.00570	0.99979	27 680	2.24
Arithmetic mean			1.5770	0.0070	0.9995	27 213	2.24
+/−			+/−	+/−	+/−	+/−	+/−
Standard deviation			0.0042	0.0008	0.0002	1 330	0.05

Data comprising the constants of the linear regression lines (slope, intercept, standard errors, and correlation coefficient) and the molecular size (molecular mass, Stokes' radius) of the proteins indicated (see Figures 2.17 and 2.18). The molecular sizes calculated according to our method are compared with the literature data and the percentage deviation determined. Additionally the molecular masses and the Stokes' radii of the enzyme carbonic anhydrase from erythrocytes of four different mammalian species are presented.

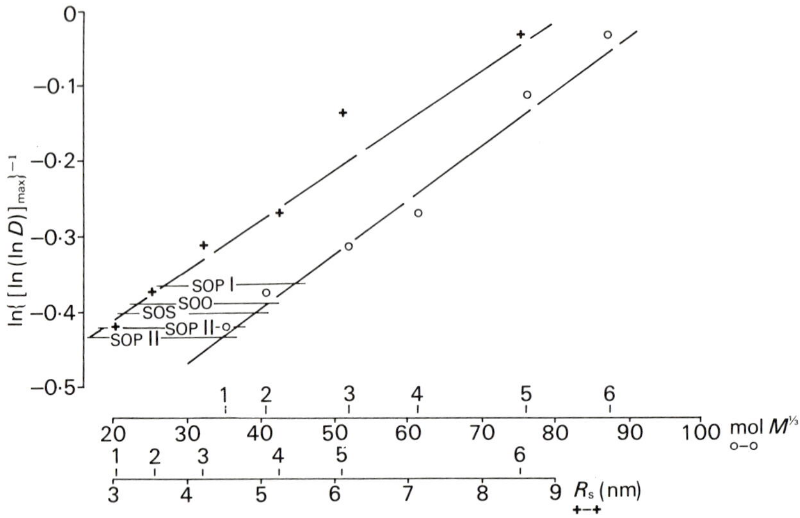

Figure 2.19 Calibration lines to estimate the molecular mass (Stokes' radius) of the enzyme shikimate oxidoreductase (EC 1.1.1.25) from pea leaves (SOP, iso-enzyme group I and II), Oscillatoria (SOO) and Scytonema (SOS) as well as the molecular mass of the chromoprotein phycoerythrin (PHY). The calibration lines were constructed by plotting the logarithm of the reciprocals of the intersecting points (found in [ln(ln D)] against $1/\sqrt{t}$ plots) against the Stokes' radii of the marker proteins or the third root of their respective molecular masses. The molecular masses and Stokes' radii were also calculated by using the regression lines presented. Deviations from the literature data are summarised in Table 2.10.

rupturing covalent bonds. In electrophoresis SDS is used:

(i) to disaggregate (enzyme) proteins into their monomeric constituents,
(ii) to estimate the mol mass of unfolded and reduced polypeptides, and
(iii) to solubilise membrane-bound proteins.

Pitt-Rivers and Impiombato (1968) found, using equilibrium dialysis, that most non-denatured proteins bind 0.9 to 1.0 g SDS per gram of protein, the protein core of some glycoproteins included. Proteins without disulphide bonds and those with reduced disulphide bridges, however, bind approximately 1.4 g SDS per gram of protein in a rather nonspecific way, solely because of the ability of SDS to unfold proteins; this is restricted only by the presence of disulphide bridges. According to Reynolds and Tanford (1970a) a concentration of SDS above 0.5 mM is sufficient for binding 1.4 g SDS per gram of protein in a primarily hydrophobic way. Binding of SDS in a micellar form was excluded. Under the influence of SDS, proteins assume the shape of rod-like particles, the length of which varies uniquely with the molecular weight of the protein moiety, occupying 0.074 nm (1.4 g SDS bound) per amino residue (Reynolds and Tanford 1970b). Certain proteins,

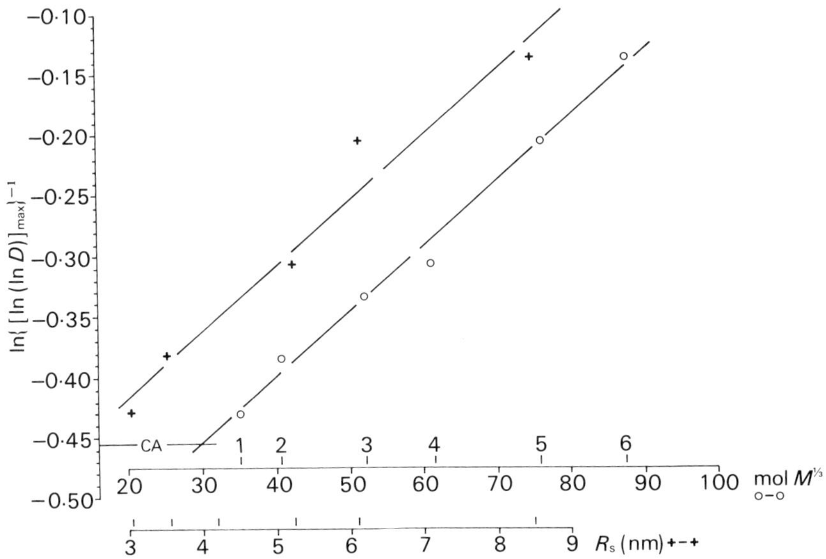

Figure 2.20 Calibration lines to calculate the molecular mass (Stokes' radius) of carbonic anhydrase (EC 4.2.1.1) from mammalian erythrocytes (CA). The calibration curves were constructed by plotting the logarithm of the reciprocals of the intersecting points (found in $[\ln(\ln D)]$ against $1/\sqrt{t}$ plots) against the Stokes' radii of the marker proteins or the third root of their molecular masses. The molecular masses and Stokes' radii of the marker proteins were also calculated by using the regression lines presented. Deviations from the literature data are summarised in Table 2.12.

such as papain, pepsin and glucose oxidase (Nelson 1971), and two different classes of proteins, the glycoproteins (Segrest *et al* 1971) and the histones (Panyim and Chalkley 1971), show an 'anomalous' behaviour towards SDS. Either they bind a relatively low amount of detergent, or SDS cannot compensate for their intrinsic charges, as is the case with the histones. But these examples are not sufficient to cast severe doubt on the validity of mol mass estimations by SDS electrophoresis (Shapiro *et al* 1967, Waehneldt 1975, Lambin 1978). It must be pointed out, however, that protein patterns in SDS electrophoresis must be tested for reproducibility either by performing separations on homogeneous gels of different pore size or by using gradient gels of different slopes (Maizel 1971).

2.3.2.2 Size estimations of denatured proteins

2.3.2.2.1 HOMOGENEOUS GEL SYSTEMS

Shapiro *et al* (1967) were the first to show that a size estimation of SDS-denatured 2-mercaptoethanol-reduced proteins is possible by PAA-gel

Table 2.13 Continuous buffer systems used in SDS electrophoresis to separate proteins and small polypeptides

Acrylamide (g/100 ml)	*Bis (g/100 ml)*	*Gel buffer*	*pH*	*Electrode buffer*	*pH*	*Mol mass range (plot) ($\times 10^3$)*	*Remarks*	*Authors*
5	0.13	0.1 M phosphate, 0.1% SDS	7.1	0.1 M phosphate, 0.1% SDS	7.1	15–200 (log mol *M* against R_f)	(i)	Shapiro *et al* 1967
10	0.27	0.1 M phosphate, 0.2% SDS	7.0	0.05 M phosphate, 0.1% SDS	7.0	15–200 (log mol *M* against *m*)	(ii)	Weber and Osborn 1969
15	0.52	0.1 M phosphate, 0.1% SDS	7.2	0.1 M phosphate, 0.1% SDS	7.2	10–60	(iii)	Dunker and Rueckert 1969
10	0.34					10–100		
5	0.17					20–350 (log mol *M* against R_f)		
12.5	1.25	0.1 M H_3PO_4, 0.1% SDS, 8 M urea, adjusted with Tris to a pH of	6.8	0.1 M H_3PO_4, 0.1% SDS adjusted with Tris to a pH of	6.8	1.2–10 (log mol *M* against *D*)	(iv)	Swank and Munkres 1971

(i) log mol *M*: logarithm of mol mass, R_f: distance of migration relative to the α and β chains of haemoglobin. Lysozyme and ribonuclease exhibit anomalous behaviour. Proteins were denatured for 3 h at 37°C in 0.1 M phosphate, pH 7.1, containing 1% SDS and 1% 2-mercaptoethanol, then they were dialysed for 16 h against 0.01 M phosphate, pH 7.1 with 0.1% SDS and 0.1% 2-mercaptoethanol.
(ii) *m*: distance of protein migration (gel length after destaining)$^{-1}$ $\times$ gel length before staining (distance of dye migration)$^{-1}$, dye, marking the moving boundary: Bromphenol blue. Proteins were denatured at 37°C for 2 h in 0.01 M sodium phosphate pH 7.0, 1% SDS and 1% 2-mercaptoethanol (0.2–0.6 mg protein/ml). Dialysis as by Shapiro *et al* (1967). Electrophoresis in gel rods (internal diameter: 6 mm, length: 100 mm). A non-linear relationship between log mol *M* and *m* was observed when doubling the amount of crosslinker.
(iii) R_f: relative mobility = migration distance of a protein, divided by that of chymotrypsinogen. log mol *M*-mobility plots undergo an inflection if mol masses fall below 15 000. Polypeptides in the mol mass range of 10 000–60 000 are better resolved in 10% than in 15% PAA gels. 10% gels require significantly less time for electrophoresis and staining than 15% gels. Proteins were dissolved at a concentration of about 2 mg per ml and denatured at 45°C for 30 to 60 min in 1% SDS, 1% 2-mercaptoethanol, and 4 M urea. Before electrophoresis they were dialysed against an electrode buffer.
(iv) *D*: migration distance (mm) relative to Bromphenol blue. 12.5% gels with 10 : 1 crosslinker and 8 M urea yield a 28-fold greater separation than 10% gels with 30 : 1 cross linker and no urea.

electrophoresis. They used homogeneous PAA gels with 5% acrylamide (acrylamide : Bis : 38 : 1) and a gel and electrode buffer of 0.1 M phosphate, pH 7.1 containing 0.1% SDS. Although the proteins which they used had widely varying intrinsic charges with isoelectric points ranging from 4 to 11, they found that all proteins migrated anodically, and that the logarithm of the mol mass is inversely related to the distance of migration. Their results were confirmed by Weber and Osborn (1969) and Dunker and Rueckert (1969) who proved, however, that proteins with mol masses below 10^4 do not fit the linear correlation between log mol M and migration distance as suggested by Shapiro *et al* (1967) (cf table 2.13).

2.3.2.2.2 DISCONTINUOUS GEL AND BUFFER SYSTEMS

A discontinuous buffer and gel system based on the original procedure suggested for the separation of native proteins (Davis 1964) was introduced by Laemmli (1970) who studied SDS solubilized structural proteins of the bacteriophage T4 (cf table 2.14). In his electrophoretic system the sample proteins are sandwiched between the leading chloride ions and the trailing glycine ions. Unstacking occurs in the separation gel when the pH changes and the protein–SDS complexes are sieved according to their apparent mol masses. Comparing continuous and discontinuous buffer and gel systems to separate SDS denatured proteins, Glossmann and Lutz (1970) came to the conclusion that discontinuous systems are superior in their resolving power. The linear relationship between log mol M and relative mobility found for homogeneous gel systems is limited to a given gel concentration: for 10% gels, the range is 1.1×10^4 to 7×10^4, for 5% gels it is 2.5×10^4 to 20×10^4, and for 3.3% gels it is 20×10^4 to 1×10^6 (Weber and Osborn 1975). Additionally, the log mol M against D relationship has two disadvantages:

(i) it cannot be applied to glycoproteins (Banker and Cotman 1972, Grefrath and Reynolds 1974, Weber and Osborn 1975), and
(ii) the expression of protein mobility for a given gel concentration is a sigmoidal rather than a linear function (Frank and Rodbard 1975, Rodbard 1976).

2.3.2.2.3 SEPARATION AND SIZE ESTIMATION OF DENATURED PROTEINS AND PEPTIDES

The methods described for the determination of denatured proteins in the range of 10×10^3 to 300×10^3 cannot be used to estimate the size of peptides in the range of 2.5×10^3 to 90×10^3 (Neville 1971, Swank and Munkres 1971, Williams and Gratzer, 1971). This failure results from the fact that peptides (below a critical size) all end up in protein–detergent complexes of the same dimension and charge (Fish *et al* 1970, Jørgensen *et al* 1982). An explanation for such observations is that if a peptide is small its complex with detergent would be smaller than the micelles formed by the

Table 2.14 Discontinuous gel and buffer systems used in SDS electrophoresis to separate proteins and polypeptides

Acrylamide (g/100 ml)	*Bis (g/100 ml)*	*Gel buffer*	*pH*	*Electrode buffer*	*pH*	*Gel strength: mol mass range (plot) (×10³)*	*Remarks*	*Authors*
3	0.2	Stacking gel: 0.0267 M H_2SO_4, 0.0541 M Tris,	6.1	Upper reservoir: 0.040 M boric acid, 0.041 M Tris, 0.1% SDS	8.6	5%: 43–200	(i)	Gainer 1971
5 10 15	0.067 0.133 0.200	Separating gel: 0.0308 N HCl 0.427 M Tris	9.2	Lower reservoir: 0.05 N HCl, 0.062 M Tris	7.5	7.5%: 25–130 10%: 12.5–90 15%: 6–55 (log mol M against R_f)		
2.5	0.625	Large pore gel: 3.125% T, 20% C 0.025% SDS, 0.20 M Tris-SO_4,	7.8	Upper buffer: 0.074 M Tris, 0.1% SDS adjusted to pH 7.8 with concentrated HCl	7.8	2.5–90 (log mol M against R_f	(ii)	Anderson *et al* 1983
7.6	0.4	Small pore gel: 8% T, 5% C, 8 M urea, 0.025% SDS, 0.20 M Tris-sulphate,		Lower buffer: 0.2 M Tris, 0.04% SDS, adjusted to pH 7.8 with H_2SO_4				
4.2	0.114	Large pore gel: 0.323 M Tris-HCl 0.1% SDS 6.3 M urea		Upper buffer: 0.213 M 4-morpholino-ethanesulphonic acid, 0.269 M pyridine, 0.1% SDS		3.7–37 ($n^{0.84}$ against ln R_f)	(iii)	Kyte and Rodriguez 1983
19	1	Small pore gel: 0.387 M HCl 0.474 M pyridine 0.1% SDS, 6.65 M urea		Lower buffer: 0.323 M Tris-HCl, 0.323 M Tris			(iii)	Kyte and Rodriguez 1983

Stacking gel: 3% (acrylamide : Bis = 30 : 0.8)	Stacking gel buffer 0.125 M Tris-HCl 0.1% SDS	6.8	0.025 M Tris 0.192 M glycine 0.1% SDS	8.3	15–150 (log mol M against R_f)	(iv)	Laemmli 1970 King and Laemmli 1971
Running gel: 8, 10 or 15% (acrylamide : Bis = 30 : 0.8)	Running gel buffer 0.375 M Tris-HCl 0.1% SDS	8.8					

(i) Electrophoresis is performed in glass capillary tubes (internal diameter, 0.58 mm; outside diameter, 1.15 mm, length: 70 mm). Before use tubes were boiled for 30 min in distilled water, dried by suction and then immersed in Column Coat (Canalco) for 5 min, dried by suction and finally stored in parafilm covered beakers. Gels were polymerised with a final concentration of 0.15% TEMED and 0.075% ammonium persulphate for the upper gel and 0.15% for the lower gel. Electrophoresis was performed at 46 μA/gel tube and the time needed to run 5, 7.5, 10, and 15% gels was 40, 50, 90, and 120 min, respectively. R_f relative mobility of protein bands to cytochrome c. The linear relationship between the log of mol mass and R_f is restricted to a certain mol mass range and depends on the gel concentration. Approximately 10^{-8} g protein per band and gel are detectable with the Coomassie blue stain. Mol mass estimations in micro gels deviated about 10% from those of macro gels.
(ii) The calibration curves obtained with this system are linear over the molecular weight range from 2.5×10^3 to 90×10^3, regardless of acrylamide concentrations in the range from 5 to 18%. Therefore, the resolving power of the system is a function of buffer composition and concentration. C = Bis. Polymerisation of the large pore gel was initiated by 0.05% ammonium persulphate (final concentration) and 0.012 ml TEMED per 23.5 ml gel solution; polymerisation of the small pore gel (8% T) was initiated by 0.0048% ammonium persulphate (final concentration) and 0.005 ml TEMED per 5.0 ml gel solution. Sample buffer, 0.139 M Tris, 0.5% SDS, 20% sucrose adjusted to pH 7.8 with glacial acetic acid. Samples were prepared at concentrations of 1 to 5 mg ml^{-1} in 0.139 M Tris, 0.5% SDS, 20% sucrose adjusted to pH 7.8 with glacial acetic acid. Power supply, 10 W for 0.75 mm flat gels. Temperature ~15°C. It takes 4–5 h for the tracking dye to migrate 15 cm. R_f = mobility of a protein band relative to Pyronin Y. Omitting urea from the separation gels gave regression lines with inflection points between 10×10^3 and 20×10^3 mol mass. The mobility of calf thymus histones was found to be completely unrelated to their mol masses.
(iii) Electrophoretic mobilities of peptides relative to cytochrome c were correlated with their chain length n by the equation $n^{0.84} = a \ln R_f + b$, where a and b are constants and $\ln R_f$ is the natural logarithm of peptide mobility relative to cytochrome c. The linear relationship between $n^{0.84}$ and $\ln R_f$ was restricted to chain lengths between 25 and 250. Standard proteins were dissolved in a 0.37 M HCl, 0.45 M pyridine, 8 M urea, 0.1% SDS buffer (peptide : SDS = 1 : 5 (w/w)). Slab gels of 14 cm × 8 cm (length) and 1 (1.5 or 2.0) mm thickness were used. The stacking gel was 1.5 cm high. The buffer volumes were 2 × 400 ml. At 80 V cytochrome c migrated ~3 cm in ~20 h.
(iv) The separating gels were polymerised by the addition of 0.025 ml TEMED and 0.3 ml of of 10% ammonium persulphate solution per 100 ml of gel solution. 10 cm gels were prepared in glass tubes of a total length of 15 cm (internal diameter: 6 mm). The length of the stacking gel was 1 cm. It was polymerised in the same way as the separation gel. The samples (0.1 to 0.2 ml) contained in final concentrations: 0.0625 M Tris-HCl (pH 6.8), 2% SDS, 10% glycerol, 5% 2-mercaptoethanol and 0.001% Bromphenol blue. The proteins were completely dissolved by immersing the sample for 1.5 min in boiling water. Electrophoresis was carried out at 3 mA per gel until the dye front had reached the bottom of the gel. The marker proteins used were: bovine serum albumin (68 000), γ-globulin, heavy chain (50 000), ovalbumin (43 000), γ-globulin, light chain (23 500) and TMV (17 000).

detergent itself, thus it simply enters a micelle. As a result the particles containing these peptides are all necessarily of the same size. In practice two different solutions have been tried to overcome this problem:

(i) addition of a solute that decreases the size of the detergent micelles (Anderson *et al* 1983, Kyte and Rodriguez 1983, Swank and Munkres 1971), and
(ii) combination of gradient gel electrophoresis and a solute that decreases the size of the detergent micelles (Hashimoto *et al* 1983).

Swank and Munkres (1971) first demonstrated that 8 M urea added to a 0.1 M phosphate buffer which contains 0.1% SDS makes polypeptides migrate according to their respective mol masses when run in a homogeneous PAA gel of 12.5% (cf table 2.13). They found a linear correlation between log mol mass and the migration distance of polypeptides

Table 2.15 Mol masses of polypeptides and peptides employed for urea–SDS gels according to Swank and Munkres (1971)

	mol mass	
Protein	*Literature value*†	*Computed*‡
Ovalbumin	46 000	
Carboxypeptidase A	34 500	§
Myoglobin	17 200	
Myoglobin I + II	14 900	
Cytochrome *c*	12 300	§
Myoglobin I	8 270	
Cytochrome *c* I	7 760	
Myoglobin II	6 420	
Bovine trypsin inhibitor	6 160	§
Adrenocorticotrophic hormone	4 550	6 500
Insulin	5 700	
Insulin B chain	3 400	
Insulin A chain	2 300	
Glucagon	3 460	1 800
Cytochrome *c* II	2 780	
Myoglobin III	2 550	
Cytochrome *c* III	1 810	
Bacitracin	1 400	
Polymyxin B	1 225	2 200

†As cited by Swank and Munkres.
‡As calculated by Swank and Munkres using least square regression analysis and assuming a linear corrrelation between log mol *M* and migration distance.
§The mol masses of these proteins also deviate considerably, if a straight line in log mol *M–D* plot is drawn through the points of carboxypeptidase A and bacitracin.

having 20 to 150 amino acid residues. For 11 peptides they calculated the standard deviation of the least squares slope to be ± 18%. Bovine trypsin inhibitor, adrenocorticotropic hormone, glucagon, and polymyxin showed larger deviations, the error in mol mass extending to 70% in the case of polymyxin (cf table 2.15). The use of uniform separation gels without a stacking gel on top and the use of a continuous buffer system however, results in the broadening of protein bands migrating in the low mol mass region (Hashimoto *et al* 1983). Moreover, there are difficulties in separating all polypeptides in such an electrophoretic system (Kadenbach and Merle 1981) because some polypeptides have very similar mol masses. These problems were circumvented by—

(i) the use of carefully adapted discontinuous gel and buffer systems containing SDS and urea (Anderson *et al* 1983, Kadenbach *et al* 1983, Kyte and Rodriguez 1983) and
(ii) SDS gradient gel electrophoresis using discontinuous gel and buffer systems containing 7 M urea in the separation gel (Hashimoto *et al* 1983).

These improved systems are summarised in table 2.11 and were used to study the polypeptide composition of cytochrome oxidase (Kadenbach *et al* 1983), the cleavage products of a cyanogen bromide digest of apohaemoglobin or $(Na^+ + K^+)$-ATPase (Kyte and Rodriguez 1983), and the detection of a trace amount of low mol mass polypeptides in body fluids such as uraemic plasma or urine samples (Hashimoto *et al* 1983).

2.3.2.2.4 GRADIENT GEL SYSTEMS TO SEPARATE DENATURED PROTEINS AND SMALL POLYPEPTIDES

PAA gradient gels, compared with homogeneous gel systems, offer greater possibilities for electrophoretic separations of proteins in the absence (Margolis and Kenrick 1968) or the presence of SDS (Lambin *et al* 1976, Lambin 1978). Protein bands are much sharper in gradient gels than in gels of one fixed PAA concentration. Moreover, such gels have a much higher resolving capacity, for example the α and β chains of haemoglobin (mol *M*: 15 126 and 15 866, respectively) can be clearly separated in a 3 to 30% PAA gradient gel, though they differ in their mol masses by less than 10^3. An 8% continuous PAA–SDS gel does not exhibit this resolving capacity (Esposito and Obijeski 1976). Lambin and co-workers (1976) demonstrated that when electrophoresis is performed in a PAA gradient gel of 3 to 30%, a linear relationship can be set up between the log of the mol mass and the log of the PAA concentration reached by a protein. Lambin (1978) confirmed the validity of the relationship log mol $M = a \log T + b$, where T is the polyacrylamide concentration reached by a protein, a is the slope and b is the intercept of the regression line, by using 34 proteins with mol masses between 13×10^3 and 950×10^3 (cf table 2.16). He demonstrated that by gradient gel electrophoresis under the influence of SDS the mol mass of

Table 2.16 SDS–PAGGE as described by Lambin (1978)†. (Taken from Rothe, 1982, with permission from the author and the publisher)

			3–30 %T† C = 8.4 %				3–30 %T‡ C = 3.8 %			
No	*Protein*	*Literature value mol M*	*D (mm)*	*± %*	*%T*	*± %*	*D (mm)*	*± %*	*%T*	*± %*
1	Prealbumin	13 745	51.5	−0.4	20.7	−4.0	56	−1.7	22.4	−6.8
2	Lysozyme	14 314	53.5	−16.5	21.4	−20.4	51.5	+16.9	20.8	+13.6
3	Ribonuclease B	14 700	52	−10.0	20.9	−14.0	55.5	−6.0	22.2	−10.3
4	Haemoglobin	15 500	51	−8.6	20.5	−11.2	55	−8.7	22	−12.3
5	Avidin	16 000	49.5	−1.7	20	−4.2	51	+7.1	20.6	+4.8
6	Soybean trypsin inhibitor	20 095	47	−6.6	19.2	−9.2	50	+10.4	20.3	−12.7
7	Papain	23 426	44.5	−4.0	18.3	−4.8	48	+15.1	19.6	−16.5
8	α-chain of IgG	23 500	—	—	—	—	—	—	—	—
9	Chymotrypsinogen A	25 666	43.5	−5.6	18	−7.0	44	+4.8	18.2	−4.8
10	Carbonic anhydrase B	28 739	41	+1.8	17.1	+2.2	42	+5.5	17.5	−4.7
11	Carboxypeptidase A	34 409	40	−8.1	16.7	−5.3	39.5	−9.5	16.7	−9.0
12	Pepsin	34 700	37.5	+11.0	15.9	+12.5	37.5	+0.4	16	+1.8
13	Glycerol 3 phosphate dehydrogenase	35 700	37.5	+7.9	15.9	+9.4	36	+6.4	15.5	+7.9
14	Lactate dehydrogenase	36 180	37.5	+6.4	15.9	+7.9	37	−1.0	15.8	+1.0
15	Aldolase	38 994	36	+11.5	15.4	+13.1	35	+3.2	15.1	+6.0
16	Alcohol dehydrogenase	39 805	35.5	+13.8	15.2	+16.4	34.5	+4.2	14.9	+7.7
17	α_1 Acid glycoprotein	40 000	35	+18.0	15	+21.7	32	+20.6	14.1	+23.9
18	Ovalbumin	43 000	35.5	+5.4	15.2	+7.8	35.5	−9.1	15.3	−7.2
19	Fibrinogen γ chain	47 000	—	—	—	—	—	—	—	—

20	Glutamate oxalacetate transaminase	50 000	32.5	+16.7	14.2	+19.2	29	+16.6	13	+21.9
21	Heavy chain IgG	50 000	—	—	—	—	—	—	—	—
22	Fibrinogen β chain	56 000	—	—	—	—	—	—	—	—
23	Catalase	57 500	32	+5.9	14	+9.1	29	+1.4	13	+6.0
24	Fibrinogen α chain	63 500	—	—	—	—	—	—	—	—
25	Albumin, monomer	66 290	31.5	+10.7	13.8	+14.9	28	+8.2	12.7	−12.3
26	Heavy chain IgM	72 000	—	—	—	—	—	—	—	—
27	Transferrin	76 000	30.5	−8.6	13.5	−6.0	24	+7.6	11.3	+12.4
28	Plasminogen	81 000	29	−1.8	13	+0.7	23	+8.5	11	+12.2
29	Phosphorylase b	96 800	25.5	+14.2	11.8	+17.1	21	+5.3	10.3	+8.9
30	Ceruloplasmin	124 000	23.5	+8.8	11.1	+11.6	17	+13.2	8.9	+16.2
31	Albumin, dimer	132 580	24	−3.3	11.2	+1.4	17.5	+1.6	9	+6.2
32	Immunoglobulin G	150 000	21.5	+10.6	10.4	+13.4	15	+11.4	8.2	+13.4
33	Immunoglobulin A	160 000	22	+1.6	10.6	+0.1	14	+14.5	7.8	+17.2
34	Reduced α_2-macroglobulin	190 000	—	—	—	—	—	—	—	—
35	Albumin, trimer	198 870	19.75	+0.8	9.8	+2.6	12	+11.8	7.1	+12.6
36	Immunoglobulin A	320 000	16	−3.1	8.5	−3.5	9	−4.0	6.1	−8.4
37	Thyroglobulin	330 000	16.5	−11.6	8.7	−12.4	11	−25.4	6.8	−26.6
38	Fibrinogen	340 000	15	+3.3	8.2	+0.4	8.5	−4.2	5.9	−8.8
39	α_2-Macroglobulin	380 000	15.5	−13.2	8.3	−13.1	8	−9.0	5.8	−16.0
40	Immunoglobulin A, trimer	480 000	13	−4.8	7.5	−9.5	6.5	−12.4	5.2	−20.6
41	Immunoglobulin M	950 000	8.75	−9.1	6	−20.1	—	—	—	—
	Average % deviation		±11.2	±7.8		±9.6		±8.7		
	Lambin (13)					±5.9				±7.4

† % Deviation of calculated mol *M* from those given in the literature. Migration distances and $\%T$ values were taken from a compilation published by Lambin (1978). Literature mol *M*s as quoted by Lambin (1978). The $\sqrt{\%T}$ scale was constructed with aid of the equation $\%T = 0.3425D + 3$.

‡ Without 2-mercaptoethanol.

both unreduced and 2-mercaptoethanol reduced proteins as well as the mol mass of glycoproteins can be determined with the same accuracy ($\pm$ 5.5%). The distances the proteins migrated and the PAA concentrations where the proteins were located are given in table 2.16. The values of the mol masses given in the literature, the values Lambin calculated by using a log mol *M*–log *T* plot, and the deviations between these two sets of values are also reported in table 2.13. According to his calculations, best results are obtained with a crosslinker concentration of 8.4% (relative to the total concentration of acrylamide plus Bis per 100 ml of solution). Papain and pepsin which are known to bind only traces of SDS (Nelson 1971, Pitt-Rivers and Impiombato 1968) were also used by Lambin and according to his observations the deviations between the expected and the observed mol *M* were small (4% and 11%, cf table 2.16). Ribonuclease and lysozyme, which bind normal amounts of SDS, migrated anomalously in SDS gels of a constant PAA concentration (Shapiro *et al* 1967); this was not the case under the conditions of SDS–PAGGE (Lambin 1978). Poduslo and Rodbard (1980) confirmed Lambin's findings that log mol *M* is significantly better correlated with log *T* than with log R_f. Furthermore, they demonstrated by computer simulations and by curve fitting that each of the two relationships is only an approximation to an underlying non-linear relationship between distance migrated and mol mass. Especially at shorter times of electrophoresis ($t < 1$ h at 150 V constant) and with proteins having mol masses below 10×10^3 or above 300×10^3 they found the curves deviated significantly from linearity.

The migration of proteins in linear PAA gradient gels and their respective mol mass can also be correlated by the equation: log mol $M = -a\sqrt{D} + b$, with *D* being the migration distance (mm), *a* the slope and *b* the intercept of the straight line (Rothe 1982). This relationship can be applied to SDS-complexed and reduced and to SDS-complexed non-reduced proteins, to glycoproteins and to carbohydrate-free proteins (Rothe 1982). Moreover, when using linear PAA gradients the following parameters do not affect the linearity of this relationship:

(i) the buffer system,
(ii) the concentration of the crosslinker within 1–8% *C*, and
(iii) the concentration range of the gradient within 3–30% at the commonly used gel length of 8–15 cm.

The constants *a* and *b*, on the other hand, are changed when the parameters (i) to (iii) are altered. These conclusions were drawn when the electrophoretic data of Lambin (1978), Lasky (1978) and Poduslo and Rodbard (1980) were re-evaluated by using log mol *M* against $\sqrt{D}$ graphs (Rothe 1982). In table 2.17 the mol masses calculated for marker proteins which migrated under the conditions of SDS gradient gel electrophoresis are compared to the corresponding literature values. It is evident that the linear rela-

Table 2.17 SDS–PAA-gel electrophoresis using high and low mol mass marker proteins (taken from Rothe 1982 with permission from the author and the publisher)

		Running time (h)													
	Literature value	*0.5*		*1*		*2*		*4*		*8*					
Protein	*mol M*	*mol M*	*± %*	*mol M*	*± %*	*mol M*	*± %*	*mol M*	*± %*	*mol M*	*± %*	*a.m.*	*σ*	*CV*	s_x
α-Lactalbumin	14 400	15 266	+6	13 882	−5	14 612	+2	16 322	+13	—	—	15 020	1 036	6.7	518
Soybean trypsin inhibitor	20 100	20 044	0	19 455	−3	19 778	−2	20 171	0	20 644	+3	20 018	445	2.2	199
Carbonic anhydrase	30 000	28 730	−4	28 281	−6	28 294	−6	29 386	−2	28 744	−4	28 687	451	1.6	202
Ovalbumin	43 000	46 717	+9	45 359	−5	45 102	+5	46 175	+7	47 692	+11	46 209	1 050	2.3	469
Bovine serum albumin	67 000	64 811	−3	64 447	−4	63 196	−6	64 100	−4	61 608	−8	63 632	281	2.0	573
Phosphorylase b	94 000	95 564	+2	91 598	−3	93 683	0	93 670	0	93 067	−1	93 516	1 425	1.5	637
Ferritin	18 500	21 225	+15	22 481	+22	22 380	+20	22 303	+20	—	—	22 097	586	2.7	293
Lactate dehydrogenase	36 000	32 728	−9	36 238	0	34 387	−4	32 887	−8	36 996	+3	34 647	1 930	5.6	863
Catalase	60 000	54 704	−9	52 406	−13	54 921	−8	48 636	−19	57 657	−4	53 665	3 372	6.3	1 508
Bovine serum albumin	67 000	59 446	−11	68 071	+2	65 538	−2	60 509	−10	65 943	−2	63 901	3 728	5.8	1 667
Fertitin	220 000	226 718	+3	231 687	+5	197 374	−10	190 050	−14	226 777	+3	214 521	19 278	9.0	8 621
Thyroglobulin	330 000	349 206	−6	332 183	+1	381 125	+15	413 479	+25	—	—	368 998	35 929	9.7	13 915
Average % deviation			6 ± 4		6 ± 6		7 ± 6		10 ± 8		4 ± 3			4.6 ± 2.9	

† % Deviation of calculated mol *M* from those given in the literature. Electrophoresis was performed for 0.5–8 h, using 6 to 27% *T* gradients ($C = 4\%$), the following electrode buffer: 0.04 M Tris, 0.02 M Na acetate, 0.002 M EDTA, pH 7.4, containing 0.2% SDS, and by applying a constant voltage of 150 V (25 °C). a.m., arithmetic mean; σ, standard deviation; CV, coefficient of variation = (σ ÷ a.m.) 100%; s_x = standard error $[\sum(x-\bar{x})^2/n(n-)]^{1/2}$.

Table 2.18 Gel and buffer systems used in SDS–PAGGE to separate denatured proteins

Gradient range (acrylamide : Bis)	*Gel dimensions*	*Current (voltage)*	*Running time (h)*	*Gel buffer system*	*Electrode buffer*	*Correlation (mol mass range)*	*Remarks*	*Authors*
3–30% (w/v) (30 : 0.8)	column, 150 × 6 mm	4 mA/gel	24	0.1 M sodium phosphate, 0.1% SDS, 5–15% (v/v) glycerol, pH 7.0	0.1 M sodium phosphate, 0.1% SDS, pH 7.0	— (12×10^3–125×10^3)	(i)	Esposito and Obijeski 1976
3–30% (w/v) (9.62 : 0.38)	length, 80 mm (slab gels)	40 V	16	10.75 g Tris, 5.04 g boric acid, 0.93 g EDTA-Na_2, pH 7.2	0.01 M sodium phosphate, 1% SDS, 1% 2-mercaptoethanol, pH 7.2	log mol *M* against log *T* (13×10^3–950×10^3)	(ii)	Lambin *et al* 1976 Lambin 1978
1.5–40% (w/v) (12.57 : 1)	micro column internal diameter, 0.43 mm; length, 15 mm	60 V	2	0.1 M sodium phosphate, pH 7.2, 0.1% SDS or 0.35 M Tris-sulphate, 0.1% SDS, pH 8.5 *or* 0.05 M Tris-glycine, 0.1% SDS, pH 8.4 *or* 0.065 M Tris-borate, 0.1% SDS, pH 9.3	29 g glycine + Tris to pH 8.4, 1 g SDS, H_2O to 1000 ml	log mol *M* against R_f (13×10^3–300×10^3)	(iii)	Rüchel *et al* 1974

1.5–40% (w/v) (12.57 : 1)	micro column internal diameter, 0.43 mm; length, 15 mm	60 V	0.33	4 g Tris and H_2SO_4 to pH 8.4, H_2O to 10 ml	see above	(13×10^3–300×10^3)		
3–30% (w/v) (28 : 1)	width 80 mm; length, 80 mm (slab gels)	150 V	0.5–8	0.04 M Tris, 0.02 M acetate Na, 0.02 M EDTA, pH 7.4, containing 0.2% SDS	same as gel buffer system	log mol M against $\sqrt{D}$ (13×10^3–950×10^3)	(iv)	Rothe 1982

(i) Before their use gels were stored at room temperature in a solution which contained 0.1 M sodium phosphate buffer, 0.01% SDS, 15% glycerol, 2 mM EDTA-Na_2 and 0.01% NaN_3. Samples were dissociated at 100°C for 3 min in 0.01 M phosphate buffer, pH 7, containing 2.5% SDS, 5% 2-mercaptoethanol, 10% glycerol and 0.005% Bromphenol blue. 20–100 μg protein were loaded onto each gel column.
(ii) T, total concentration of acrylamide and Bis per 100 ml solvent. Preparation of proteins, proteins (0.5 mg ml^{-1}) were incubated in 0.01 M phosphate buffer, containing 1% SDS, pH 7.2 for 3 min in a 100°C bath; for cleavage of disulphide bridges 1% 2-mercaptoethanol was added. The PAA concentration reached by each protein after electrophoresis was determined and log T plotted against log mol mass.
(iii) Resolution was found to be better in discontinuous than in continuous buffer systems. For discontinuous electrophoresis 0.1% SDS was added to the electrophoresis buffer only because the SDS always passed through the gels prior to the sample. Samples (1 mg protein/ml) were treated for 2 min at 100°C with 1% SDS and 1% 2-mercaptoethanol in 0.035 M Tris-sulphate, pH 8.6, 0.35 M Tris-sulphate, pH 8.6 or 0.1 M phosphate. Complete removal of SDS from proteins can be achieved with SDS-free electrode buffers. The activity of beta-galactosidase denatured with SDS and separated on an SDS-free gradient gel could be restored to 10%.
(iv) $\sqrt{D}$: square root of migration distance. Re-evaluation of the data from Lambin (1978), Lasky (1978) and Poduslo and Rodbard (1980) confirmed the validity of the log mol M–$\sqrt{D}$ relationship, found when evaluating time dependent SDS–PAA-gradient gel electrophoresis using marker proteins in the range of 10^3 to 33×10^3.

Table 2.19 Gel and buffer systems used in SDS–PAGGE to separate small polypeptides

Acrylamide (g/100 ml)	*Bis (g/100 ml)*	*Gel buffer*	*pH*	*Electrode buffer*	*pH*	*Mol mass range (plot)* $(\times 10^3)$	*Remarks*	*Authors*
10–18	0.5–0.9	Stacking gel: 5% acrylamide, 0.13% Bis, 0.067 M Tris-HCl, pH 6.8, 0.1% SDS, 0.067% ammonium persulphate and 0.067% TEMED; separation gel: 0.45 M Tris-HCl, pH 8.8, 0.1% SDS, 0.05% ammonium persulphate, 0.05% TEMED, 7 M urea, and 0–10% linear gradient sucrose	6.8	0.05 M Tris, 0.38 M glycine, 0.1% SDS	8.5	1.4–15 (log mol *M* against *D*)	(i)	Hashimoto *et al* 1983, Laemmli 1970

(i) The system can also be used to separate lipopolysaccharides and phospholipids. The addition of iodoacetamide to samples prior to electrophoresis eliminates artifacts currently observed in silver staining of protein bands. Log mol masses correlated non-linearly with migration distance (*D*) in the mol mass range of 1.4×10^3–17×10^3. A linear correlation between log mol *M* and *D* in the range of 1.410×10^3–17.2×10^3 can be used to calculate unknown mol *M*s. Bands were stained using the silver staining method of Merril *et al* (1981). PAA concentration increased linearly from 10–18% acrylamide. Flat gels of the dimension 150×140 (height) $\times$ 1 mm^3 were used. Gels were run for at least 15 h at 120 V. Samples were heated for 2 min at 100°C in a sample buffer containing 10% sucrose, 0.0625 M Tris-HCl, pH 6.8, 2% SDS, 10 mM dithiothreitol and 0.0025% Bromphenol blue (if necessary they were treated with iodoacetamide).

tionship between log mol M and $\sqrt{D}$ is practically time-independent. This means that mol mass estimations can be made when the best resolution for a given set of proteins is obtained; in other words one need not wait until the proteins have reached their so-called 'stacking point'. Moreover, caution has to be exercised in as much as protein–SDS complexes can reach a pore size under prolonged electrophoresis in which they may be partially or totally stripped of the complexing SDS molecules. This is particularly true when the upper electrode buffer contains no detergent (Waehneldt 1975). Also, when using gradient gels continuous buffer systems are preferable. With discontinuous buffer systems partial deloading of SDS–protein complexes has been observed, leading to a confusing multitude of bands (Rüchel *et al* 1974, Waehneldt 1975, cf table 2.18).

Polypeptides in the range of 1.4×10^3 to 10×10^3 cannot be resolved with SDS gradient gels ranging from 3 to 30%. Separation is possible, however, if a gradient of 10–18% PAA is used and 7 M urea is included in the separation gel system (cf table 2.19, Hashimoto *et al* 1983).

2.3.2.3 SDS Electrophoresis on a Micro and Macro Scale

SDS electrophoresis with homogeneous or heterogeneous gel and buffer systems is performed in glass tubes of 4–6 mm inner diameter and a length of 100–150 mm or in flat gels of 100×100 (length) $\times$ 1 mm^3 and requires at least 10^{-6} g of an individual protein (Weber and Osborn 1969). Microelectrophoresis in glass capillaries needs only 5×10^{-10} g protein (Condeelis 1977, Gainer 1971, Neuhoff 1973, Neville 1971, Stephens 1975, Wilson 1977) and can be performed with homogeneous or gradient gels (Rüchel *et al* 1974).

SDS–PAA-gel electrophoresis can also be scaled up to preparative dimensions. To name just a few, macro-scale preparations have been performed for example for rat brain myeline proteins (Waehneldt 1971), for proteins of sarcoplasmic reticulum membranes (Martonosi and Halpin 1971), for pigmented protein from *Rhodopseudomonas* (Fraker and Kaplan 1972), for the major proteins of the human erythrocyte membrane (Tanner and Boxer 1972), for human serum orosomucoid (Svendsen 1979), and for human serum transferrin (Svendsen 1979).

2.3.2.4 The Effect of SDS on Functional Proteins

With SDS present in bound and unbound form (Helenius and Simons 1975), functional proteins are in most cases subject to irreversible denaturation. An exception are some of the proteolytic enzymes which appear to be rather stable in the presence of this detergent. Therefore the degradation of biologically active proteins by proteolytic enzymes in complex samples has to be prevented. For several commercial preparations of yeast hexokinase (EC 2.7.1.1) which are contaminated with at least one proteolytic activity, Pringle (1970) found the three following procedures effective in minimising proteolysis prior to SDS-electrophoresis:

(i) exposing the sample to heating immediately after the addition of SDS and 2-mercaptoethanol (ME);
(ii) treating the sample with phenylmethanesulphonyl fluoride (PMSF) prior to the addition of SDS and ME;
(iii) initially dissolving the enzyme preparation in hot guanidinium hydrochloride instead of buffer only, followed by reduction with ME and alkylation by iodoacetic acid, with dialysis then into a SDS solution.

These methods seemed to inactivate the proteolytic enzyme(s) present in the preparations sufficiently rapidly before performing SDS electrophoresis, since they resulted in equivalent values for the hexokinase subunit upon SDS electrophoresis (mol mass $\sim 50 \times 10^3$); additionally they are in good accordance with the result obtained by gel filtration in 6 M guanidinium hydrochloride (Pringle 1970). Because of their easy practicability, the first two techniques have been used predominantly for the protection of many other enzymes so far. Denaturing proteins by heating with SDS present generally has to be carried out under strictly controlled conditions, for the time of exposure to heating or alterations of the pH during the boiling period might induce the cleavage of proteins as demonstrated for the enzyme β-galactosidase (EC 3.2.1.23) from different sources (Kowit and Maloney 1982). The cleaving mechanism is assumed to be of a chemical nature because of the formation of dehydroamino acid residues at serine, threonine, and cysteine.

Generally, membrane-bound enzymes are gradually inactivated by increasing amounts of SDS, e.g. p-nitrophenylphosphatase activity (EC 3.1.3.41) is almost immediately lost upon the solubilisation of mycoplasm membranes (Ne'eman *et al* 1971). On the other hand, the enzyme NADH oxidase (EC 1.6.99.2) retains about 25% of its original activity in 0.4% SDS (Bishop *et al* 1967). Glucose oxidase (EC 1.1.3.4), papain (EC 3.4.22.2) and pepsin (EC 3.4.23.1) bind only traces of SDS and retain most of their activity for a prolonged period of time (Nelson 1971). Membrane-bound ATPases (EC 3.6.1.8) are activated at low SDS concentrations, whereas an excess of the detergent causes inactivation (Marchesi and Palade 1967, Wallach 1969) (for further details about membrane proteins see §2.3.3).

2.3.2.5 Effect of Commercial SDS Preparations on the Electrophoresis and Subsequent Renaturation of Functional Proteins

As can be taken from table 2.20 commercially available SDS preparations differ significantly in their concentration of smaller and longer chain

Table 2.20 Composition of alkyl sulphates of commercial SDS preparations as determined by gas–liquid chromatographic analysis. Data in percent of total detergent according to Lacks *et al* (1979) (I) and Dohnal and Garvin (1979) (II)

			% C_n						
Company	*Product no*	*Lot no*	C_{10}	C_{12}	C_{14}	C_{16}	C_{18}	*Reference*	*Remarks*
British Drug Houses (BDH)	30 176	17 995 60	0.0	100.0	0.0	0.0	†	I	
	red label	23 712 00	0	100	0	0	0		
	green label	23 009 80	0	100	0	0	0		
Gallard–Schlessinger	44 244	11 672/46	0.4	98.7	0.9	0.0	†	I	repacking of BDH product 44 244
Bio–Rad		16 782	0.0	99.4	0.6	0.0	†	I	
Serva		E 3192	0.0	97.7	1.5	0.7	†	I	
Fisher Scientific		7 215 05	0.0	67.4	26.4	6.2	†	I	
		7 664 41	0.5	80.1	16.8	2.6	0	II	
Pierce	98 365	03026-5	0.0	65.9	27.1	7.0	†	I	
	28 365	01248.28	0	78.2	18.8	0	0	II	unidentified 30%
	28 364	11 087.12	0	77.6	20.2	0	0	II	unidentified 2.2%
Sigma		16 000 72	0	78.2	18.8	3.0	0	II	
Schwarz–Mann		9 026 96	0	72.0	24.3	3.7	0	II	
Matheson, Coleman and Bell (MCB)	DX 2490	27	0.0	68.4	31.0	0.6	†	I	
	DX 2490	12G21	0.0	63.3	34.9	1.8	†	I	

† not determined.

length alkyl sulphate impurities. The use of any of these products markedly influences both the resolution (Matheka *et al* 1977) and the banding pattern (Margulies and Tiffany 1984) of proteins in SDS-electrophoresis. Additionally the purity of the SDS preparation used strongly influences the renaturability of denatured proteins. SDS analogues bind more tenaciously to proteins, with tetradecyl sulphate (Dohnal and Garvin 1979) and hexadecyl sulphate (Lacks *et al* 1979) exhibiting increasing affinities with growing chain lengths. SDS preparations containing longer alkyl sulphates appear to inhibit enzyme activities at lower detergent concentrations as compared to pure SDS. Besides these major impurities, compounds other than SDS analogues contaminate most of the electrophoresis grade SDS preparations. These lipophiles are present in varying proportions and together at less than 0.4% (w/w) include dodecyl alcohol (DDA), 1-dodecene, didodecyl ether (DDE), didodecyl sulphate (DDS), and other unidentified long chain hydrocarbons which also bind tightly to proteins, thus seriously interfering with the renaturing process of enzymes (Blank *et al* 1982, 1983, Thelen *et al* 1982). It could be demonstrated for RNase (EC 3.1.27.5) that its detection was prevented by DDE, DDS, and other long-chain hydrocarbons at 0.1% (w/w), while DDA at the same concentration reduced sensitivity but did not prevent detection (Thelen *et al* 1983). By treating the commercial products with charcoal and recrystallising them, SDS can be at least partially freed of these inhibitory substances. With respect to recovery of enzyme activity, it is the purity of the SDS preparation used during electrophoresis which severely affects renaturation, as the detergent present in the sample is replaced by the SDS present in the electrophoretic buffer system. To improve the yield in restored enzyme activity Lacks *et al* (1979) suggested running the gels in electrophoretic buffer systems with lower concentrations of contaminating longer chain alkyl sulphates. Thus, similar recoveries were obtained for several nucleolytic enzymes if for example the SDS preparation DX 2490 containing 31% tetradecyl sulphate was used at 0.05%, while the SDS preparations from BDH were required at 0.2% (Lacks *et al* 1979).

2.3.2.6 Renaturation of Enzymes and Other Functional Proteins Following SDS Electrophoresis

2.3.2.6.1 GENERAL ASPECTS

The properties which make SDS unique among detergents in the isolation and electrophoretic separation of proteins, namely disintegrating membranes and aggregated protein structures as well as the ability to maintain the resulting fragments in a solitary state, appear to be seriously disadvantageous for their biological functions. Generally, water soluble proteins are markedly susceptible to inactivation by SDS, but hydrophobic membrane proteins are likewise denatured even if to a lesser degree. An ever increasing number, mainly of enzymes, reportedly regain—at least partially—their activity after SDS electrophoresis. This means that the use of this detergent does not necessarily bring about the irreversible loss of biological activity. Factors which affect the renaturability of SDS denatured functional proteins are essentially the same for water soluble and hydrophobic membrane proteins and are discussed together here. It is evident that renaturation will be most effective if no disulphide bonds are essential for activity, and if the native protein is not composed of subunits of different sizes (Lacks and Springhorn 1980). Manrow and Dottin (1982) observed that the renaturation process of many enzymes is favoured by high concentrations of sodium chloride, glycerol, substrates, or dithiothreitol. Sulphydryl reducing agents, however, should be kept at a minimum when staining solutions contain phenazine methosulphate (PMS) and tetrazolium salts, as these compounds are non-enzymatically reduced (Dottin *et al* 1979). Any denaturing actions of ammonium persulphate or feasible polymerisation products of the gelling procedure are avoided during electrophoresis if the gels are pre-electrophoresed (Weber and Kuter 1971). Cofactors which are essential for activity and might have been removed during electrophoresis, e.g. Zn^{2+} in alcohol dehydrogenase (EC 1.1.1.1) (Lacks and Springhorn 1980), must be included in the renaturation medium (see below). Protease inhibitors like phenylmethanesulphonyl fluoride (PMSF) eliminate proteolytic degradation of the proteins during the denaturation process (Rosenthal and Lacks 1977). Casting substrates and/or bovine serum albumin into PAA gels may increase effectively the recovery of biological activity (Lacks and Springhorn 1980). The significant influence of commercial SDS preparations on the renaturability has been discussed above (see §2.3.2.5).

2.3.2.6.2 REMOVAL OF SDS

The key step in the renaturing process is the elimination of SDS in bound and unbound form from the sample. Reactivation, especially of water soluble proteins, is best if the denaturant is removed completely. In samples containing hydrophobic proteins, SDS must be replaced by a less denaturing (non-ionic or zwitterionic) detergent to keep them solubilised. Traces of residual SDS may interfere with the chemical or immunological assay systems (Lacks *et al* 1979, Zaman and Verwilghen 1979), but by some very sensitive methods, e.g. by the spectrophotometric analysis using acridine orange, as little as 0.001% SDS can be rapidly detected in the sample (Sokoloff and Frigon 1981). As pointed out by Hager and Burgess (1980), the removal of SDS must proceed slowly so that the protein molecules have sufficient time to refold into their native conformation. In order to recover maximum activity, many functional proteins—including several nucleolytic enzymes (Lacks *et al* 1979) and the enzyme fructosyltransferase (EC

2.4.1.9) (Russell 1979)—necessitated renaturation times exceeding 12 h. On the other hand, activity staining of the membrane enzymes phosphodiesterase (EC 3.1.4.1) and alkaline phosphatase (EC 3.1.3.1) directly following SDS electrophoresis was completed within half an hour, probably due to their resistance to SDS (Dulaney and Touster 1970). In principle, renaturation of functional proteins after electrophoresis under denaturing conditions is practicable in two ways:

(i) proteins are renatured directly on the PAA gels, and
(ii) renaturation is performed after eluting the proteins from the gels.

Different techniques for the removal of SDS have been developed accordingly.

2.3.2.6.3 RENATURING *IN SITU*

If SDS denatured proteins are renatured *in situ*, the PAA gels are incubated into aqueous media in which the detergent can diffuse prior to activity staining. The renaturation media used comprise buffer solutions only (e.g. Lacks *et al* 1979, Rosenthal and Lacks 1977), or they consist of a buffer containing the non-ionic detergents Triton X-100 (Russell 1979), Nonidet P-40 (Manrow and Dottin 1982), or G3707 (Hjertén 1983). When an isopropanol concentration of 25% is included in the renaturation medium the impurities present in commercial SDS preparations are washed out of the PAA gels (Blank and Dekker 1982, Blank *et al* 1982, 1983, cf §2.3.2.5). Relatively high yields of activity could be obtained for many functional proteins, even after prolonged washing procedures. As put forward by Lacks and Springhorn (1980), this phenomenon is presumably attributable to the behaviour of a considerable number of polypeptide chains that fold around the PAA matrix or form covalent bonds with the gel during the renaturation process. These linkages might result from transamidation reactions catalysed by SDS or from the addition of non-essential cysteine residues to unreacted alkene groups of the gel matrix (Lacks and Springhorn 1980).

A special variant of renaturation *in situ* can be used if the native enzyme consumes substrates of high mol masses. By entrapping the appropriate compounds in the PAA gels prior to SDS electrophoresis, several hydrolytic enzymes, including amylases, DNases, proteases (Lacks and Spinghorn 1980) and RNases (Blank and Dekker 1980) were effectively renatured. Since proteins are separated according to their molecular size in SDS electrophoresis, it is evident that renaturation *in situ* is only practicable for monomeric and oligomeric proteins composed of identical subunits. Reassociation to the native structure of heteromeric proteins is clearly impossible, as the polypeptide chains are at different locations on the PAA gel after SDS electrophoresis. Renaturation and activity staining *in situ* is also possible if the SDS–protein complexes are subjected to isoelectric focusing in the presence of 6 M urea (Manrow and Dottin 1980).

2.3.2.6.4 RENATURATION FOLLOWING ELUTION

Renaturation of functional proteins eluted from PAA gels following SDS electrophoresis was first reported for oligomeric proteins from *E. coli* with identical subunits, including the lac repressor $[4\times(38\times10^3)]$ and the enzyme β-galactosidase (EC 3.2.1.23) $[4\times(135\times10^3)]$. To restore biological activity, Weber and Kuter (1971) first exchanged SDS for the less denaturing agent urea by using the ion exchange resin Dowex 1X2. Urea transforms the protein into a random coil structure (Tanford 1968). In the following step urea was removed by dialysis, and the proteins refolded into their native conformations. Further methods used successfully to remove SDS from the protein samples include the use of the anion exchange resin, Bio-Rad 1X10 (Lenard 1971), and of inert polystyrene beads (Fox *et al* 1978). SDS anions were extracted as ion pairs with triethylammonium or tributylammonium cations into appropriate solvents (Henderson *et al* 1979). Hager and Burgess (1980) used acetone which precipitated the sample proteins and left SDS dissolved in the supernatant. Clarke (1981) observed the direct renaturation of the cytoplasmic enzymes L-lactate dehydrogenase (EC 1.1.1.27) and malate dehydrogenase (EC 1.1.1.37) by adding an excess of Triton X-100. Proteins which modulate the activity of DNA polymerase (EC 2.7.7.7) were recovered from PAA–SDS gels by electrophoresis in the presence of 6 M urea (Hanaoka *et al* 1979). Ziola and Scraba (1976) separated proteins from SDS by electrophoresis into hydroxyapatite. By electrophoresis through a layer of Lubrol PX-agarose (Converse and Papermaster 1975) or Triton X-100-agarose (Lee *et al* 1978), several proteins were renatured. In specifically designed apparatus proteins were separated electrophoretically (Tuszynski *et al* 1977) or by electrodialysis (Tuszynski and Warren 1975). Most of the methods listed here include the use of urea and guanidinium hydrochloride, or non-ionic detergents in a transient step to the restoration of the native protein conformation. Before activity staining, these detergents are removed by dialysis.

2.3.2.6.5 CONCLUSIONS

The comparison of renaturation *in situ* with the renaturing procedures for proteins eluted from PAA–SDS gels makes the former appear to be the alternative which is easier to handle. Without underestimating their respective renaturation potency, the latter procedures are somehow laborious and tend to be susceptible to errors that result from dissecting the gels into equal parts, extracting proteins quantitatively from the gel sections, or recovering them completely after a step to remove SDS etc. These obvious disadvantages become less decisive, however, when proteins are renatured for preparative purposes, even if some of the above listed methods were designed for micro-scale application (e.g. Henderson *et al* 1979).

On the other hand, renaturation of proteins *in situ* following SDS electrophoresis appears to be an important analytical technique for complex samples far too small to be subjected to classical methods of fractionation. By considering its restricted applicability to monomeric and homomeric proteins only, particular functional proteins can be easily located on the PAA gels, their mol masses rapidly estimated, and feasible isozymic forms semiquantitatively determined.

By applying these different renaturation techniques, a wide variety of SDS denatured enzymes and other functional proteins like nucleases, proteases, dehydrogenases as well as histones and RNA binding proteins successfully have been able to reconstitute their native conformation (cf literature already cited and Babczinski *et al* 1980, Bjerrum *et al* 1975, Hanspal *et al* 1983, Huet *et al* 1978, Kelleher and Juliano 1984, Manrow and Dottin 1980, Rosenthal and Lacks 1978, Schieven *et al* 1982, Spanos *et al* 1981, Sugiyama *et al* 1981).

2.3.3 Separation of Membrane Proteins

2.3.3.1 Solubility of Membrane Proteins

According to their characteristics of solubility, functional membrane proteins may be put into two classifications:

(i) (strongly) hydrophobic lipoproteins; these are often constitutents of large aggregates of proteins and phospholipids. They require a lipophilic environment to be separated in an active state from a membrane (Tzagaloff and Penefsky 1971).
(ii) Proteins which acquire the properties of water-soluble proteins after they have been removed from the membranes (Penefsky and Tzagoloff 1971).

Such proteins are frequently anchored in membranes by protein–protein interactions through complexes with divalent cations such as Mg^{2+} and Ca^{2+}, or electrostatic forces.

Some considerations are presented here which may facilitate the selection of an appropriate technique applicable to solubilise membrane proteins in their functional state.

Buffer or salt solutions are suitable means for releasing proteins which are retained on membranes by electrostatic forces. Thus, cytochrome *c* is solubilisable by a 0.015 M (isotonic) solution of KCl (MacLennan *et al* 1966). Buffers of alkaline pH values apparently bring about alterations in the charges of membrane components. King (1963) used a 60 mM Tris-buffer of pH 8.9 to separate active succinic dehydrogenase (EC 1.3.99.1) from the inner mitochondrial membrane. By adding EDTA to the extraction medium, McCarthy and Racker (1966) isolated an active ATPase (EC 3.6.1.8) from spinach chloroplasts. Metal ions most probably retain this enzyme at its membrane locations. A variety of membrane proteins can be extracted by using solutions with extremely high concentrations of salts, e.g. 2 M NaCl for ATPase (EC 3.6.1.8) from bovine heart submitochondrial electron transport particles (MacLennan *et al* 1968). Other agents used comprise the chaotropic salts, e.g. sodium perchlorate at a concentration of 2 M, 6 M guanidinium hydrochloride (for a compilation see Hatefi and Hanstein 1974) and 8 M urea. For many functional proteins, however, these high concentrations may be denaturing, although renaturation might be possible in some instances if the denaturant can be removed (see §2.3.2.6). In the solubilisation of microsomal enzymes and enzymes from the plasma membrane of mammalian tissues, the use of lipolytic enzymes was found to be a very potent tool. Treating membranes with lipase (EC 3.1.1.3) and/or phospholipase A_2 (EC 3.1.1.4) causes the degradation of membrane lipids and brings about the release of enzymes like NADPH-cytochrome *c* reductase (EC 1.6.2.4) (Masters *et al* 1967). Sonic oscillations may also separate some proteins from membranes, as has been demonstrated for mitochondrial ATPase (EC 3.6.1.8)(Datta and Penefsky 1970).

Some strongly hydrophobic proteins can be isolated from membranes when organic solvents are used. For example, rhodopsin can be isolated by ether or hexane (Darszon *et al* 1977), and a large number of specific enzymes, including phosphatases, peptidases, dehydrogenases, esterases, and a transaminase, can be isolated from intestinal mucosa by *n*-butanol (Morton 1950). The majority of membrane proteins, however, can be solubilised with the aid of detergents. These amphiphilic agents can be split into three groups; the first includes ionic detergents, e.g. deoxycholate (anionic) or cetyltrimethylammonium bromide (cationic) and the non-ionic detergents like the Brij, Triton or Tween series. Their respective solubilising potencies on membranes and membrane protein constituents have been thoroughly reviewed by Helenius and Simons (1975) and Tanford and Reynolds (1976). A new class of non-ionic detergents, composed of bile acid moieties and polar gluconamide groups, was recently introduced by Hjelmeland *et al* (1983a). The third category comprises the zwitterionic detergents such as dodecyl-*N*-betaine or the recently synthesised series of *N*-alkyl-sulphobetaines (Zwittergents) (Gonenne and Ernst 1978), and improved modifications thereof (Hjelmeland *et al* 1983b).

It may be desirable in some instances to use as many detergent species as are available for comparative purposes, e.g. to draw conclusions about the nature of interaction between a certain protein constituent and its membrane location, or to study their influence on the biological activities of functional proteins. Thus Womack *et al* (1983) tested more than fifty detergents from all three categories in their effectiveness to solubilise some hydrolytic enzymes, including sulphatase (EC 3.1.6.1), alkaline phosphatase (EC 3.1.3.1) and an esterase preparation from wheat, without denaturing them.

For initial solubilisation tests, it appears practicable, however, to apply only a restricted number of agents. Hjelmeland and Chrambach (1984) pointed out that there are only eight chemical categories to be considered when solubilising tests are to be performed. These include the bile salts, CHAPS (3-[(cholamidopropyl)dimethylammonio]-1-propanesulphonate), deoxyBIGCHAP (*N*, *N*-bis(3-D-gluconamidopropyl)deoxycholamide), digitonin, the zwittergent series, the Triton-X (polyoxyethylene p-t-octyl phenol) and the Lubrol (polyoxyethylene alcohol) series. If specific properties of these agents are considered the selection becomes even more limited. The removal of the detergent from the sample protein was only possible when the bile salts, CHAPS, deoxyBIGCHAP and octyl glucoside were used as solubilising agents. The bile salts should be omitted for initial solubilisation, especially with respect to the electrophoretical separation of membrane proteins according to the differences between their proper charges. (Chrambach and Hjelmeland 1983). These agents also precipitate divalent metal ions which may be essential to preserve biological activity. On the other hand, these species as well as SDS solubilise membranes and protein constituents more efficiently than the non-ionic detergents do. It must be emphasised that SDS should not be completely excluded from considerations about solubilising membrane proteins because an ever increasing number of enzymes have demonstrated their respective stability in its presence (see Dulaney and Touster 1970, Hjertén, 1983). Feasible losses in activities might be reversed by the renaturation procedure discussed above (see §2.3.2.6).

2.3.3.2 A General Procedure to Solubilise Membrane Proteins using Detergents

It is suggested by Hjelmeland and Chrambach (1984) that octyl glucoside or CHAPS are used to start the solubilising trials. The following is a brief summation of their strategy on how to optimise the solubilisation of membrane proteins in an active state by using detergents. In their two-step procedure, the suitable buffer system and appropriate solubilisation agent is determined first, followed by the optimum detergent-to-protein ratio being found out (for more detailed information, see original paper).

Step 1

(i) Suspend membranes in 50 mM buffer containing 0.15 M KCl at a protein concentration of 10 g l^{-1}.

(ii) Prepare stock solutions of detergent at 10% (w/v) (digitonin at 4% (w/v)) in buffer.

(iii) Dilute detergent stock solutions to 0.01, 0.03, 0.1, 0.3, 1.0, and 3.0% (w/v).

(iv) Mix stock solutions and the dilution series with the suspended protein and buffer to a final protein concentration of 5 g l^{-1}.

(v) Stir the mixtures at 4°C for 1 h.
(vi) Centrifuge them at 105 000 × g for 1 h at 4°C.
(vii) Remove the supernatants from the respective pellets, and resuspend the pellets in an equal volume of buffer with the identical detergent concentration.
(viii) Perform assays for activity and protein concentration both on the supernatant liquids and the resulting pellets.
(ix) Plot percentage of the particular activity against detergent concentration for supernatants and pellets.
(x) Take from the plots either an optimum value of activity at some detergent concentration, or the fact that the detergent is either ineffective or even deleterious on the activity.
(xi) If all detergents failed to be useful singly, test combinations or add stabilising agents to buffer and start again.

Step 2
(i) Dilute protein solution of 10 $g\,l^{-1}$ to 7.5 $g\,l^{-1}$ and 5 $g\,l^{-1}$, then follow instructions (ii)–(viii) in *Step 1*.
(ii) Plot activity against detergent-to-protein ratio.
(iii) Take from the plot the optimum value of detergent-to-protein ratio.

2.3.3.3 Electrophoresis of Functional Membrane Proteins

Contrary to the numerous reports on electrophoresis of water soluble proteins under non-denaturing conditions, comparatively few functional membrane proteins have been subjected to this method so far. This is most probably due to the problems arising from the fact that electrophoresis of hydrophobic proteins has to be carried out in the presence of detergents to keep them solubilised. Suitable detergents have to be selected taking into account—

(i) their feasible effects on the respective biological activities,
(ii) their specific properties as amphiphiles in aqueous solutions, and
(iii) their interactions with the sample constituents.

It should be borne in mind that detergents above a concentration which is specific for each of them (critical micelle concentration, *CMC*) are aggregated in well-defined stable micelles. In the case of the non-ionic detergents these are large and consist of many monomers which aggregate at fairly low concentrations, whereas only a few anionic detergent molecules combine to form small-sized micelles at relatively high concentrations. For example, 2–4 molecules of deoxycholate form aggregates of micellar masses of 900–1800 above the *CMC* value of 13–15 mM, while the voluminous Triton X-100 micelles, with masses of 90 000, are composed of 140 monomers and are formed when the detergent is used at concentrations exceeding 0.24 mM (for more details see Helenius and Simons 1975). These micelles bring about the entrapment of membrane protein constituents and other lipids to form artefacts which differ significantly from the net charge and the original size of a native protein. It is advisable, therefore, to use a detergent with a concentration as low as possible in order to avoid the formation of artefactual aggregates, but which will still keep the hydrophobic proteins in a solubilised state on the PAA gels.

It is most convenient if the same detergent or detergent mixture can be used at the same concentration both for initially isolating the membrane protein constituents and for subsequently subjecting them to electrophoresis. On the other hand, several techniques can be applied to remove an excess of detergent in the protein sample (Furth 1980) or to exchange one solubilising agent for another (Horigome and Sugamo 1983) prior to electrophoresis.

Because of their non-denaturing potential on most functional hydrophobic proteins, it is mainly non-ionic detergents, such as Triton X-100, which are used in PAA-gel electrophoresis. Problems arise because the large non-ionic detergent aggregates penetrate poorly into the PAA gel and migrate with low electrophoretic mobilities leading in many instances to diffuse protein bands. Possibly the large pore matrix of agarose is the more suitable support medium for electrophoresing membrane proteins in the presence of Triton X-100 (cf Helenius and Simons 1977). On the other hand, alkaline phosphatase isozymes (EC 3.1.3.1) from human sources are separated electrophoretically when Triton X-100 is present in the sample at a concentration of 0.5% (w/v), and both the gels ($T = 7\%$) and the electrode buffer (0.1 M Tris-borate of pH 9.5) contain the detergent at 0.5% (w/v) (Fishman 1974). In the electrophoresis of a variety of enzymes from rat liver membranes, Dewald *et al* (1974) used a discontinuous system containing detergents at a final concentration of 0.1% (w/v). Among the detergents Triton X-100, deoxycholate, and SDS, Triton X-100 was the best at retaining enzyme activities.

In the presence of a mixture of non-ionic and ionic detergents, the resulting mixed detergent–protein complexes are charged and exhibit increased mobilities as compared to the corresponding non-ionic detergent–protein aggregates. This 'charge shift' technique, originally applied in agarose electrophoresis (Helenius and Simons 1977), appears to be practicable also in PAA-gel electrophoresis. To characterise the enzyme adenylate cyclase (EC 4.6.1.1) from rat liver membranes, Newby and Chrambach (1979) performed PAA gel electrophoresis in the presence of 0.1% (w/v) Lubrol plus 0.03% (w/v) deoxycholate. In the presence of this mixture, the molecular radius of the enzyme decreased from 4.9 nm—as determined in the presence of Lubrol only—to 4.0 nm, with greater than 90% recovery of enzymatic activity. Bordier *et al* (1978) used a mixture consisting of 0.5% (w/v) Triton X-100 plus 0.25% (w/v) deoxycholate to analyse the molecular structure of a developmentally regulated protein complex in membranes from *Dictyostelium discoideum*.

At present the use of zwitterionic detergents has only been reported for isoelectric focusing (Hjelmeland *et al* 1979) and not for PAA gel electrophoresis. These detergents have the advantage that they solubilise membrane proteins more efficiently and do not form unwanted artefacts. The recently synthesised derivatives of *N*-alkylsulphobetaines appear to be absolutely non-denaturing on all functional membrane proteins and thus might be especially suitable for the electrophoresis of hydrophobic proteins (Hjelmeland *et al* 1983b).

2.4 Affinity Electrophoresis

2.4.1 Introduction

Affinity electrophoresis combines electrophoretic techniques with the biospecific binding of proteins to certain biological molecules or their derivatives. Some modifications of affinity electrophoresis are comparable to affinity chromatography, whereas others resemble certain immunoelectrophoretic techniques (cf reviews of Bøg-Hansen and Han 1983, Hořejší *et al* 1979, Takeo 1984, Takeo *et al* 1978). The method is used mainly for analytical purposes, which are:

(i) the detection of specific ligand-binding properties of proteins,
(ii) checking the homogeneity of purified preparations,
(iii) quantitation of interacting components,
(iv) evaluation of the strength of the interaction (Bøg-Hansen and Han 1983, Hořejší *et al* 1982; Takeo 1984).

So far most experience with affinity electrophoresis has been obtained with lectins, carbohydrate binding glycoproteins of plant origin (cf table 2.21). It is remarkable that SDS does not interfere with the binding of a lectin to a protein containing carbohydrate (Bøg-Hansen and Han, 1983). Therefore lectins can also be used to identify serum and other glycoproteins following SDS electrophoresis.

Molecules which have a specific affinity towards a certain protein or a certain group of proteins are called ligands. Ligands may be macromolecules such as the lectins or they may be of low mol mass such as a competitive inhibitor of an enzyme. Macromolecular affinity ligands such as lectins, starch or glycogen are merely mixed into a gel like agarose or PAA during the polymerisation process, while low mol mass ligands are attached to a reactive spacer arm which in turn is covalently bound to agarose and then dissolved in the polymerisation solution, or which is copolymerised into a PAA gel.

Table 2.21 Sources and specificities of some lectins (cf Roth 1978)

Name	*Source*	*Sugar-specificity*	*mol M* ($\times 10^3$)	*Sub units*	*Me^{2+}*	*Blood group spec.*
Anti A-lectin	Dolichos biflorus	α-*N*-acetyl-D-galactosamine 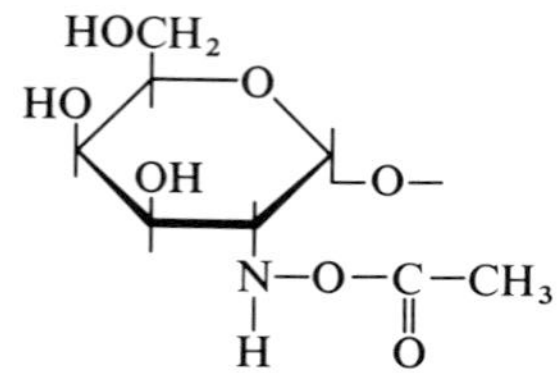	135	4		A
Anti A + B-lectin	Sophore japonica	α-*N*-acetyl-D-galactosamine and galactose	133	—		A,B
Anti H-lectin, type I	Ulex europaeus	L-fucose (= 6-deoxy-L-galactose 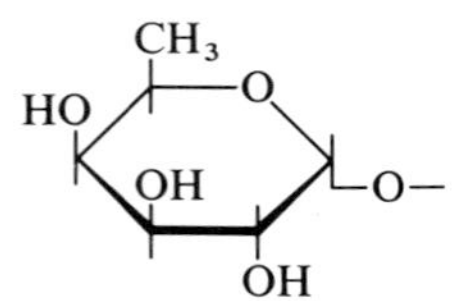	170	—		H(O)
Anti H-lectin, type II	Ulex europaeus	di-*N*-acetyl-chitobiose 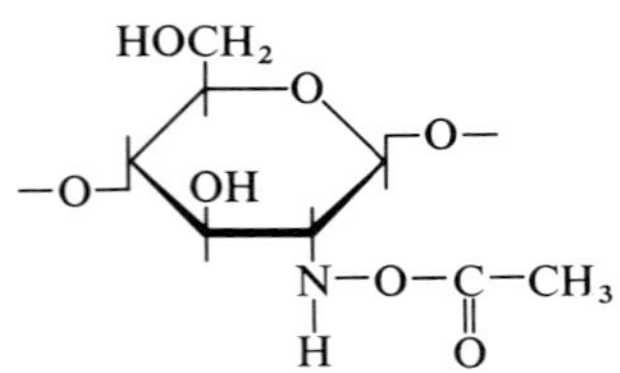	170	—		H(O)
[illegible]	[illegible]	[illegible]	60	2		—

lectin-120	communis					
Concanavalin A	Canavalia ensiformis	D-glucose 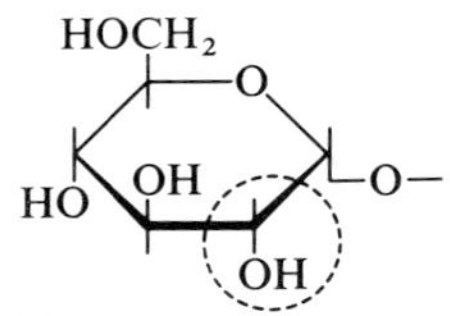 D-mannose 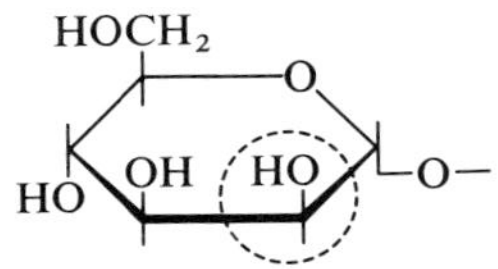	105 pH 7	4	$Ca2^{+}$ Mg^{2+}	
Lentil-lectin	Lens culinaris	D-glucose, D-mannose and sterically related sugars	60 40	2 2	Ca^{2+} Mg^{2+}	—
Limulin	Limulus polyphemus	*N*-acetyl-neuraminic acid (=derivative of mannosamine) 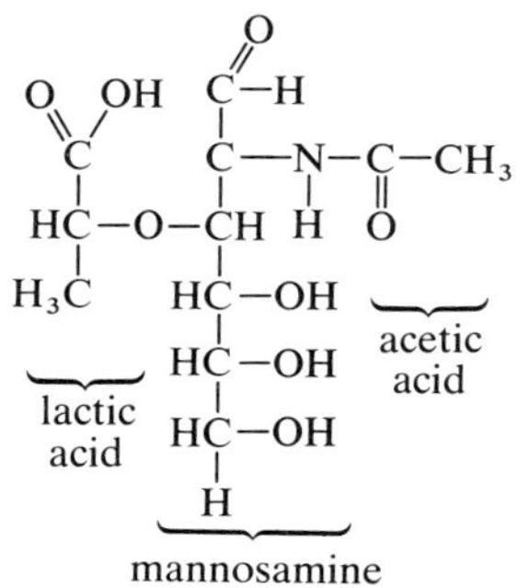	335	18	Mn^{2+}	—

(*Continued*)

Table 2.21 Continued

Name	*Source*	*Sugar-specificity*	*mol M* ($\times 10^3$)	*Sub units*	*Me^{2+}*	*Blood group spec.*
Peanut lectin	Arachis hypogea	galactose (β1,3) *N*-acetyl-D-galactosamine	106	4		—
Soybean lectin	Glycine max.	*N*-acetyl-D-galactosamine and D-galactose	120	4		—
Wheat germ lectin	Triticum vulgare	Di-*N*-acetyl-D-glucosamine	35 pH7	2		—
	Vicia faba	D-mannose and D-glucose	47	2		—
	Solanum tuberosum	β-1,4-*N*-acetyl-glucose	105	2		—
	Pisum sativum	D-mannose, I D-glucose, II	54 53	2	Mn^{2+} Ca^{2+}	—

To study a ligand-induced reaction in affinity electrophoresis, any macromolecular or particle-bound interacting substance may be included in the electrophoretic support medium. The ligand need not be charged. The main advantages of affinity electrophoresis at an analytical scale are (Bøg-Hansen and Han 1983):

(i) it is not necessary to purify the interacting components,
(ii) a multitude of proteins reacting with the same ligand may be studied simultaneously,
(iii) macromolecules which interact with a specific ligand can be separated from those that do not,
(iv) the method can be used for the prediction of preparative experiments.

2.4.2 *Preparation of Affinity Gels*

Affinity gels are prepared from 1% agarose or 5% PAA. In general these gels contain the ligand in an immobilised form. Immobilisation is either caused by coupling it to an uncharged macromolecule or by copolymerising it with the PAA matrix.

2.4.2.1 *Incorporation of Macromolecular Substrates into Gel Matrices*

Polysaccharides complexing with proteins can be incorporated into PAA gels without any modification (Borrebaeck and Etzler 1980, Stegemann 1967, Takeo and Nakamura 1972). The lectin concanavalin A (Con A) is specifically retarded by a Dextran-T-2000 containing matrix, α-amylase at a critical concentration decreases in mobility when starch or glycogen are embedded in a PAA gel, and phosphorylase from mammalian tissues or potato tubers migrates at slower rates when separated on a glycogen containing PAA gel (cf Takeo 1984).

When α-amylase or phosphorylase are separated on a PAA gel containing starch or glycogen, they start to digest their substrate. If the amount of enzyme is relatively high in comparison to the substrate concentration, then both enzymes migrate practically unhindered. This is the case for example if decreasing amounts of human salivary amylase are separated on a constant amount of starch incorporated into an agarose or PAA gel. It loses migration velocity progressively until a limiting amount of enzyme has been reached. At that point a maximum of retardation occurs and further reduction of enzyme concentration leaves the migration velocity uninfluenced. Under the latter conditions the calculation of binding constants is possible (Takeo 1984). Table 2.22 summarises the dissociation constants calculated for various α-amylases, phosphorylases and other enzymes.

An easy way to study the carbohydrate residues of glycoproteins is to mix a sugar-complexing protein (lectin) to melted agarose of low electroendosmosis, to gelatinise the agarose, apply the sample and run the gel at

Table 2.22 Dissociation constants calculated for lectins and enzymes by use of affinity electrophoresis

Protein	*Source*	*Ligand*	*Dissociation constant (mol l^{-1})*	*Authors*
Alcohol dehydrogenase	Rape, seed	Blue dextran†		Ticha *et al* 1978
isozyme 1			2.2×10^{-6}	
isozyme 2			7.9×10^{-6}	
Aldolase	Rabbit, muscle	Blue dextran†	2.2×10^{-5}	Ticha *et al* 1978
α-amylase	Salivary, human	soluble potato starch	2.6×10^{-6} (g ml^{-1})	Takeo 1984
	EDTA treated human salivary		1.9×10^{-4} (g ml^{-1})	Inoue 1978, 1980
Antitrypsin	Human	Concanavalin A		Bøg-Hansen and Takeo 1980
component 1			3.1×10^{-5}	
component 2			1.3×10^{-5}	
Carboxylesterase	Murine	Conanavalin A		Bøg-Hansen *et al* 1981
component 1			1.5×10^{-5}	
component 2			1.3×10^{-5}	
component 3			0.9×10^{-5}	
Creatine kinase	Rabbit, muscle	Blue dextran†		Ticha *et al* 1978
isozyme 1			8.6×10^{-5} to	
isozyme 2			1.8×10^{-4}	
α-foetoprotein	Human	Concanavalin A		Bøg-Hansen and Han 1983
component 1			5.0×10^{-6}	
α-foetoprotein	Murine	Concanavalin A		
component 1			4.5×10^{-6}	
component 2			3.3×10^{-6}	
Glucose-6-phosphate dehydrogenase	Rabbit, liver	8-substituted adenosine 2′-monophosphate-polyacrylamide	8.0×10^{-4}	Nakamura *et al* 1980
Haptoglobin	Human	Concanavalin A	3.2×10^{-5}	Bøg-Hansen and Takeo 1980
α_2-HS-glycoprotein	Human	Concanavalin A	4.0×10^{-5}	Bøg-Hansen and Takeo, 1980

Table 2.22 continued

Protein	*Source*	*Ligand*	*Dissociation constant (mol l^{-1})*	*Authors*
Isocitrate dehydrogenase	Rabbit, liver	8-substituted adenosine 2′-monophosphate-polyacrylamide	1.6×10^{-2}	Nakamura *et al* 1980
Lactate dehydrogenase	Rabbit, brain	8-substituted adenosine 5′-monophosphate-polyacrylamide		Nakamura *et al* 1980
isozyme 1 (H_4)			3.2×10^{-3}	
isozyme 2 (H_3M)			1.9×10^{-3}	
isozyme 3 (H_2M_2)			1.2×10^{-3}	
isozyme 4 (HM_3)			4.1×10^{-4}	
isozyme 5 (M_4)			8.6×10^{-5}	
Lactate dehydrogenase	Soybean	Blue dextran†		Ticha *et al* 1978
isozyme 1			3.0×10^{-6}	
isozyme 2			1.3×10^{-6}	
Lactate dehydrogenase	Bovine, heart	Blue dextran†		Ticha *et al* 1978
isozyme 1A			1.6×10^{-5}	
isozyme 1B			5.8×10^{-6}	
isozyme 2			1.9×10^{-6}	
isozyme 3			7.5×10^{-7}	
isozyme 4			4.0×10^{-7}	
Lactate dehydrogenase	Bovine, heart	5′-AMP covalently linked to 6-amino-hexanoic acid through the 6-amino group of adenine; copolymerised with hydroxy-propylmethacryl-ate polymer		Ticha *et al* 1980
isozyme 1			2.7×10^{-3}	
isozyme 2			2.3×10^{-3}	
isozyme 3			1.9×10^{-3}	
α_1-lipoprotein-associated aryl-esterase	Human	Concanavalin A	1.1×10^{-5}	Bøg-Hansen and Takeo 1980
Malic enzyme	Rabbit, liver	8-substituted adenosine 2′-monophosphate-polyacrylamide	4.5×10^{-4}	Nakamura *et al* 1980
	Carrot		7.3×10^{-3}	

(*Continued*)

Table 2.22 continued

Protein	*Source*	*Ligand*	*Dissociation constant (mol l^{-1})*	*Authors*
Orosomucoid	Human	Concanavalin A		Bøg-Hansen and Takeo 1980
component 2			1.5×10^{-5}	
component 3			1.1×10^{-5}	
6-Phosphogluconate dehydrogenase	Rabbit, liver	8-substituted adenosine 2′-monophosphate-polyacrylamide	8.5×10^{-5}	Nakamura *et al* 1980
Phosphorylase	Rabbit, brain	Glycogen	1.3×10^{-3}	Takeo and Nakamura 1972
	Rabbit, muscle		6.1×10^{-4}	
	Rabbit, liver		2.2×10^{-3}	
	Uterine carcinoma	Glycogen	2.1–2.7×10^{-3}	Takeo *et al* 1974
	Vesicular carcinoma	Glycogen	1.4×10^{-3}	
isozyme 1	Potato tubers	Glycogen	5×10^{-5} (g/100 ml)	Gerbrandy and Doorgeest 1972
isozyme 2			2×10^{-3}	
Pyruvate kinase	Rabbit, muscle	Blue dextran†	2.6×10^{-5}	Ticha *et al* 1978

† Blue dextran, Cibacron Blue F3GA bound to agarose.

a pH close to the pH of the lectin so that it is practically immobile (Bøg-Hansen *et al* 1977).

Since concanavalin A (con A) binds to α-(1 → 4) and α-(1 → 6) linked glucans, the addition of Con A to starch gels immobilises the protein (Swallow *et al* 1976). Provided free binding sites remain on the Con A molecule when entrapped in the gel, the modified starch gel can be used as an affinity gel to separate glycoproteins with α-(1 → 4) and α-(→ 6) glucosyl residues. The isozyme C of β-D-N-acetylglucosaminidase (EC 3.2.1.30) could be separated from the isozymes A and B by using this technique, as could the acidic from the neutral form of α-mannosidase (EC 3.2.1.24) from human tissues (Swallow *et al* 1976). On the other hand, some of the enzymes which bind to Con A-Sepharose in affinity chromatography, such as alkaline phosphatases, failed to bind to the modified starch gel (5 ml of loosely packed Con A-Sepharose per 150 ml of starch gel)(Swallow 1977).

2.4.2.2 Coupling of Spacer-attached Ligands to Agarose

A generally applicable method of preparing affinity gels is to mix melted agarose to which a ligand has been bound covalently via a spacer arm, to either melted agarose or a heated polymerising mixture of acrylamide and Bis. Hořejší *et al* (1982) tested a number of differently activated agarose beads (Sepharose) and found periodate-oxidised beaded agarose to have a number of advantages as a general ligand carrier:

(i) the activation reaction is simple and safe,
(ii) the activated product can be stored for a long time,
(iii) coupling of a ligand covalently attached to a reactive amino-hexyl or amino-phenyl spacer is easy and efficient when cyanoborohydride is used,
(iv) periodate-oxidation leaves the agarose un-crosslinked so that the substituted beads are still fully meltable when heated.

Hořejší *et al* (1982) described the following procedure to bind ligands having a free amino group to beaded agarose:

5 ml of washed Sepharose 4 B are added to 5 ml of 0.5 M $NaIO_4$ and the suspension is agitated for 2 h at room temperature. The oxidised gel is thoroughly washed with water and stored at 4°C. 2 ml of oxidised gel are washed with 0.5 M phosphate buffer, pH 6.0, and added to 2 ml of the same buffer containing 20 mM ligand and 20 mM $NaBH_3CN$ and agitated for 24 h at room temperature. After washing, the substituted beads are stored in 0.1 M Tris–HCl buffer, pH 8.2, containing 0.1% NaN_3 at 4°C.

To prepare an affinity gel the beads are equilibrated in the separating gel buffer so that a 2% agarose solution results. The mixture is heated until the agarose melts and then cooled down to 50°C. One volume of ligand-substituted agarose and one volume of 10% acrylamide + Bis (acrylamide: Bis = 30 : 1), containing 0.08% TEMED and 0.004% ammonium persulphate heated to 50°C are mixed together, poured into parafilm-sealed glass tubes (70 (length) × 2.5 mm), overlayered with water, and maintained at 50°C until polymerisation has come to an end. The concentration of the immobilised ligand is ~1–1.5 μmol ml^{-1} of packed gel. Finally the water overlay is replaced by the buffer contained in the gel, and the tubes are left to stand for at least 1 h at 4°C.

The following buffer-systems were used to perform affinity electrophoresis:

(i) the discontinuous system of Laemmli (1970) without SDS,
(ii) the discontinuous system of Reisfeld *et al* (1962),
(iii) the discontinuous system of Williams and Reisfeld (1964) (except that reversed polarity was used to separate trypsin),
(iv) the continuous buffer system of 0.02 M ammonium acetate, pH 5.2 which served for the separation of ribonuclease.

When discontinuous systems were used a stacking gel was omitted. 5–20 g of the protein in 10–20 μl of 10% glycerol in tenfold diluted gel buffer were applied on top of the gel rods. Electrophoresis was run at 60 or 100 V for 1 to 3 h.

Table 2.23 indicates the immobilised ligands and the buffer systems used to separate specific binding proteins.

Table 2.23 Immobilised ligands coupled to periodate-oxidised Sepharose 4B and buffer systems as used by Hořejší *et al* (1982) to separate specific binding proteins in affinity electrophoresis

Immobilised ligand	*Binding protein*	*Electrophoretic buffer system*
p-Aminophenyl-α-D-mannopyranoside	Con A, pea seed lectin	Laemmli (1970) without SDS, Reisfeld *et al* (1962)
Uridine 3′,5′-diphosphate 5′-p-aminophenyl ester	Ribonuclease	0.02 M ammonium acetate, pH 5.2
p-Aminobenzamidine	Trypsin	Williams and Reisfeld (1964), but letting the proteins migrate from anode to cathode

2.4.2.3 Copolymerisation of Allyl- or Acryloyl Substituted Ligands

Another method uses unsaturated, polymerisable derivatives of a ligand. These substances are added to a solution of acrylamide and Bis and the solution is polymerised to give a ligand substituted PAA gel. The ligand is incorporated into the gel due to its ability to co-polymerise with the monomers. Before electrophoresis can be performed the unreacted part of the unsaturated derivative must be removed by extensive washing of the gel. PAA gels are usually used to study the binding properties of enzymes and the types of ligands incorporated into the matrix are substrates or inhibitors, haptens interacting with lectins or antibodies (Hořejší 1984).

α-D-mannosyl-allylamine-acrylamide was co-polymerised with acrylamide and Bis and used to study the binding properties of pea seed lectin and Con A (Hořejší *et al* 1982). 8-substituted adenosine 2′-monophosphate-polyacrylamide and 8-substituted adenosine 5′-monophosphate-polyacrylamide gels were used to study the binding of plant and animal dehydrogenases (see table 2.19). The ligand was prepared from 8-bromo-2′-AMP and allylamine and reacted to 8-allyl-2′-AMP. 8-substituted 2′-AMP was then copolymerised with acrylamide and then incorporated into a solution of acrylamide and Bis (Nakamura *et al* 1980) (see Figure 2.21).

Copolymerisation of *N*-(2-hyroxypropyl)methacrylamide with 4-nitrophenyl esters of *N*-methacroyl derivatives of ω-amino acids leads to copolymers to which AMP can be bound with its 6-amino group via the 4-nitrophenyl group (Ticha *et al* 1980). AMP spaced from the PAA matrix by a C_6 spacer arm, using this technique, was applied to separate bovine heart lactate dehydrogenase isozymes (cf table 2.22).

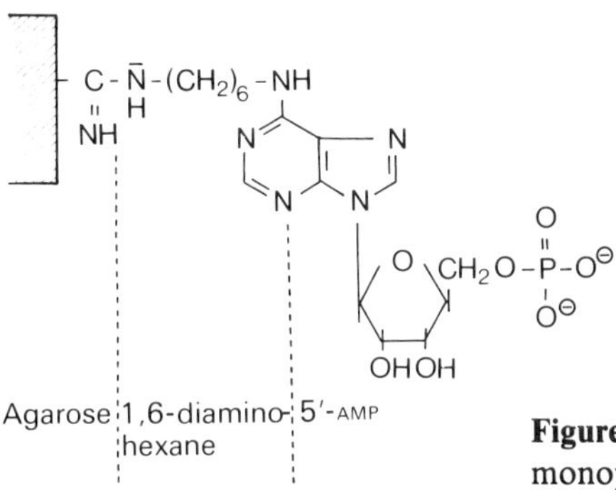

Figure 2.21 Formula of 5′-adenosine-monophosphate (AMP) bound to the N_6-atom of the adenine ring which is attached to a 6C spacer arm which is covalently attached to agarose (cf Craven *et al* 1974).

2.4.3 Experimental Modifications

Affinity gels may be prepared homogeneously, layered on top of a ligand-free gel, or placed between the first and second dimension gel in two-dimensional electrophoresis (Hořejší 1984, Bøg-Hansen and Han 1983).

In immuno-electrophoresis-like techniques, lectins may be incorporated homogeneously in agarose or PAA gels, choosing the lectin concentration so that lectin–glycoprotein precipitates are formed during electrophoresis which are comparable to antigen–antibody precipitates. If it is the aim to sieve out a certain glycoprotein from a crude protein mixture then it is sufficient to bind a lectin to agarose beads and apply these—mixed into an agarose gel—on top of a ligand-free gel (Raftell 1977, Horejsi *et al* 1982).

To study the interaction of two proteins such as the binding of glycoproteins to a lectin it is sufficient to distribute the lectin evenly in an agarose gel before electrophoresis. The pH in the gel is chosen so that the lectin has practically no mobility. PAA gels on the other hand are used to study the interaction of enzymes with immobilised inhibitors or to study the interaction of haptens with immobilised lectins or antibodies (Hořejší 1984).

If affinity gels of low electro-endosmosis can be prepared (e.g. by coupling an amino-containing ligand to periodate-oxidised agarose) then these gels may be used for affinity isoelectric focusing (Horejsi and Tichá 1981; Hořejší *et al* 1982). In this technique the immobilised ligand is incorporated into approximately 20% of the total gel just under the site where the sample is applied. A ligand-binding protein present in the sample solution is captured in the ligand-containing part of the gel, while the non-interacting proteins are sharply resolved according to their isoelectric points in the ligand-free part of the gel.

Bøg-Hansen suggested that electro-immunoprecipitation was used as an analytical reference method because proteins remain in their native state when this method is applied (Bøg-Hansen and Han 1982). Cholinesterase (EC 3.1.1.8) for example could be stained histochemically with 1-naphthyl acetate and fast red TR salt after immuno-electrophoresis or affinity electrophoresis with immobilised lectins (Bøg-Hansen *et al* 1977).

Another variant is to react specific binding proteins with immobilised lectins before electrophoresis and to study the supernatant fluid by crossed immuno-electrophoresis (Bøg-Hansen and Han 1983). The typical reaction is the disappearance of precipitates from the pattern. Several glycoproteins appear as multipeak precipitates when electrophoresed in 1% agarose (Tris–Veronal, pH 8.6) containing Con A. These are for instance: human serum α-1-antitrypsin, orosomucoid (α-1-acid glycoprotein), α-foetoprotein, haptoglobin and high-density lipoprotein (Bøg-Hansen and Han 1983).

α-glucosidase (maltase EC 3.2.1.20) from human lysosomes shows a two-banded pattern when separated on starch gel. The cause of the separation is the difference in affinity that the two isozymes have for starch: while isozyme 1 is bound to starch, isozyme 2 has no affinity for α-(1 → 4) and α-(1 → 6) polyglucans under the conditions applied. Mixing maltose with the gel at a concentration of ~2% releases isozyme 1 from the starch gel (Swallow 1977).

2.4.4 Affinity Chromatography Followed by Electrophoresis

It is possible to separate enzymes by affinity chromatographic techniques and then study the eluants by using electrophoretic techniques. When doing so, Swallow (1977) found that most enzymes from human tissues which bind to Con A-Sepharose or wheat germ lectin-Sepharose belong to the hydrolase class of enzymes and originated from lysosomes. The hydrolases found to have affinity for Con *A* (A) or wheat germ lectin (*W*) were: esterase (EC 3.1.1.1; *A*), cholinesterase (EC 3.1.1.8; *A*, *W*), alkaline phosphatase (EC 3.1.3.1; *A*, *W*), acid phosphatase (EC 3.1.3.2; *A*, *W*), aryl sulphatase

(EC 3.1.6.1; *A*), acid α-glucosidase (EC 3.2.1.20; *A*, *W*), neutral α-glucosidase (EC 3.2.1.20; *A*), α-galactosidase (EC 3.2.1.22; *A*, *W*), β-galactosidase (EC 3.2.1.23; *A*, *W*), α-mannosidase (EC 3.2.1.24; *A*, *W*), β-D-*N*-acetylglucosaminidase (EC 3.2.1.30; *A*, *W*), β-glucuronidase (EC 3.2.1.31; *A*), α-L-fucosidase (EC 3.2.1.51; *A*) and adenosine deaminase (EC 3.5.4.4; *A*, *W*). The fact that all the lysosomal acid hydrolases bind to Con A and in part to wheat germ lectin is consistent with the widely held view that most of these enzymes are glycoproteins.

2.4.5 Quantitative evaluations

The binding of a protein to a ligand is mainly influenced by the following:

(i) the mass–action law,
(ii) the size of the protein–ligand complex,
(iii) the buffer system,
(iv) the reaction temperature.

Provided the parameters (ii)–(iii) are kept constant, three different situations can be postulated:

(i) no protein is bound to the ligand; this is the case if a relatively large amount of protein faces a small concentration of ligand.
(ii) the protein is completely retarded; this happens if a relatively small amount of protein is run through a gel containing a large amount of ligand.
(iii) the protein is partially bound to the ligand; this happens at a critical relation of protein to ligand.

The latter situation can be used to calculate the dissociation constant of the protein–ligand complex. The dissociation constant is the reciprocal value of the binding constant. It can be calculated for the case where—

(i) the protein binds to an immobilised ligand and the protein–ligand complex has no electrophoretic mobility,
(ii) the case where a protein is partially desorbed from an immobile ligand by a competitive inhibitor (determination of the inhibition constant),
(iii) the case where a protein forms a complex with a high mol mass and uncharged substrate (see figure 2.22) and where the complex still has an electrophoretic mobility, although this mobility is diminished in comparison to the free protein.

For each of these three situations a special equation has been formulated to calculate the dissociation constant of the protein–ligand complex or the inhibition constant of the protein–inhibitor complex (cf Takeo 1984).

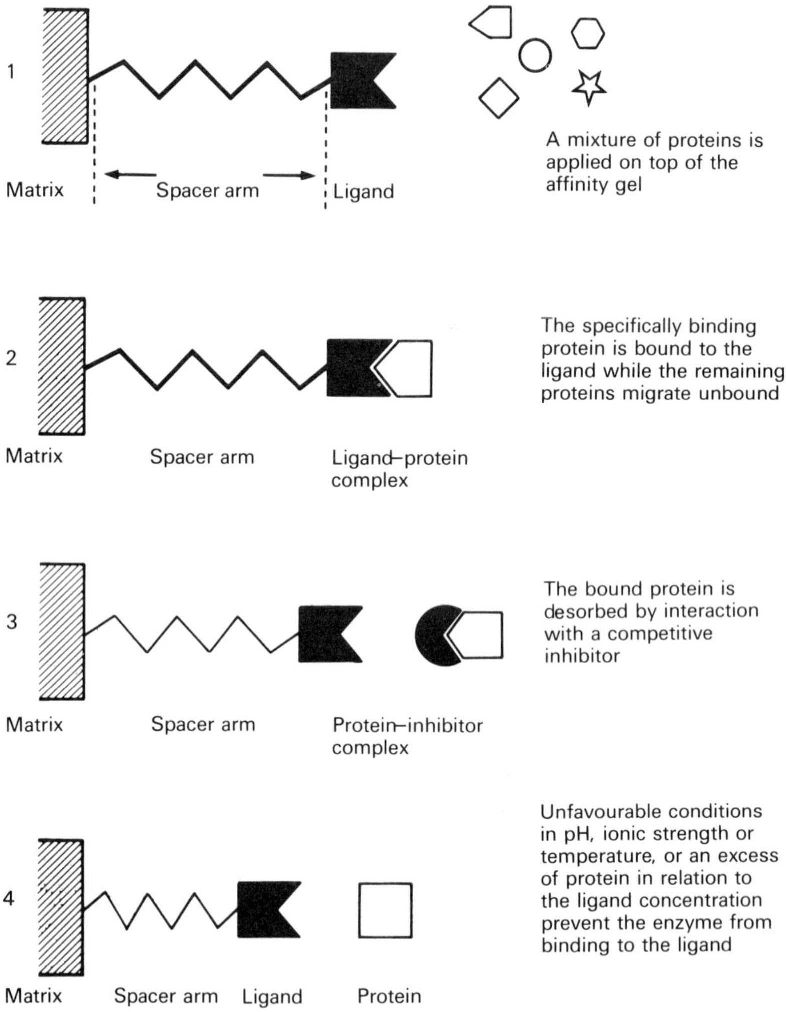

Figure 2.22 Scheme of the specific binding of a biologically active protein to an immobilized ligand covalently attached to the PAA matrix and desorption of the protein by a competitive inhibitor.

2.4.5.1 Determination of the Dissociation Constant if Protein and Ligand Form an Immobile Complex

A decrease in electrophoretic mobility is observed when a constant amount of rabbit phosphorylase is separated on a 5% PAA gel containing 0.005–0.32% rabbit liver glycogen (Takeo 1984). The interaction between the phosphorylase enzyme (E) and the ligand (L) glycogen leads to the formation of an enzyme–ligand complex (EL). The adsorption of the enzyme (phosphorylase) on to the ligand (glycogen) can be considered as an equilibrium between enzyme (E), ligand (L), and enzyme–ligand complex (EL) (Gerbrandy and Doorgeest 1972)

$$E + L \rightleftharpoons EL \tag{2.18}$$

and by applying the law of mass–action it follows that the dissociation constant K is equal to

$$K = \frac{[L]([E_t] - [EL]}{[EL]} = \frac{[L([E] + [EL] - [EL])}{[EL]} = \frac{[L][E]}{[EL]} \tag{2.19}$$

where $[E_t]$ is the total amount of enzyme (which is equal to the amount of free enzyme $[E]$ plus the amount of ligand bound enzyme $[EL]$, and $[L]$ is the concentration of ligand in the gel.

To describe the effect of the ligand on the mobility of the enzyme relative to a dye-marked protein, the quotient D_o/D_L is used, where D_o is the maximal migration distance without ligand in the gel and D_L is the migration in the presence of a ligand. At low enzyme concentrations the quotient D_o/D_L is equal to the ratio of the free enzyme concentration to the total enzyme concentration (Gerbrandy and Doorgeest 1972, Takeo, 1984)

$$\frac{D_o}{D_L} = \frac{[E]}{[E] + [EL]} = K. \tag{2.20}$$

Rearranging and combining equations (2.17) and (2.18) yields

$$\frac{1}{D_o - D_L} = \frac{K}{D_o} \times \frac{1}{[L]} + \frac{1}{D_o}. \tag{2.21}$$

This equation represents a linear relationship between $(D_o - D_L)^{-1}$ and $1/[L]$. The intercept on the ordinate corresponds to $1/D_o$, the intercept on the abscissa with $1/(-K)$.

2.4.5.2 Determination of the Inhibitor Constant

The binding of a macromolecule to an immobilised ligand can be hindered by the inclusion of a competitive inhibitor of low mol mass. For example the oligosaccharides maltotriose or β-cyclodextrin included in an affinity gel containing glycogen inhibit the binding of phosphorylase to it. The increase in phosphorylase mobility depends on the concentration of the oligosaccharide included and the inhibitor concentration dependent change in migration can be used to calculate the inhibitor constant (Reisfeld *et al* 1962, Takeo and Nakamura 1978, Takeo, 1984).

Under the requirements that—

(i) the mobility of the enzyme–inhibitor complex is the same as that of free phosphorylase, and that

(ii) the total concentration of immobilised ligand (glycogen) and inhibitor maltotriose, or β-cyclodextrin) are much higher than that of the enzyme, the inhibitor constant K can be calculated by applying the following equation

$$\frac{D_i}{D_o - D_i} = \frac{K}{cK_i} i + \frac{K}{c} \tag{2.22}$$

with D_i, migration of the enzyme in the presence of a constant amount of substrate (glycogen) and a varying amount of inhibitor (maltotriose); D_o, migration of the enzyme in the presence of the same amount of substrate but without inhibitor; K, the dissociation constant of the enzyme–substrate complex; c, the concentration of the substrate; K_i, the inhibitor constant; and i, the inhibitor concentration (cf Takeo 1984). A plot of $(D_i/D_o - D_i)$ against i (mM) leads to a straight line. The intercept with the ordinate gives K/c and the intercept of the line with the abscissa gives the negative value of the dissociation constant for maltotriose $(-K_i)$. Equation (2.22) is also applicable to the interaction between both a lectin and an oligosaccharide and an antibody and a hapten.

2.4.5.3 *Determination of the Dissociation Constant when Protein and Ligand Form a Mobile Complex and the Mol Mass of the Complex is Larger Than That of the Free But Still Mobile Protein*

When a protein forms a complex with a relatively large and uncharged ligand which is not covalently attached to the matrix, the sieving properties of the PAA matrix must be taken into consideration. The interaction of Con A with dextran molecules with mol masses in the range of 10^4–10^6 gives an example for this situation (Bøg-Hansen and Takeo 1980, Hořejší 1979, Takeo 1984). Con A binds specifically α-(1 → 4) and α-(1 → 6)-D oligo- and polysaccharides (Poretz and Goldstein 1970, So and Goldstein 1967). Goldstein *et al* (1974) and Takeo (1984) found identical K_i values for a number of maltose-type oligosaccharides having an increasing number of glucosyl residues. The mobilities of the Con A–maltose complexes were the same as the mobility of free Con A. But when using dextran molecules of a mol mass of 10^4–2×10^6 the mobility of the Con A–dextran complexes decreased concomittantly, while the K values of the various Con A–dextran complexes were found to be the same, regardless of the size of the dextran used (Takeo 1984).

To calculate the dissociation constant K for a protein interacting with a ligand of a relatively large mol mass that is not covalently attached to the matrix and where the protein–ligand complex has a slower migration velocity than the free protein, Bøg-Hansen and Takeo (1980) suggested that the following equation was used (for details see Takeo 1984)

$$\frac{1}{D_o - D_L} = \frac{K}{D_o - D} \frac{1}{c} + \frac{1}{D_o - D} \tag{2.23}$$

with D_o, migration distance of the protein in absence of any macromolecular affinity ligand in the gel; D_L, migration distance of the protein in the presence of a macromolecular ligand; D, migration distance in the presence of an excess amount of affinity ligand with all protein molecules binding the ligand, and c, the concentration of the ligand. When plotting $(1/D_o - D_L)$ against $1/c$ a straight line is obtained. Its intercept with the abscissa gives $-K$. The slope equals $(K/D_o - D)$.

Equation (2.23) can also be used to calculate K values when proteins interact with charged ligands and the protein–ligand complex has a mobility different from that of the free protein (Hořejší 1979, Takeo 1984).

Acknowledgments

The authors would like to thank Mrs Inge Milderberger for her consistently cheerful secretarial assistance and Mrs Holly Maurer for her proofreading.

References

Alligood J P 1983 *Anal. Biochem.* **134** 284–7

Alpers D H and Glickman R 1970 *Anal. Biochem.* **35** 314–20

Ameyama M, Shinagawa E, Matsushita K and Adachi O 1981 *J. Bacteriol.* **145** 814–23

Andary T J and Dabich D 1974 *Anal. Biochem.* **57** 457–66

Anderson L O, Borg H and Mikaelson M 1972 *FEBS Letters* **20** 199–202

Anderson B L, Berry R W and Telser A 1983 *Anal. Biochem.* **132** 365–75

Anker H S 1970 *FEBS Letters* **7** 293

Babczynski P 1980 *Anal. Biochem.* **105** 328–333

Bahr G F, Engler W F and Mazzone H M 1976 *Quart. Rev. Biophys.* **9** 459–89

Balakrishnan C V, Ravindranath S D and Appaji Rao N 1974 *Arch. Biochem. Biophys.* **164** 156–64

Banker G A and Cotman C W 1972 *J. Biol. Chem.* **247** 5856–61

Baumann G and Chrambach A 1976 *Anal. Biochem.* **70** 32–38

Bieber A L 1974 *Anal. Biochem.* **60** 206–12

Bishop D H L, Claybrook, J R and Spiegelman S 1967 *J. Mol. Biol.* **26** 373–87

Bispink G and Neuhoff V 1977 *Electrofocusing and Isotachophoresis* ed. B J Radola and B Graessling (Berlin: de Gruyter) pp 135–46

Bjerrum O J, Bhakdi S, Bøg-Hansen T C, Knüfermann H and Wallach D F H 1975 *Biochim. Biophys. Acta* **406** 489–504

Blakley R L 1960 *Nature* **188** 231–2

Blank A and Dekker C A 1982 Biochem. **20** 2261–7

Blank A, Silber J R, Thelen M P and Dekker C A 1983 *Anal. Biochem.* **135** 423–30

Blank A, Sugiyama R H and Dekker C A 1982 *Anal. Biochem.* **120** 267–75

Blattler D P, Garner F, Van Slyke K and Bradley A 1972 *J. Chromatogr.* **64** 147–155

Bøg-Hansen T C 1973 *Anal. Biochem.* **56** 480–8

Bøg-Hansen T C, Bjerrum O J and Brogren C H 1977 *Anal. Biochem.* **81** 78–87
Bøg-Hansen T C and Han J 1983 in: *J. Chromatogr. Library vol. 18B* ed. Z Deyl, (Amsterdam: Elsevier) pp 219–52
Bøg-Hansen T C, Jensen P, Hinnerfeldt F and Takeo K 1981 *Lectins, Biology Biochemistry and Clinical Biochemistry* ed. T C Bøg-Hansen vol. 1 (Berlin: de Gruyter) pp 241–58
Bøg-Hansen T C and Takeo K 1980 *Electrophoresis* **1** 67–71
Bordier C, Loomis W F, Elder J and Lerner R 1978 *J. Biol. Chem.* **253** 5133–9
Borrebaeck C and Etzler M E 1980 *FEBS Letters* **117** 237–40
Brewer G J and Sing Ch F 1970 *An Introduction to Isozyme Techniques* (New York: Academic) p 17
Bryan J K 1977 *Anal. Biochem.* **56** 480–8
Cain D F and Pitney R E 1968 *Anal. Biochem.* **22** 11–20
Campbell W P, Wrigley C W and Margolis J 1983 *Anal. Biochem.* **129** 31–6
Choules G L and Zimm B H 1965 *Anal. Biochem.* **13** 336–344
Chrambach A and Hjelmeland L M 1983 *Electrophorese-Forum '83* ed. B J Radola *4. Diskussionstagung TU München* 24 26.10.1983 pp 9–20
Clarke J T 1964 *Ann. NY Acad. Sci.* **121** 428–436
Clarke S 1981 *Biochim. Biophys. Acta* **670** 195–202
Clavert P 1975 *Nature* **254** 104
Colombo G and Marcus F 1973 *Biol. Chem.* **248** 2743–5
Condeelis J S 1977 *Anal. Biochem.* **77** 195–207
Content J and Duesberg P H 1971 *J. Mol. Biol.* **62** 273–285
Converse C A and Papermaster D S 1975 *Science* **189** 469–472
Cottolico R A and Jones R F 1975 *Anal. Biochem.* **66** 35–46
Craven D B, Harvey M J, Lowe C R and Dean P D G 1974 *Eur. J. Biochem.* **41** 329–33
Darszon A, Montal M and Phillip M 1977 *FEBS Letters* **74** 135–8
Datta A and Penefsky H S 1970 *J. Biol. Chem.* **245** 1537–44
Davies R C and Neuberger A 1973 *Biochem. J.* **133** 471–92
Davies R C and Neuberger 1979 *Biochem. J.* **177** 649–59
Davies B J 1962 *Disc Electrophoresis* Distillation Prod. Div. Eastman Kodak Co. Rochester NY
Davies B J 1964 *Ann. NY Acad. Sci.* **121** 404–27
De Issaly S M 1968 *Biochim. Biophys. Acta* **151** 473–483
De Rosa G, Duncan D S, Keen C L and Hurley L S 1979 *Biochim. Biophys. Acta* **566** 32–9
Dewald B, Dulaney J T and Touster O 1974 *Meth. Enzymol. 32*, eds S Fleischer and L Packer pp 82–91
Dohnal J C and Garvin J E 1979 *Biochim. Biophys. Acta* **576** 393–403
Dottin R P, Manrow R E, Fishel B R, Aukerman S L and Culleton J L 1979 *Meth. Enzymol.* **68** 513–29
Dulaney J T and Touster O 1970 *Biochim. Biophys. Acta* **196** 29–34
Dunker A K and Rueckert R R 1969 *J. Biol. Chem.* **244** 5074–80
Esposito J J and Obijeski J F 1976 *Prep. Biochem.* **6** 431–442
Fan L L and Masters B S S 1974 *Arch. Biochem. Biophys.* **165** 665–671
Fawcett J S and Morris C J O R 1966 *Separation Sci.* **1** 9–26
Faye L 1981 *Anal. Biochem.* **112** 90–95

Felgenhauer K 1969 *Protides Biol. Fluids* **17** 505–509
Felgenhauer K 1974 *Hoppe-Seyler's Z. Physiol. Chem.* **355** 1281–90
Ferguson K A 1964 *Metabolism* **13** 985–1002
Fish W W, Reynolds J H and Tanford C 1970 *J. Biol. Chem.* **245** 5166–8
Fishman L 1974 *Biochem. Med.* 309–315
Flory P J 1953 *Prin. Pol. Chem.* Cornell Univ. Press, NY
Fox J L, Steven S E and Taylor C P 1978 *Anal. Biochem.* **87** 253–6
Fraker P J and Kaplan S 1972 *J. Biol. Chem.* **247** 2732–7
Frank R N and Rodbard D 1975 *Arch. Biochem. Biophys.* **171** 1–13
Furth A J 1980 *Anal. Biochem.* **109** 207–15
Gainer H 1971 *Anal. Biochem.* **44** 589–605
Gainer H 1973 *Anal. Biochem.* **51** 646–50
Gerbrandy S J J and Doorgeest A 1972 *Phytochemistry* **11** 2403–7
Glossmann H and Lutz F 1970 *Hoppe-Seyler's Z. Physiol. Chem.* **351** 1583–5
Goldstein I J, Reichert C M and Misaki A 1974 *Annals NY Acad. Sci.* **234** 283–96
Gonenne A and Ernst R 1978 *Anal. Biochem.* **87** 28–38
Gonenne A and Lebowitz J 1975 *Anal. Biochem.* **64** 414–24
Goren R and Huberman M 1976 *Anal. Biochem.* **75** 1–8
Graft H and Gaafer H A 1974 *Am. Rev. Resp. Disease* **110** 320–3
Grefrath S P and Reynolds J A 1974 *Proc. Natl. Acad. Sci. USA* **71** 3913–16
Gregory E M and Fridovich I 1974 *Anal. Biochem.* **58** 57–62
Grossbach U 1965 *Biochem. Biophys. Acta* **107** 180–2
Grove T H and Levy H R 1975 *Anal. Biochem.* **65** 458–65
Hager D A and Burgess R R 1980 *Anal. Biochem.* **109** 76–86
Hallock R O and Yamada E W 1973 *Anal. Biochem.* **56** 84–90
Hanaoka F, Shaw J L and Mueller G C 1979 *Anal. Biochem.* **99** 170–174
Hansen N J 1976 *Anal. Biochem.* **76** 37–44
Hansl R 1964 *Ann. NY Acad. Sci.* **121** 391–403
Hanspal J S, Bushell G R and Ghosh P 1983 *Anal. Biochem.* **132** 288–93
Harris H and Hopkinson D A 1976 *Handbook of Enzyme Electrophoresis in Human Genetics* (Amsterdam: North Holland)
Hashimoto F, Horigome T, Kanbayashi M, Yoshida K and Sugano H 1983 *Anal. Biochem.* **129** 192–9
Hatefi Y and Hanstein G 1974 *Meth. Enzymol.* **31** ed. S Fleischer and L Packer pp 770–90
Hedrick J L and Smith A J 1968 *Arch. Biochem. Biophys.* **126** 155–64
Helenius A and Simons K 1975 *Biochim. Biophys. Acta* **415** 29–79
Helenius A and Simons K 1977 *Proc. Natl. Acad. Sci. USA* **74** 529–32
Henderson E J, Nagano H, Zalkin H and Hwang L H 1970 *J. Biol. Chem.* **245** 1416–23
Henderson L E, Oroszlan S and Konigsberg W 1979 *Anal. Biochem.* **93** 153–7
Hjelmeland L M and Chrambach A 1984 *Meth. Enzymol.* **104** ed. B Jacoby pp 305–318
Hjelmeland L M, Klee, W A and Osborne J C 1983a *Anal. Biochem.* **130** 485–90
Hjelmeland L M, Klee, W A and Osborne J C 1983a *Anal. Biochem.* **95** 201–8
Hjelmeland L M, Nebert D A and Osborne J C 1983b *Anal. Biochem.* **130** 72–82
Hjertén S 1962 *Arch. Biochem. Biophys. Suppl. 1* pp 147–151
Hjertén S 1983 *Biochim. Biophys. Acta* **736** 130–6

Honold G R, Farkas G L and Stahmann M A 1966 *Cereal Chem.* **43** 517–29

Honold G R 1967 *Cereal Chem.* **44** 373–82

Honold G R and Stahmann M A 1968 *Cereal Chem.* **45** 99–108

Hořejší V 1979 *J. Chromatogr.* **178** 1–13

Hořejší V 1984 *Meth. Enzymol.* **104** ed. B Jacoby pp 275–81

Hořejší V and Kocourek J 1974 *J. Biochim. Biophys. Acta* **336** 338–43

Hořejší V and Ticha M 1981 *Anal. Biochem.* **116** 22–26

Hořesjší V, Ticha M and Kocourek J 1979 *Trends Biochem. Sci.* **4** N6–N7

Hořejší V Ticha M, Tichy P and Holy A 1982 *Anal. Biochem.* **125** 358–69

Horigome T and Sugano H 1983 *Anal. Biochem.* **130** 393–6

Hrkal Z 1979 *J. Chromatogr. Library* **18** ed. Z Deyl, F M Everaerts, Z Prusik and P J Svendsen, Part A (New York: Elsevier) pp 299–324

Hubert E and Marcus F 1974 *FEBS Letters* **40** 37–40

Hucklesby D P and Hageman R H 1973 *Anal. Biochem.* **56** 591–2

Huet J, Sentenac A and Fromageot P 1978 *FEBS Letters* **94** 28–32

Hyden H, Bjurstam K and McEwen B 1966 *Anal. Biochem.* **17** 1–15

Ichihara K, Kusunose E and Kusunose M 1973 *Eur. J. Biochem.* **38** 463–72

Inoue T 1978 *Physico-Chem. Biol* **22** 135–8

Inoue T 1980 *Physico-Chem. Biol.* **24** 1–10

Jørgensen P L, Karlish S J D and Gitler G 1982 *J. Biol. Chem.* **257** 7435–42

Kadenbach B, Jarausch J, Hartmann R and Merle P 1983 *Anal. Biochem.* **129** 517–521

Kadenbach B and Merle P 1981 *FEBS Letters* **135** 1–11

Kavavolas H J, Baedecker M L and Engel L L 1970 *J. Biol. Chem.* **245** 4948–52

Kelleher P J and Juliano R C 1984 *Anal. Biochem.* **136** 470–5

Kimura K, Miyakawa A, Imai T and Sasakawa T 1977 *J. Biochem.* **81** 467–76

Kinard F E 1957 *Rev. Sci. Instr.* **28** 293–4

King J and Laemmli U K 1971 *J. Mol. Biol.* **62** 465–477

Kopperschläger G, Diezel W, Bierwagen B and Hofmann E 1969 *FEBS Letters* **5** 221–4

King T E 1963 *J. Biol. Chem.* **238** 4037–51

Kowit J D and Maloney J 1982 *Anal. Biochem.* **123** 86–93

Kyte J and Rodriguez H 1983 *Anal. Biochem.* **133** 515–22

Lacks S A, Springhorn S S and Rosenthal A L 1979 *Anal. Biochem.* **100** 357–363

Lacks S A and Springhorn S S 1980 *J. Biol. Chem.* **255** 7467–73

Laemmli U K 1970 *Nature* **227** 680–685

Lambin P 1978 *Anal. Biochem.* **85** 114–25

Lambin P and Fine J M 1979 *Anal. Biochem.* **98** 160–8

Lambin P, Rochu D and Fine J M 1976 *Anal. Biochem.* **74** 567–75

Lasky M 1978 *Electrophoresis '78* ed. N Catsimpoolas (Amsterdam: North Holland) pp 195–210

Laurent T C and Killander J 1964 *J. Chromatogr.* **14** 317–30

Lee L T, Deas J E and Howe C 1978 *J. Immunol. Methods* **19** 69–75

Lenard J 1971 *Biochem. Biophys. Res. Comm.* **45** 662–8

McCarthy R E and Racker E 1966 *Brookhaven Symp. Biol.* **19** 202–13

McLellan T 1982 *Anal. Biochem.* **126** 94–9

MacLennan D H, Lenaz G and Szarkowska L 1966 *J. Biol. Chem.* **241** 5251–59

MacLennan D H, Smoley J M and Tzagoloff A 1968 *J. Biol. Chem.* **243** 1589–97

Mahadik S P 1976 *Anal. Biochem.* **76** 615–33
Maizel J V 1971 *Methods in Virology* **5** 179–246
Manrow R E and Dottin R P *1980 Proc. Natl. Acad. Sci. USA* **77** 730–4
Manrow R E 1982 Anal. Biochem. **120** 267–75
Marchesi V T and Palade G E 1967 *J. Cell Biol.* **35** 385–404
Margolis J and Kenrick K G 1968 *Anal. Biochem.* **25** 347–62
Margolis J and Wrigley C W 1975 *J. Chromatogr.* **106** 204–9
Margulies M M and Tiffany H L 1984 *Anal. Biochem.* **136** 309–13
Markert C L and Møller F 1959 *Proc. Natl. Acad. Sci. USA* **45** 753–63
Martelli G P, Quacquarelli A and Russo M 1971 *CMI/AAB Descriptions of Plant Viruses* **69**
Martin R G and Ames B N 1961 *J. Biol. Chem.* **236** 1372–1374
Martonosi A and Haplin R A 1971 *Arch. Biochem. Biophys.* **144** 66–77
Masters B S S, Williams C H and Kamen H 1967 *Meth. Enzymol.* **10** ed. R W Estabrook and M E Pullman pp 565–573
Matheka H D, Enzmann P J, Bachrach H L and Migl B 1977 *Anal. Biochem.* **81** 9–17
Maurer H R 1968 Disk-Elektrophorese, *Theorie und Praxis der diskontinuierlichen Polyacrylamidgel-Elektrophores* (Berlin: de Gruyter) pp 1–9
Merril C R, Goldman D, Sedman S A and Ebert M H 1981 *Science* **211** 1437–8
Meyer J M and Stadtman E R 1981 *J. Bacteriol.* **146** 705–12
Miller R E, Shelton E and Stadtman E R 1974 *Arch. Biochem. Biophys.* **163** 155–71
Morton R K 1950 *Nature* **166** 1092–5
Nakamura S, Takeo K, Tanaka K and Ueta T 1960 *Hoppe–Seyler's Z. Physiol. Chem.* **318** 115–28
Nakamura K, Kuwahara A, Ogata H and Takeo K 1980 *J. Chromatogr.* **192** 351–62
Ne'eman Z, Kahane I and Razin S 1971 *Biochim. Biophys. Acta* **249** 169–76
Nelson C A 1971 *J. Biol. Chem.* **246** 3895–901
Neuhoff V 1968 *Arzneim. Forsch.* **18** 35–9
Neuhoff V 1973 *Micromethods in Molecular Biology* (Berlin: Springer) pp 4–83
Neuhoff V and Lezius A 1968 *Z. Naturforsch.(B)* **23** 812–19
Neuhoff V and Schill W B 1968 *Hoppe-Seyler's Z. Physiol. Chem.* **349** 795–800
Neville D M 1971 *J. Biol. Chem.* **246** 6328–34
Newby A C and Chrambach A 1979 *Biochem, J.* **177** 623–30
Nimmo H G and Nimmo G A 1982 *Anal. Biochem.* **121** 17–22
Nixon P F and Blakley R L 1968 *Biol. Chem.* **243** 4722–31
O'Connell PBH and Brady CJ 1976 *Anal. Biochem.* **76** 63–73
Öfverstedt L G, Sundelin J and Johansson G 1983 *Anal. Biochem.* **134** 361–7
Ogston A G 1958 *Trans. Faraday Soc.* **54** 1754–7
Ogston A G and Phelps C F 1961 *Biochem. J.* **78** 827–33
Ornstein L 1964 *Ann. NY Acad. Sci.* **121** 321–349
Ornstein L B and Davis J 1962 *Disc Electrophoresis* Distillation Products Ind. Rochester and CANALCO brochures Bethesda, Unpublished
Otto M and Šnejdárková M 1981 *Anal. Biochem.* **111** 111–4
Panyim S and Chalkley R 1969 *Arch. Biochem. Biophys.* **130** 337–46
Panyim S 1971 *J. Biol. Chem.* **246** 7557–60
Patterson M S and Greene R C 1965 *Anal. Chem.* **37** 854–7
Paus P N 1971 *Anal. Biochem.* **42** 372–6

Penefsky H S and Tzagoloff A 1971 *Meth. Enzymol.* **22** ed. W Jacoby pp 219–230
Pharmacia 1980 *Polyacrylamide Gel Electrophoresis–Laboratory Techniques* **June** Uppsala, Sweden)
Pierce M, Cummings R C and Roth S 1980 *Anal. Biochem.* **102** 441–9
Pitt-Rivers R and Impiombato F S A 1968 *Biochem. J.* **109** 825–30
Poduslo J F and Rodbard D 1980 *Anal. Biochem.* **101** 394–406
Poretz R D and Goldstein I J 1970 *Biochemistry* **9** 2890–6
Porter A L G 1981 *Anal. Biochem.* **117** 28–31
Pringle J R 1970 *Biochem. Biophys. Res. Comm.* **39** 46–52
Purkhanbaba H and Rothe G M 1984 *Electrophoresis '84* ed. V Neuhoff (Weinheim: Verlag-Chemie) pp 252–5
Raftell M 1977 *Immunochemistry* **14** 787–92
Raymond S 1962 *Clin. Chem.* **8** 455–70
Raymond S and Nakamichi M 1962 *Anal. Biochem.* **3** 23–30
Raymond S and Weintraub L 1959 *Science* **130** 711
Reisfeld R A, Lewis U J and Williams D E 1962 *Nature* **201** 281–3
Reynolds J A and Tanford C 1970a *Proc. Natl. Acad. Sci. USA* **66** 1002–7
Reynolds J A 1970b *J. Biol. Chem.* **245** 5161–5
Risi S, Höckel M, Hulla F W and Dose K 1977 *Eur. J. Biochem.* **81** 103–9
Robdard D 1976 *Methods of Protein Separation* ed. N Catsimpoolas vol. 2 (New York: Plenum) pp 145–218
Rodbard D and Chrambach A 1970 *Proc. Natl. Acad. Sci. USA* **65** 970–7
Rodbard D and Chrambach A 1971a *Anal. Biochem.* **40** 95–134
Rodbard D and Chrambach A 1971b *Science* **172** 440–51
Rodbard D, Kapadia G and Chrambach A 1971 *Anal. Biochem.* **40** 135–57
Rodbard D, Levitor C and Chrambach A 1972 *Sep. Sci.* **7** 705–23
Rosenthal A and Lacks S A 1977 *Anal. Biochem.* **80** 76–90
Rosenthal A 1978 *J. Biol. Chem.* **253** 8674–6
Rosenthal A 1980 *J. Mol. Biol.* **141** 133–46
Roth J 1978 *The Lectins* (Jena: VEB Fischer)
Rothe G M 1982 *Electrophoresis* **3** 255–62
Rothe G M 1983 *Electrophoresis '82* ed. D Stathakos (Berlin: de Gruyter) pp 103–8
Rothe G M and Purkhanbaba H 1982a *Electrophoresis* **3** 33–42
Rothe G M 1982b *Electrophoresis* **3** 43–8
Rüchel R, Mesecke S, Wolfrum D I and Neuhoff V 1974 *Hoppe-Seyler's Z. Physiol. Chem.* **355** 997–1020
Rüchel R, Steere R L and Erbe E F 1978 *J. Chromatogr.* **166** 563–75
Russell R R B 1979 *Anal. Biochem.* **97** 173–5
Satoh K, Imai F and Sato K 1978 *FEBS Letters* **95** 239–42
Schäfer H J, Scheurich P and Rathgeber G 1978 *Hoppe-Seyler's Z. Physiol. Chem.* **359** 1441–2
Schieven G L, Blank A and Dekker C A 1982 *Biochem.* **21** 5148–55
Schuerch A R, Mitchell W R and Joklik W K 1975 *Anal. Biochem.* **65** 331–45
Segrest J P, Jackson R L, Andrews E P and Marchesi V T 1971 *Biochem. Biophys. Res. Comm.* **44** 390–5
Shaik-M M B, Guy A L and Pancholy S K 1980 *Anal. Biochem.* **103** 140–3
Shapiro A L, Viñuela E and Maizel J V 1967 *Biochem. Biophys. Res. Comm.* **28** 815–20

Shepherd R J 1971 *CMI/AAB Descriptions of Plant Viruses* **57**
Shuler J K and Tryfiates G P 1977 *Enzyme* **22** 262–5
Shuster L 1971 *Meth. Enzymol.* **22** ed. W Jacoby p 412
Siegel L M and Monty K J 1966 *Biochim. Biophys. Acta* **122** 346–62
Slater G G 1968 *Anal. Biochem.* **24** 215–7
Slater G G 1969 *Anal. Chem.* **41** 1039–41
Smith A E and Yamada E W 1971 *J. Biol. Chem.* **246** 3610–17
Smithies O 1959 *Biochem. J.* **71** 585–7
So L L and Goldstein I J 1967 *J. Biol. Chem.* **242** 1617–22
Sober H A, Hartley R W, Carroll W R and Peterson E A 1965 *The Proteins* **3** 2
Sokoloff R L and Frigon R P 1981 *Anal. Biochem.* **118** 138–41
Solomon S S, Palazzolo M and King L E 1977 *Diabetes* **26** 967–972
Spanos A, Sedgwick S G, Yarranton G T, Hübscher U and Banks G R 1981 *Nucl. Acid Res.* **9** 1825–39
Späth P J and Koblet H 1979 *Anal. Biochem.* **93** 275–285
Stegemann H 1967 *Hoppe-Seyler's Z. Physiol. Chem.* **348** 951–2
Stephens R E 1975 *Anal. Biochem.* **65** 369–79
Strålfords P and Belfrage P 1983 *Anal. Biochem.* **128** 7–10
Strongin A Y, Azarenkova N M, Vaganova T I, Levin E D and Stepanov V M 1976 *Anal. Biochem.* **74** 597–9
Sugiura M, Ito Y, Hirano K and Sawaki S 1977 *Anal. Biocehm.* **78** 481–4
Sugiyama R H, Blank A and Dekker C A 1981 *Biochem.* **20** 2268–74
Svendsen P J 1979 *Electrophoresis, a survey of techniques and application, part A, Journal of Chromatography Library vol. 18,* ed. Z Deyl, F M Everaerts, Z Prusik and P J Svendsen (New York: Elsevier) pp 307–324
Swallow D M 1977 in: *Isozymes: Current topic in biological and medical research* ed. M C Rattazzi, J G Scandalios and G S Whitt (New York: Alan R Liss) pp 159–87
Swallow D M, Evans L, Saha N and Harris H 1976 *Ann. Hum. Genet. Lond.* **40** 55–6
Swank R T and Munkres K D 1971 *Anal. Biochem.* **39** 462–77
Takeo K 1984 *Electrophoresis* **5** 187–95
Takeo K, Fugimoto M, Suzuno R and Kuwahara A 1978 *Physico—Chem. Biol.* **22** 139–44
Takeo K and Nakamura S 1972 *Arch. Biochem. Biophys.* **153** 1–7
Takeo K 1978 *Affinity Chromatography* ed. O Hoffmann-Ostenhof, M Breitenbach, F Koller, D Kraft and O Schreiner (Oxford: Pergamon) pp 67–70
Takeo K, Nitta K and Nakamura S 1974 *Clin. Chim. Acta* **57** 45–54
Tanford C 1968 *Adv. Protein Chem.* **23** 121–282
Tanford C and Reynolds J A 1976 *Biochim. Biophys. Acta* **457** 133–70
Tanner M J A and Boxer D H 1972 *Biochem. J.* **129** 333–47
Tas J, De Vries A C J and Berndsen R G 1979 *Anal. Biochem.* **100** 264–70
Thelen M P, Blank A, McKeon T A and Dekker C A 1982 *Fed. Proc.* **41** 1203
Thomas J M and Hodes M E 1981 *Anal. Biochem.* **118** 194–6
Ticha M, Barthova J, Labsky J and Semansky M 1980 *J. Chromatogr.* **194** 183–9
Ticha M, Horejsi V and Barthova J 1978 *Biochim. Biophys. Acta* **534** 58–63
Tombs M P 1965 *Amal. Biochem.* **13** 121–32
Tsou K C, Lo K W and Yip K F 1974 *FEBS Letters* **45** 47–9

Tuszynski G P, Damsky C H, Fahrer J P and Warren L 1977 *Anal. Biochem.* **83** 119–29

Tuszynski G P and Warren L 1975 *Anal. Biochem.* **67** 55–65

Tzagoloff A and Penefsky H S 1971 *Meth. Enzymol.* **22** ed. W Jacoby pp 204–19

Uehara K, Fujimoto S and Taniguchi T 1974 *J. Biochem.* **75** 627–38

Ueng S T H, Hartanowicz P, Lewandoski C, Keller J, Holick M and McGuiness E T 1976 *Biochemistry* **15** 1743–9

Vollmert B 1962 *Grundlagen der makromolekularen Chemie* (Berlin: Springer) p 388

Waehneldt T V 1971 *Anal. Biochem.* **43** 306–12

Waehneldt T V 1975 *Bio-Systems* **6** 176–87

Wallach D F H 1969 *J. Gen. Physiol.* **54** 3s

Weber K and Kuter D J 1971 *J. Biol. Chem.* **246** 4504–9

Weber K and Osborn M 1969 *J. Biol. Chem.* **244** 4406–12

Weber K 1975: *The Proteins* 3rd edn, vol. 1 ed. H Neurath and R L Hill (New York: Academic) pp 179–223

White M L 1960 *J. Phys. Chem.* **64** 1563–5

Willhardt I and Wiederanders B 1975 *Anal. Biochem.* **63** 263–6

Williams D E and Reisfeld R A 1964 *Ann. N Y Acad. Sci.* **121** 373–81

Williams J G and Gratzer W B 1971 *J. Chromatogr.* **57** 121–5

Wilson D L 1971 *J. Gen. Physiol.* **57** 26–40

Womack M D, Kendall D A and MacDonald R C 1983 *Biochim. Biophys. Acta* **733** 210–15

Wright G L, Farrell K B and Roberts D B 1973 *Biochim. Biophys. Acta* **295** 396–411

Yonezawa S and Hori S H 1975 *J. Histochem. Cytochem.* **23** 745–51

Young R B, Orcutt M and Blauwiekel P B 1980 *Anal. Biochem.* **108** 202–6

Young Y P L and Young R J 1974 *Anal. Biochem.* **58** 286–93

Zaman Z and Verwilghen R L 1979 *Anal. Biochem.* **100** 64–9

Ziegler A and Hengartner H 1977 *Eur. J. Immunol.* **7** 690

Zink M W and Katz J S 1973 *Can. J. Microbiol.* **19** 1187–1196

Ziola B R and Scraba D G 1976 *Anal. Biochem.* **72** 366–371

Zöllner E J, Müller W E G and Zahn R K 1973 *Z. Naturforsch.* **28c** 376–8

Zwaan J 1967 *Anal. Biochem.* **21** 155–68

Chapter 3

Conventional Isoelectric Focusing and Immobilised pH Gradients

by
P G Righetti, C Gelfi and E Gianazza

3.1 Introduction

We have written many reviews on isoelectric focusing (IEF) over the years. An extensive book has recently appeared on this topic (Righetti 1983a) and the modern explosion of books dealing with electrophoretic techniques in general makes it look as though all laboratories using or developing electrophoresis want to write their own 'state of the art' account. We will have to struggle hard to give a modern and fresh view of the two techniques (IEF and immobilised pH gradients, IPG) with which we are involved.

Conventional IEF

3.2 The Principles of IEF

It took the brilliant theoretical work of Svensson (1961, 1962a, b) to lay the foundations of modern IEF and to introduce the subtle concept of 'carrier ampholyte' (CA) (figure 3.1). The buffers used for IEF had to be amphoteric, so that they would seek a stationary position in the system. Moreover, they had to be a 'carrier' of buffering power, so that they would prevent local pH changes, and also of conductivity, so as to ensure unhindered flow of current through the focusing cell. With Vesterberg's (1969) elegant synthesis of such buffers, present day IEF was born (figure 3.1). Basically, a mixture of oligoamines (the more heterogeneous the better, e.g. triethylene triamine, tetraethylene pentamine, pentaethylene hexamine) is reacted with an α–β unsaturated compound (the best being acrylic acid) forming a highly complex mixture of aliphatic oligo-amino, oligo-carboxylic acids. These have pIs in the pH range 3–10, with small (pI–pK proximal) values and are thus able to buffer and conduct in this pH region. Over the years several synthetic approaches have been described (Charlionet *et al* 1979, Grubhofer and Borja 1977, Just 1980, Righetti *et al* 1975, Vinogradov *et al* 1973, even high molecular weight (M_r) species for fractionation of peptides, Righetti and Hjerten 1981), but all have been based on the classical synthetic approach of Vesterberg (1969). In his patent Vesterberg stated that CAs should not contain peptide bonds. This was a valid statement at the time but it subsequently allowed Pharmacia Fine Chemicals to patent what were essentially Vesterberg's ideas by using dito tetra-peptides in the synthesis of CAs (Williams and Sonderberg 1979). The Pharmalytes have one unique property, though: when focusing proteases, the Pharmalytes are digested by these enzymes on their path to the pI position, with concomitant destruction of the pH gradient.

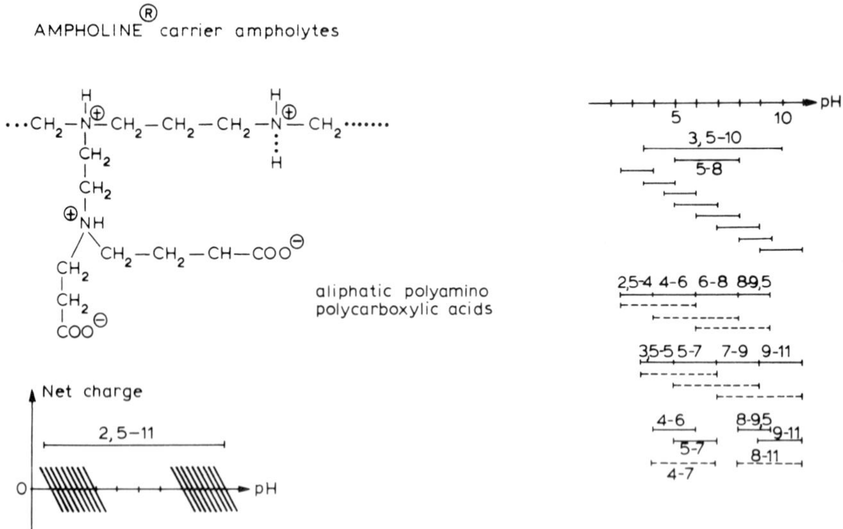

Figure 3.1 Composition of Ampholine. On the upper left side a representative chemical formula is shown (aliphatic oligo-amino-oligo-carboxylic acids). On the lower left side, portions of hypothetical titrations curves of Ampholines are depicted. On the right different pH cuts for wide and narrow range carrier ampholytes. (Courtesy: LKB Produkter AB.)

The establishment of a pH gradient is illustrated in figure 3.2. Any narrow or wide mixture of carrier ampholytes is a random concoction of possibly hundreds of species having pI's in the stated pH interval. It is as

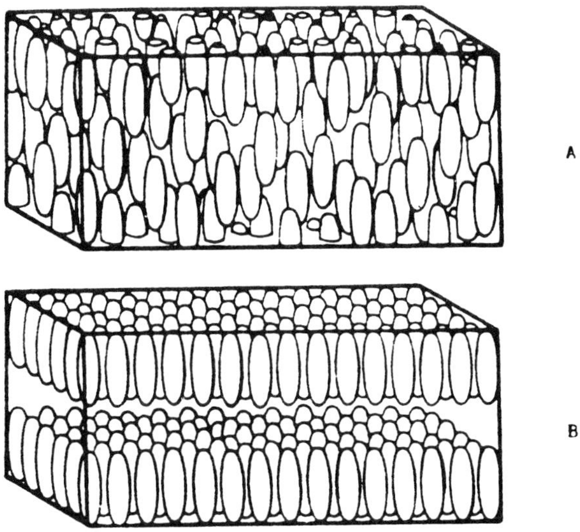

Figure 3.2 Molecular arrangements of liquid crystals. States, A = nematic; B = smectic. In the nematic state, molecules may freely rotate around their own axis or move along in three space directions (this is taken to represent non-focused carrier ampholytes). In the smectic state, molecules can rotate along their longitudinal axis and two adjacent molecular strata are organised into parallel layers (this is taken to represent focused carrier ampholytes) (from Valerio *et al* 1983).

if they are in a 'nematic' state, as in liquid crystals (figure 3.2(*a*)). Once a voltage gradient is applied, the current sweeps them, at different velocities, to their pI position, thus arranging them in a stack of contiguous zones which automatically dictate and stabilise a pH gradient. This could be envisaged as the passage to a 'smectic' state in liquid crystals (figure 3.2(*b*)) (in the latter case however, the force driving the two different states is the temperature) (Valerio *et al* 1983). The parallel we have drawn will hold with the proviso that figure 3.2(*b*) has many layers of adjacent, isoelectric CA species and by considering contiguous layers as not physically separated, but slightly interdigitating. It is this enormous complexity of CAs that ensures production of smooth pH gradients (it has been calculated that at least 30 carrier ampholytes/pH unit are needed for production of a stepless pH gradient) (Almgren 1971). It came as a shock to us to learn that we should perform IEF in a mixture of only 47 buffers (Cuono and Chapo 1982). This approach was tried and discarded in 1962 by Svensson (who gave a list of 41 amphoteric buffers as possible candidates for IEF). Are we assisting in the 'great leap backward'? We do not even want to mention here the incredible concept of IEF in non-amphoteric buffers (BEF) (Chrambach and Hjelmeland 1984), where the buffering ions are fully stripped of counter-ions (*sic*!): this is a 'gigantic leap backward' to some unheard of prehistoric time.

3.3 IEF in Agarose Matrices

Agarose is a particularly attractive support for IEF , because it can be gelled with an extremely large pore size, unmatched by any other matrix. Righetti *et al* (1981a) have titrated the porosity of agarose gels and found that the most diluted gel which can be cast with agarose for IEF (0.16%) has an average pore diameter of 500 nm. Serwer (1980), using Miles HGT(P) agarose, has described gels containing as little as 0.075% solids, which should have an average porosity of about 800 nm. How can agarose present such an open pore structure? It has been demonstrated (Arnott *et al* 1974) that this polysaccharide in solution exists as a double helix (figure 3.3(a)) and therefore it is considerably more rigid than a polyacrylamide strand. Moreover, seven to eleven such helices form bundles which extend as long rods (figure 3.3(b)) thus further strengthening the architectural framework of the gel. It does not seem possible to cast hydrophilic gels with porosities greater than the 500–800 nm range reported, since their structure would collapse under 1 *g* gravity.

Agarose suitable for IEF is now available from Marine Colloids (FMC Corp. Rockland, Mn.) as Isogel and from Pharmacia as Agarose IEF (the latter containing positive charges to balance the negative ones of the matrix, thus exhibiting residual electro-endosmosis at pH extremes). Agarose IEF has been independently described by Rosen *et al* (1979) and by Saravis *et al* (1979, 1980). Excellent resolution of very high M_r proteins such as α-2-macroglobulin (780 000 d), 19S immunoglobulin M (900 000 d), keyhole limpet haemocyanin (3×10^6 d) and zinc glycinate human tumour marker (2×10^6 d), could be achieved by IEF in agarose. Besides its much larger pore size, agarose also presents other significant advantages over other supports, such as polyacrylamide:

(i) it is non-toxic;
(ii) it is gelled without the aid of catalysts, which in polyacrylamide produces long-lived radicals;
(iii) it is fully compatible with subsequent immunofixation and crossed electrophoretic techniques;
(iv) it allows quick and efficient staining and destaining and it is easily dried for permanent records.

A simple gel casting procedure has been made possible by the use of clear, flexible, polyester-based plastic film (Gel Bond, from Marine Colloids)

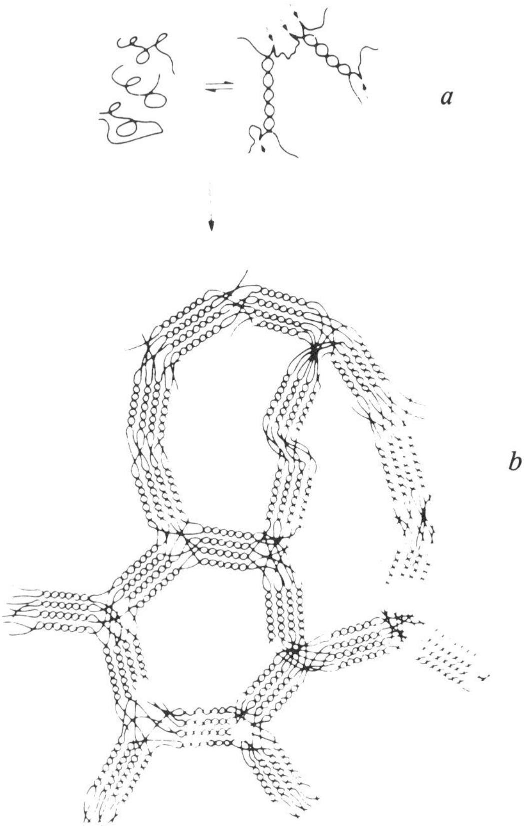

Figure 3.3 Gelation mechanism of agarose: (*a*) melted strands, with free-draining ends, in equilibrium with double strands; (*b*) formation of large-pore gel by lateral aggregation of seven to eleven helices (from Serwer 1983).

rendered hydrophilic on one side for adherence of gels. A piece of Gel Bond film (usually $12.5 \times 12.5\ cm^2$) is placed, hydrophobic side down (the side where a drop of water does not spread, but stays in a bead shape), on a levelling table and, if needed, rolled flat with the aid of a rubber roller (figures 3.4(*a*), (*b*)). In order to avoid too rapid a heat dissipation due to contact with the metal table, which could result in uneven gelling, thermal insulation is suggested by placing a sheet of polystyrol foam beneath the Gel Bond film. Any concentration of agarose between 0.5% and 1.25% (w/v) can be used for IEF; however, since the higher concentrations retain water better than the lower ones, a good compromise is to cast a 0.8% agarose gel, which still has an excellent porosity (170 nm). The desired volume of agarose solution (or a bigger volume, which can then be dispensed in aliquots in individual glass tubes and stored at 4°C until remelted when

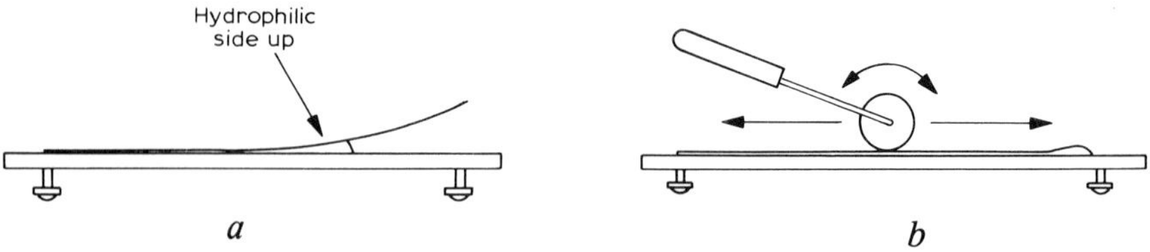

Figure 3.4 (*a*) Placing and (*b*) rolling flat a Gel Bond film on a levelling table for casting agarose gels (from Saravis and Cook 1979. Courtesy: Marine Colloids.)

needed) is boiled in a water bath for 10 min. The flask should be covered with a cold finger or inverted small beaker to prevent evaporation. When agarose is fully dissolved, the flask is transferred to a water bath at 56–58°C, and allowed to equilibrate with the added pre-heated carrier ampholytes to a final concentration of 2.5% and then poured directly onto the Gel Bond horizontal sheet. The volume of agarose solution is chosen so

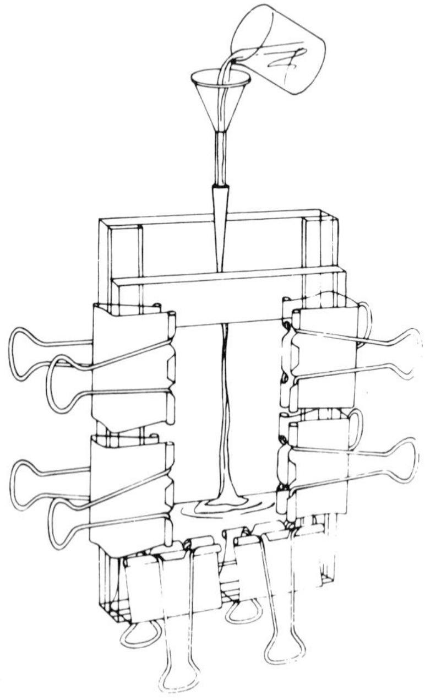

Figure 3.5 Cassette for casting agarose gels thinner than 1 mm. The casting apparatus should be pre-warmed at about 50°C and should contain the Gel Bond film on one face (from Saravis and Cook 1979. Courtesy: Marine Colloids.)

as to have a thickness of about 1 mm of gelled matrix. If thinner gels are required, it is better to let the agarose gel in a vertical cassette, as regularly used for polyacrylamide gel polymerisation (figure 3.5), in order to ensure uniform gel thickness, particularly along the border. The casting assembly will have to be prewarmed and should contain, on one face, the Gel Bond film. The gelled plates are usually aged overnight at 4°C to increase the mechanical strength of the gels. However, it has been suggested (Saravis *et al* 1980) that it is preferable to keep them at room temperature to eliminate the syneresis (gel collapse and water exudation) observed upon storage at 4°C. The samples are applied as 3 μl droplets onto the gel surface, or soaked into Paratex filter pieces (LKB), or with the aid of a 52 μm thick Mylar sheet containing loading slits, or, in the case of biopsies, by direct tissue application onto the matrix (Saravis *et al* 1979). The focusing process is usually over within 60 min at a constant wattage of 10 W (1100 V at steady state) (Vesterberg 1980). At the end of the IEF run, the agarose plate is quickly fixed in 33% methanol, 5% TCA and 3.5% sulphosalicylic acid in water. After the fixation step, the agarose gel is dried onto the Gel Bond film by the sequence illustrated in figure 3.6. It is covered with filter paper and blotting towels and left for 30 min under a 0.5 to 1 kg weight, then fully dried with a hair dryer and freed from the gel-adhering filter paper by a quick wetting step (steps 3 and 4 in figure 3.6). At this point, the dried agarose film can be stained and destained by classical Coomassie Blue methods or, as suggested by Saravis *et al* (1980), by using Crowle and Cline (1977) stain (2.5 g Crocein scarlet, 150 mg Coomassie brilliant blue R-250, 50 ml glacial acetic acid, 30 g TCA in 1 litre final volume).

When running proteins under denaturing and/or reducing conditions (8–9 M urea and 2-mercaptoethanol) it would be ideal to use an open pore matrix like agarose, since upon unfolding the Stokes' radius of a protein is increased (Creighton 1979). Unfortunately agarose gels and urea are not quite compatible, since this matrix is extensively hydrogen bonded and urea is a typical H-bond breaking agent. This problem has been solved by Olsson and Läas (1981) by simply increasing the agarose level from 0.8% to 2% (w/v), adding 10% sorbitol to the gelling solution and letting the 8 M urea, 2% agarose and 10% sorbitol mixture gel in a cassette for 15–20 hrs at 21°C. If there is a danger that cyanate ions, in equilibrium with urea (their concentration is 20 mM in 8 M urea at 20°C and pH > 6), will carbamylate amino groups in proteins, 1% 2-mercaptoethanol (an effective scavenger of cyanate ions) can be gelled in the agarose matrix.

Agarose IEF is also suitable for a variety of applications. Saravis *et al* (1980) have described an immunoperoxidase labelling by which focused antigens can be detected in agarose gels at an antibody dilution of <1 : 12 500. If this sensitivity is not high enough, a 100-fold amplification can be obtained by exploiting the strong binding between avidin and biotin. A complex is formed among biotinylated secondary antibody, avidin and biotinylated horseradish peroxidase: since this is much more stable than the anti-primary (goat–antimouse)-peroxidase conjugate, the focused antigen can be detected at antibody dilutions of up to 1 : 250 000, a remarkable increment in sensitivity.

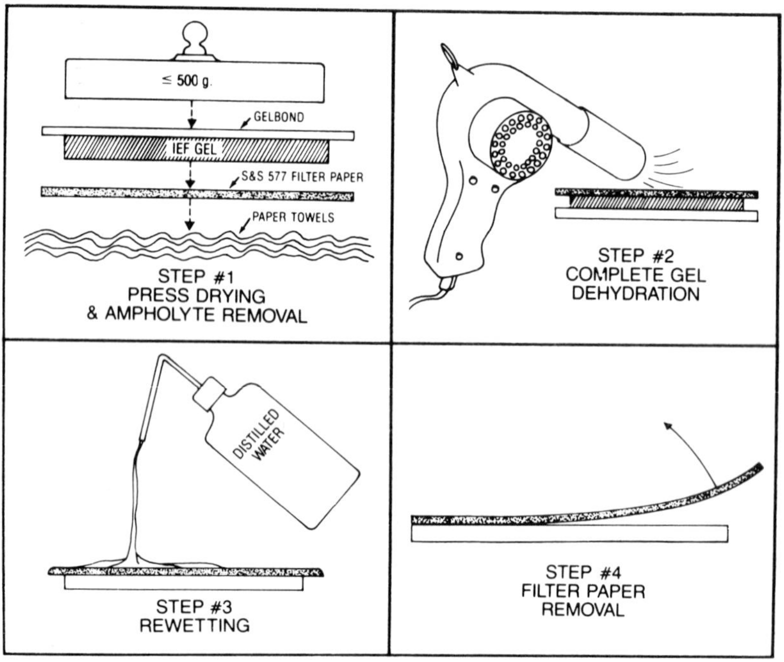

Figure 3.6 Procedure for drying an agarose gel prior to staining. After fixing in TCA, the gel is pressed for 30 min against filter paper and paper towels with a weight of 0.5 to 1 kg. The dehydration is completed with a hair-dryer (step 2) and then the filter paper is removed from the agarose film by a quick re-wetting step (3 and 4) (from Saravis and Cook 1979. Courtesy: Marine Colloids.)

At this point, we would normally have to discuss the other three anti-convective media compatible with IEF: cellulose acetate, granulated gels and polyacrylamide matrices. The first one, notwithstanding recent reports (Ambler and Walker 1979, Harada 1975, Harada *et al* 1980, Janik and Dane 1981), has always given us less than satisfactory results, and we frankly would not dare to suggest it for routine use. Granulated gels (Sephadex, polyacrylamide beads) are excellent supports for IEF but they are more useful for preparative runs than for analytical applications (Radola 1973), therefore they will not be dealt with here. We will thus proceed into the most versatile matrix discovered so far, polyacrylamide layers.

3.4 IEF in Polyacrylamide Matrices

3.4.1 Gel Polymerisation

We will describe in this section a general methodology valid for both gel cylinders and slabs. For detailed information on polymerisation kinetics as a function of different crosslinkers, temperatures and catalysts, the reader is referred to Gelfi and Righetti (1981a, b) and Righetti *et al* (1981b).

Reagents required: recrystallised acrylamide and Bis (or any other appropriate crosslinker except DATD); TEMED; ammonium (or potassium) persulphate, carrier ampholytes; optional gel additives, e.g. sucrose, sorbitol, glycerol, urea, neutral or zwitterionic detergents.

The following stock solutions should be prepared:

Monomer solution: 28.8 g acrylamide + 1.2 g Bis in 100 ml distilled water (30% *T*, 4% *C*);

Carrier ampholytes: 40% solutions (except for pH 2.5–4.0 and pH 9–11, which are 20% (w/v));

TEMED: 1 in 10 dilution in distilled water NH_4-persulphate: 40% (w/v) prepared fresh weekly.

Care should be taken in handling acrylamide since it is a neurotoxin in its monomeric form. Contact with skin should be prevented. There is usually no need to maintain separate solutions of acrylamide and Bis except when using highly-crosslinked gels. It is best to prepare stocks which will last no longer than a month, as acrylamide in solution tends to hydrolise very slowly producing free acrylic acid.

3.4.2. Procedures

(i) Add the desired volume of gel cocktail (a 5% *T*, 4% *C* (Bis) or 4% C (DHEBA) or 4% *C* (BAC) and 2% carrier ampholyte is a universal recipe) to a conical flask with a side arm.

(ii) Stopper the flask and degas thoroughly on a vacuum or water line for 5 min. For reproducible operation, we prefer a mechanical pump giving a vacuum of about 0.1 mm Hg and the Erlenmayer to be connected to a nitrogen tank via a three-way tap, so that, once the vacuum is interrupted, the flask is filled back with nitrogen instead of air. Degassing, in order to be efficient, should be done while warming the gel mix and not while keeping it in ice.

(iii) Add 5 $\mu l\,ml^{-1}$ of TEMED and 1 $\mu l\,ml^{-1}$ of ammonium persulphate (in this order) and mix rapidly.

(iv) Immediately pour the gel cocktail in the slab cassette or in the gel tubes. In the former case, the mould is filled to the brim, while in the latter the cylinder is filled to about 0.5 to 1 cm from the rim and then the gel solution is overlayed with a dilute catalyst mix to ensure uniform gel polymerisation and to avoid a gel meniscus.

(v) Set the gels aside to polymerise (as adopted for IPGs, we now prefer 1 h at 50°C as universal, standard polymerisation procedure).

3.4.3 *Choice of Electrode Solutions*

Suggested formulations for anode and cathode solutions are given in table 3.1 (as suggested in LKB, Pharmacia and Bio Rad technical bulletins). As seen from the table, there is a general agreement to use strong acids and bases only in the wide pH ranges (pH 3–10) while, in narrow ranges, progressively weaker anolytes and catholytes, or mixtures of ampholytes, are recommended.

Table 3.1 Suitable electrode solutions for IEE in gel slabs (from Righetti 1983a)

pH range	*Cathode(−)*	*Anode(+)*
LKB		
3.5–9.5	1 M NaOH	1 M H_3PO_4
2.5–6	0.5% Ampholine, pH 5–7	1 M H_3PO_4
5–8.5	0.1–1 M NaOH or 1% Ampholine, pH 8–10	0.1–1.0 M H_3PO_4 or 1% Ampholine, pH 5–7
7.5–10.5	1 M NaOH	0.1% Ampholine, pH 7–9
Pharmacia		
3–10	1 M NaOH	40 mM aspartic acid
2.5–5	0.2 M histidine or 0.1 M NaOH	0.1 M H_2SO_4
4–6.5	0.2 M histidine	40 mM glutamic acid
5–8	1 M NaOH	40 mM glutamic acid
6.5–9	1 M NaOH	0.25 M HEPES†
8–10.5	1 M NaOH	0.25 M HEPES
Bio Rad		
3–10	1 M NaOH	1 N H_3PO_4
2.9–5.0	2% Biolyte, pH 6–8	1 N H_3PO_4
3.7–6.2	1 M NaOH	1 N H_3PO_4
4.4–7.2	1 M NaOH	1 N H_3PO_4
5.6–8.2	1 M NaOH	2% Biolyte, pH 4–6
7.3–9.5	1 M NaOH	1 N HEPES
8.2–10.0	1 M NaOH	1 N HEPES

† HEPES: *N*-2-hydroxyethylpiperazine-*N'*-2-ethane-sulphonic acid.

3.4.4 *Casting of Ultrathin Polyacrylamide Layers*

The method has been developed by Görg *et al* (1978). The matrix has to be supported by a plastic foil (Gel Bond PAG, the light sensitive brand from Marine Colloids, being the most popular). Gel layers of 120, 240 or 360 μm thickness can be cast by using as gaskets in the gel cassette either one, two or three U-shaped frames cut out from Parafilm sheets. The chamber is assembled as shown in figure 3.7: a thick glass plate (3 mm) is sprinkled with some water, the polyester foil lowered onto it and pressed flat with a rubber roller (see figure 3.4), the U-gasket placed next and finally the cover lid lowered onto it. The cover, if made of glass, should be coated with Repelsilane (LKB) for ease of removal. Since filling of the chamber is not an easy operation, especially with 120 and 240 μm thin gel, the two upper clamps are removed and two paper clips introduced between the glass plates, thus allowing the insertion of the needle of the syringe containing the polymerisation mixture (figure 3.8). When the clips are removed and the clamps fastened in place, the liquid level in the chamber rises to the desired height (this casting procedure, with a vertically standing cassette, is the only operative method for preparing IPG gels!)

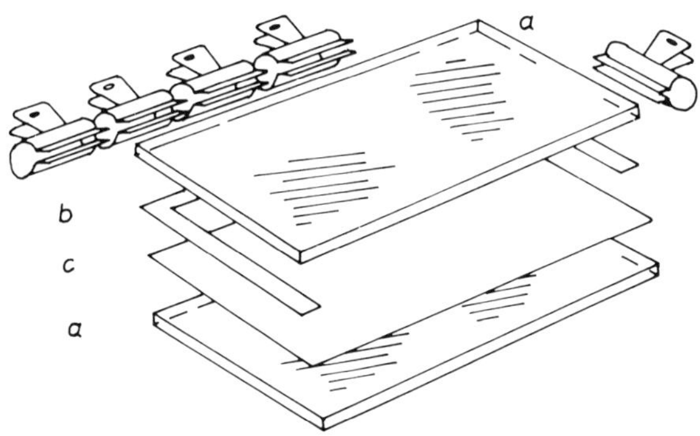

Figure 3.7 Procedure for casting ultrathin polyacrylamide gels onto polyester foils. Assembly of the gel cassette: (*a*) 3 mm thick glass; (*b*) U-shaped Parafilm gasket (1 layer = 120 μm); (*c*) polyester foil (from Görg *et al* 1978).

Several advantages are inherent in this technique:

(i) heat transfer is much more efficient, thus allowing higher field strengths than in 1–2 mm thick slabs;
(ii) as a result, band sharpness and resolution are markedly improved as compared with conventional gels;
(iii) by adhering to the polyester foil during all the operation steps, the gels can be handled very conveniently and are protected from fracture;
(iv) staining, destaining and drying are completed in a fraction of the time needed for thicker gels;
(v) the demand for carrier ampholytes and other reagents is drastically lowered;

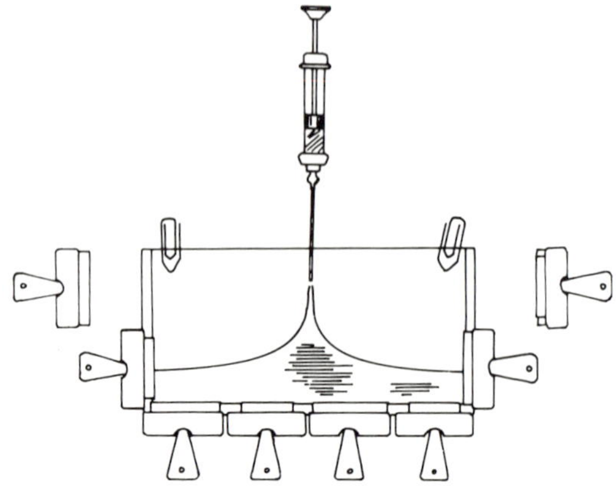

Figure 3.8 Moulding chamber for ultrathin gels. The two upper clamps are removed and the two paper clips inserted between the glass walls, for easier pouring of the polymerisation solution. After the gel mixture has been completely poured in the cassette, the paper clips are removed and the two clamps placed back in the frame, so that the liquid rises to fill up the chamber (from Görg *et al* 1978).

(vi) zymograms can be developed within a few minutes, thus retaining the high band sharpness of the IEF dimension.

For even thinner gels, in the 50–100 μm range, a modification of the above method, called the flap technique, has been described by Radola (1980). Since it would be impossible by the conventional cassette technique to cast such ultrathin gels, the problem has been solved by using a horizontal glass plate with appropriate spacer strips, onto which the total volume of polymerisation mixture is poured. The chamber is sealed by slowly lowering the cover plate onto it, so that the gel mixture is spread evenly between the two plates (figure 3.9). The thin gel veil can be made to adhere to polyester films, or can be covalently bound to the glass surface by using a silane coupling agent (γ-methacryl oxypropyltrimethoxysilane) having the following formula (Bianchi Bosisio *et al* 1980):

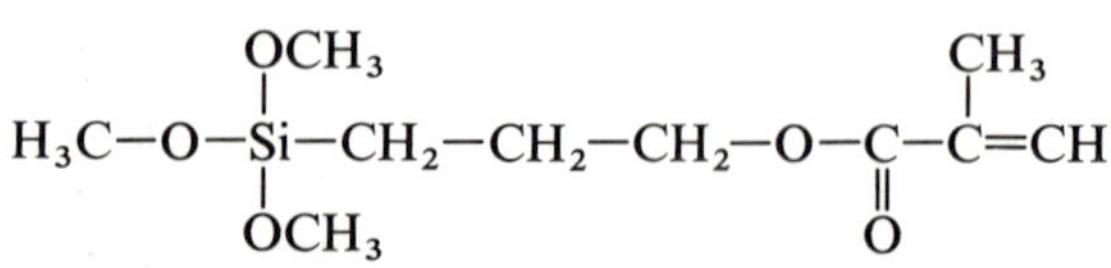

In this last case a true covalent bond is formed since the silane bridge forms a:

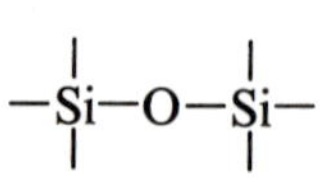

link with the glass surface, while at the other extremity of the molecule, methacryl groups are available for copolymerisation into the polyacrylamide gel. This reagent is available as Silane A-174 from Pharmacia, as Polyfix 1000 from Desaga or as Bindsilane form LKB.

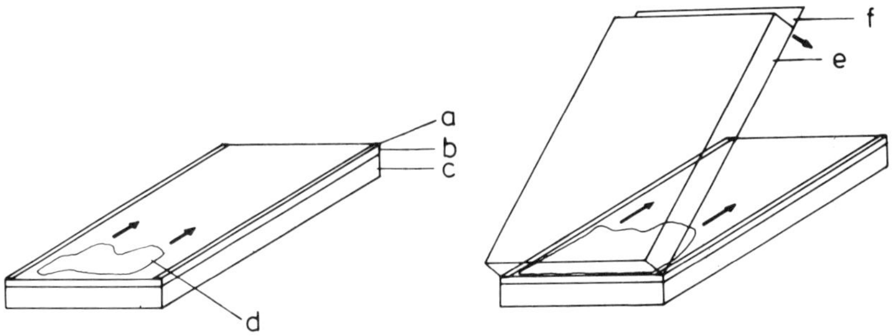

Figure 3.9 Gel preparation with the flap technique: (*a*) 50 or 100 μm spacer strips; (*b*) silanised glass plate; (*c*) glass base plate; (*d*) polymerisation mixture; (*e*) glass cover plate; (*f*) cover film, 100 μm hydrophilic polyester film protruding from the upper plate for easier removal of the polymerised gel (from Radola 1980).

We recommend the following procedure: make up 0.2% Silane A-174 in anhydrous acetone. Just before use, dip a clean, thin glass plate in this solution for 30 s and let it drip to dryness. The silane solution, kept in the dark and at 4°C, is stable for several months. Silanised glass plates are a must when using detergent-laden gels, as this additive will prevent the binding of polyacrylamide layers to Gel Bond PAG. If use of polyester foils is needed even in presence of detergents, we suggest the following process: polymerise an empty gel (e.g. 4 or 5% *T*, 4% *C*) in the absence of additives, including CAs; wash it in 500 ml distilled water for 30 min, then in 100 ml of 2% glycerol; dry it with a fan at room temperature (the gel will thus dry onto the PAG foil without cracking or peeling); re-swell it in 2% CAs plus detergent plus any other additive needed (e.g. 8 M urea, etc) (when using expensive additives, you can slip the dried gel back into the polymerisation cassette and just fill it up with the exact volume needed for swelling back to original size).

Another alternative to Görg's technique is the 'sliding' method of Garoff and Ansorge (1981) (figure 3.10). Three plates of specular glass, 4 mm thick, are used, in one of which ((2) of figure 3.10(*a*)) a notch is made (as in the Studier apparatus, 1973). Plates (1) and (3) are placed on an even support one after the other with a gap of a few cm in between. Along both long sides of the plates teflon spacers, 1.5 cm wide and 0.2 mm thick (4) (or any other suitable thickness) are placed. They are fixed at their ends by adhesive tape. The gap between plates (1) and (3) is covered with a rectangular teflon band of the same thickness as the spacers, having a width matching the distance between the two spacers. One side of this teflon band is made to protrude so as to cover the rim of plate (1), thus completing the framing of this edge. Plate (3) is a dummy and serves as a sliding guide for the cover plate (2). Plate (2) is now placed on top of the dummy plate (3), so that the notched side is about 5 mm in front of the teflon band edge laying on the lower rim of plate (1). The gelling solution is now slowly poured in front of the notch letting the capillary force fill the space between plates (1) and (2). The latter is now slid completely over plate (1) (see figure 3.10(*b*)). The teflon band is now removed, a well former inserted, and the glass plates clamped till polymerisation is completed.

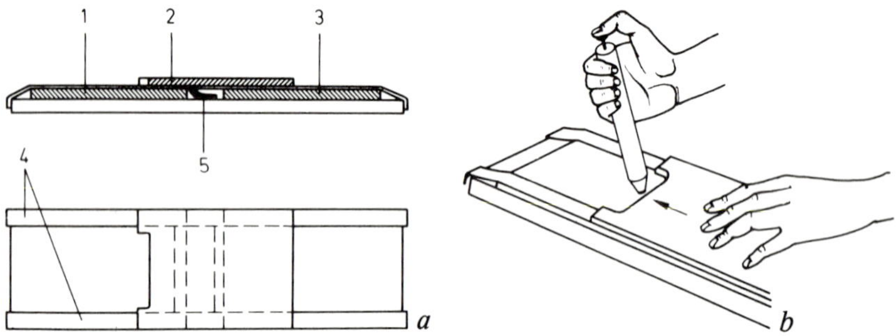

Figure 3.10 (*a*) The layout of the mould for polyacrylamide gel preparation by the sliding technique: (1) bottom glass plate of the chamber; (2) sliding, notched glass cover; (3) dummy glass plate, used as a guide for the cover plate (2); (4) teflon spacers, 1.5 cm wide and of variable thickness; (5) rectangular teflon band covering the gap between plates (1) and (3). (*b*) Moulding of a 0.2 mm thick polyacrylamide gel using the sliding technique (from Garoff and Ansorge 1981).

3.4.5 Sample Application

A small volume of sample solution (10–15 μl) can be applied as a drop or a strip directly onto the gel, the concentration of protein being in the range 1–10 mg ml^{-1}. Thus the total load of each lane may vary from 15 μg to 150 μg (with silver stain this load can be lowered by about 100-fold; see chapter 7 by C Merril). Up to 150 μl may be applied using a plexiglass frame. A piece of artificial sponge (e.g. polyurethane foam) soaked in sample solution can also be employed. A rectangular piece of filter paper impregnated with a sample solution is often used to apply discrete amounts of protein; a 1.5×10 mm^2 strip of Whatman 3MM paper absorbs about 10 μl of liquid. A cotton thread can absorb up to 2 μl. Accurate loading can be made by placing a dry filter paper strip on the gel and pipetting a known amount of sample solution onto it (figure 3.11(*a*)). Alternatively, a superimposable sample applicator, such as a rigid foil with evenly distributed slits can be used. Often wells for sample application are made in gels during casting, though their depth should not exceed $\frac{2}{3}$ of the gel thickness so as not to disturb conductivity too severely (figure 3.11(*b*)).

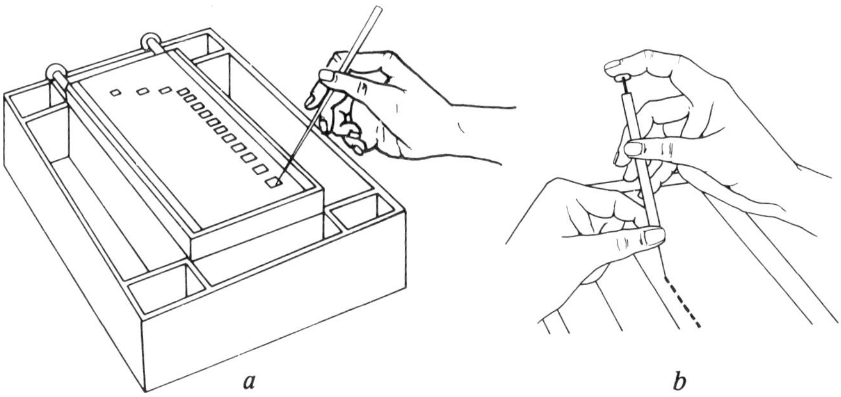

Figure 3.11 (*a*) Sample application onto a surface filter paper. The paper can be placed at different positions between anode and cathode and can be loaded with known sample volumes with the aid of a microsyringe. (*b*) Sample application into slots precast in the gel layer. (Courtesy: LKB Produkter AB.)

The application filter paper can be removed from the gel 1 h after the start of the run. When large original volumes are applied, e.g. with plastic feeders, the total duration of IEF is increased. Some proteins from the original protein mixture can be adsorbed by the filter paper and the degree of this adsorption may be pH dependent. This means that the region of the pH gradient where the paper is placed may be of some importance and, when selecting conditions for a run, several points over the gradient should be checked. Cellulose is known to possess some negative charges and thus the adsorption is as a rule more pronounced at neutral to mildly alkaline pHs where the carboxyls are fully dissociated. Losses are of the order of about 1 μg/10 mm of paper strip, though possible selectivity for minor protein components must be taken into account. Dilute proteins in body fluids (urine, cerebrospinal fluid) can be concentrated and desalted by adsorption onto phenyl Sepharose; elution is affected by directly applying the hydrophobic resin to the gel surface (Vesterberg and Hansen 1978). This is also an excellent method for eluting proteins tenaciously bound to gels for affinity chromatography (Haff *et al* 1979).

3.5 Titration Curves

Figure 3.12 shows how pH mobility curves are generated. A polyacrylamide gel slab is cast with a trench in the middle, $9 \times 0.1\ \text{cm}^2$, which can be loaded with up to 150 μl of sample (kits for generating titration curves are available from LKB Produkter and Bio Rad). The first dimension consists of sorting out electrophoretically the carrier ampholytes contained in the gel, thus generating a stationary pH gradient. No sample is applied at this stage. At this point, the electrode strips, with the respective gel layers underneath, are chopped away with a long knife (figure 3.12(*b*)). This step is essential, otherwise much heat will be generated in these regions in the second dimension, due to the presence of the strong acid and base. New electrode strips are then applied perpendicular to the first run and the trench filled with the sample to be analysed (figure 3.12(*c*)). Now electrophoresis perpendicular to the stationary pH gradient is run, usually at 50 V cm^{-1} and for periods from 10 up to 45 min (figure 3.12(*d*)). Here are examples of fields of application for this technique:

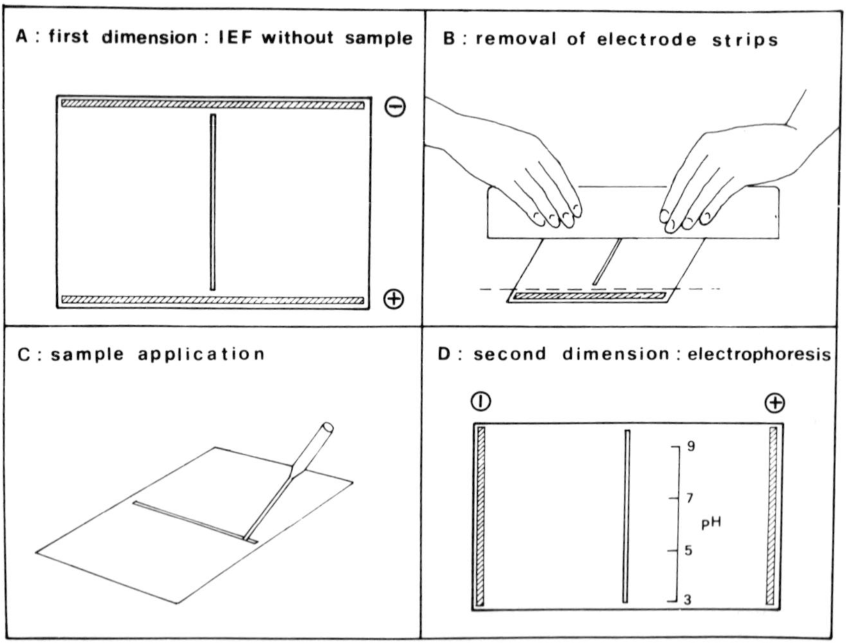

Figure 3.12 Experimental procedure for generating titration curves by IEF electrophoresis. (*a*) The pH gradient is first formed by focusing the carrier ampholyte mixture; (*b*) the electrode strips are then removed; (*c*) the sample is applied in a trench cut perpendicular to the pH gradient; (*d*) the second dimension run is started perpendicular to the first dimension axis (from Righetti and Gianazza 1979).

(i) Genetic Mutants

It is possible to perform 'differential' titration curves by running a protein and its genetic mutants in a mixture. The shape of the respective titration curves should reveal which charged amino acid has been substituted in the mutant phenotype (figure 3.13). For instance, in the case of Lys mutants, the two curves should meet around pH 11, while for Glu or Asp mutants the confluence point should be around pH 3. Double charge mutations (e.g. Lys → Glu) or same charge replacements (e.g. Arg → His) can also be detected. These theoretical titration curves were experimentally verified by running normal human adult haemoglobin (HbA) mixed with any of the following mutants: HbS, HbC, Hb Suresnes and Hb Philadelphia (Righetti *et al* 1978c, Righetti and Gianazza 1979).

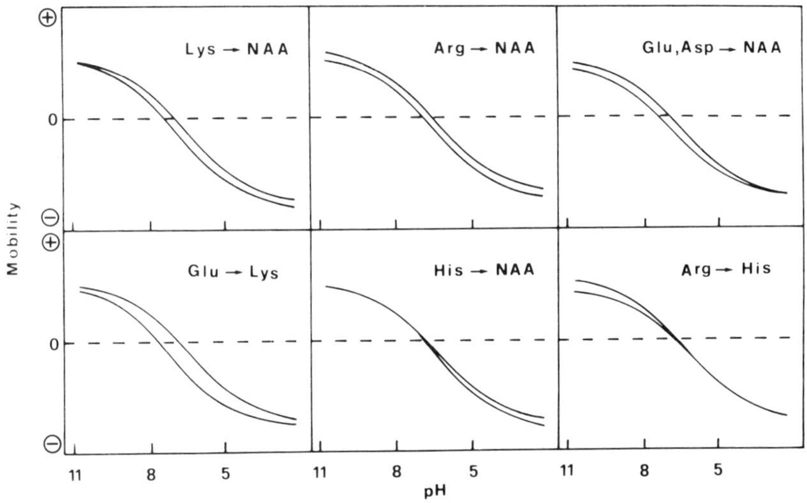

Figure 3.13 Theoretical titration curves of haemoglobin and its genetic mutants in the presence of a given amino acid substitution. NAA: neutral amino acid. The broken lines represent the zero mobility plane (application trench). The arrows indicate the direction of amino acid mutation (from Righetti and Gianazza 1979).

(ii) Macromolecule–Ligand Interaction

Krishnamoorthy *et al* (1978) have been able to run titration curves of met-Hb-inositol hexaphosphate (IHP) and met-Hb-inositol hexasulphate (IHS) complexes. The pH ranges of stability of these ligated states are pH 4.5–6.0 for met-Hb-IHP and pH 3.7–6.0 for met-Hb-IHS. Both complexes appear to have a dissociation constant (K_d) of the order of μm, and survive under these experimental conditions up to 8–10 min of electrophoresis. When the Hb-2,3-diphosphoglycerate complex (known to have a K_d greater than 1 mM) was titrated, only the titration curve of free Hb could be demonstrated within the pH 3–10 range, indicating that the complex was immediately split by the current. It should be possible, however, to detect very labile intermediates by performing the experiment at −10 to −20°C, as described by Perrella *et al* (1978, 1979, 1980). By using the same 2D technique Constans *et al* (1980) have demonstrated the binding of vitamin D3, and its derivatives, to the human serum vitamin D binding protein. It has been shown that the binding of ligand induces conformational changes in the apoprotein. The stoichiometry of the complex, and the pH range of stability of the holoprotein could also be investigated.

(iii) Macromolecule–Macromolecule Interactions

Righetti *et al* (1978b) have exploited this 2D technique to study the interaction between cytochrome b5 (Cyt b5) and met Hb. While the two proteins, when run singly in the gel, develop the classical sigmoidal shape of a titrated macromolecule, when run in a mixture they exhibit strongly distorted patterns, above the pI in the case of Cyt b5 and both above and below the pI for met Hb. The maximum interaction appears to occur in the pH 8.0–8.3 range, and is consistent with a predominant role of Lys residues of met Hb in the binding to acidic amino acids of Cyt b5. As an extension of this work, Lostanlen *et al* (1980) have demonstrated the interaction of NADH-cytochrome b5 reductase and Cyt b5 by a direct enzyme titration curve. Direct detection of titration curves by zymograms or immunoprints has been independently reported by Gianazza *et al* (1980). The interaction between Hb and haptoglobin (Hp), as well as the binding of intact α- and β-globin chains to Hp, have also been studied. While Hp–Hb complexes were completely stable in the pH range 3–10 indicating a predominant role of non-ionic interaction, free β-chains did not appear to bind to Hp while free α-globins exhibited an intriguing, pH dependent pattern, indicating a mixed type of interaction (Righetti and Gianazza 1981).

(iv) Direct pK Determinations from Titration Curves

Righetti *et al* (1979) and Valentini *et al* (1980) have developed a mathematical theory which would allow direct pK determination of ionisable groups from the shape of the pH-mobility curves. Equations have been derived linking cationic or anionic mobilities to the degree of ionisation of simple cations and anions and of uni-uni-valent amphoteric molecules. For non-amphoteric ions, a direct determination of either pKc or pKa can be made by measuring the pH ($pH\frac{1}{2}$) corresponding to $\frac{1}{2}$ mobility in the cathodic or anodic direction, respectively. For amphoteric species, the $pH\frac{1}{2}$ values will have to be corrected by a term accounting for the influence of the degree of ionisation of the opposite charge ion on the mobility curve of the ion being measured.

(v) K_d Determinations from Titration Curves

One of the most recent extensions of pH mobility curves is the possibility of determining dissociation constants (K_d) of ligands to proteins and their pH dependence. This is achieved by extension of techniques developed for affinity electrophoresis (Horejsi 1979). If the ligand is a macromolecule it is simply entrapped in the gel matrix. If it is a small molecule it is covalently bound to the gel fibres. In the presence of increasing concentrations of ligands, the titration curve of the protein is progressively retarded in a pH-dependent fashion. The mobility decrements, when plotted against the ligand molarity in the gel, can be used to calculate K_d values at any pH value. Ek *et al* (1980) and Ek and Righetti (1980) have developed this technique for studying the binding of glycogen to phosphorylases a and b of blue dextran to several dehydrogenases and of sugars to lectins.

As a general conclusion, it can be stated that titration curves hold a great potential for investigating several physico-chemical parameters of proteins and dynamic aspects of their interaction with ligands.

3.6 Production of Narrow pH Gradients

It is well known, from Rilbe's equation on the resolving power in IEF (Svensson 1962a), that good resolution is obtained with compounds having a low diffusion coefficient and a high pH mobility slope at their pI, as well as by increasing the field strength (V/cm^{-1}) and by decreasing the slope of the pH gradient (dpH/dx). This last proposition is, in practice, achieved by utilising the so-called 'narrow pH ranges', which create pH gradients of 1.5 to 2 pH units over the separation distance. When higher resolution is needed, even shallower pH ranges are available, such as pre-cast gel slabs in the pH 4–5 range for the phenotyping of α-1-AT. Available pH ranges, or home-made batches, can be further subfractionated by the IEF process (Righetti and Drysdale 1971, Gianazza *et al* 1975), down to narrow cuts encompassing about 0.4 to 0.5 pH units. Another way of modifying the pH gradient slope in IEF is the 'separator' technique of Caspers *et al* (1977) and Brown *et al* (1977). By adding amphoteric substances (separators) to an ampholine pH gradient, its slope is altered and usually flattened in the pH region corresponding to, or in the proximity of, the pI of the added separator. The following chemicals have been used as separators: tetraglycine (pI 5.2), proline (pI 6.3), β-alanine (pI 6.9), 5-amino-valeric acid (pI 7.5), 4-aminobutyric acid, 7-amino caprylic acid, histidine (pI 7.6) and 6-amino caproic acid (pI 8.0). All of them, except His, are poor carrier ampholytes (pI–pK values of 3 or 4) and thus focus as broad plateaus in the ampholine gradient. Therefore, in order to be able to modify the pH gradient, they have to be used at rather high concentrations, usually about 0.3–0.6 M. On the other hand, good ampholytes such as His (or the dipeptide His–Gly, successfully used for the separation of HbA from HbA1c, a minor glycosylated Hb) (Beccaria *et al* 1978) are already effective at concentrations of 10–50 mM and probably act by interposing themselves between the two species to be separated. 'Separator' IEF adds another dimension to IEF by further improving separations which are at the resolution limit of the technique. Mixtures of different 'separators' can also be used to improve the resolution in a given pH region. Thus Cossu *et al* (1982) have used an equimolar mixture of β-alanine and 6-amino caproic acid for the separation of HbA, HbF and HbFac in umbilical cord blood for thalassemia screening in newborns. Separator IEF has also been recently applied to agarose matrices. Hence Qureshi and Punnett (1981) have described the phenotyping of α-1-AT in agarose gels containing *N*-(2-acetamido)2-aminoethane sulphonic acid (ACES) (Good's buffers for α-1-AT fractionation have also been reported by Frants *et al* 1978).

Yet another variant for pH gradient flattening is the 'strip' technique reported by Altland and Kaempfer (1980). It is based on the assumption that modifying the profile of the gel along the pH axis should be accompanied by changes of pI distance along that axis. Such a modification of the gel thickness along the pH axis can be achieved simply by overlaying a gel strip of a defined thickness and width on the gel at any desired location. In the electric field the overlaid gel strip delivers ampholytes of a distinct pI interval into the strip-free part of the base gel and removes some ampholytes from the latter which were present in the system; this process results in a flattening of the pH gradient in the IEF gel slab. As an example, a pH range 6–8 is transformed into a 7.2–8.0 pH gradient with an anodic strip, whilst with a cathodic strip overlay it will generate a 6–7 pH gradient: with a symmetrical, anodic and cathodic overlay, the central portion of the gradient (pH 6.5–7.5) will be expanded. For best results, the volume of the overlay gel strip should be at least equivalent to the total volume of the base gel slab, while the area occupied on its surface should not exceed 20–25% of the electrode distance. For these reasons, ultra-thin base slabs are preferred (200–300 μm) so that the thickness of the overlay gel can be kept in the range 1.5–2.0 mm. This technique has also been subsequently reported by Läas and Olsson (1981), who have named it 'thickness modified pH gradients'. These last authors have described another way of altering the course of a pH gradient in IEF: 'concentration modified pH gradients'. The principles of these two methods are shown in figure 3.14(*a*), (*b*), (*c*), (*d*).

The latter technique is a bit more cumbersome, since it requires polymerising different gel sections containing variable amounts of carrier ampholytes. In any event, the mechanism of both methods is similar in that over the gel discontinuity there exists a much higher voltage gradient as compared with the surrounding gel regions, which also results in a sharpening of the protein bands over this gel zone. It might also be possible to combine these three different methods ('separators', thickness and concentration modified pH gradients) to achieve even greater resolution for difficult separation problems.

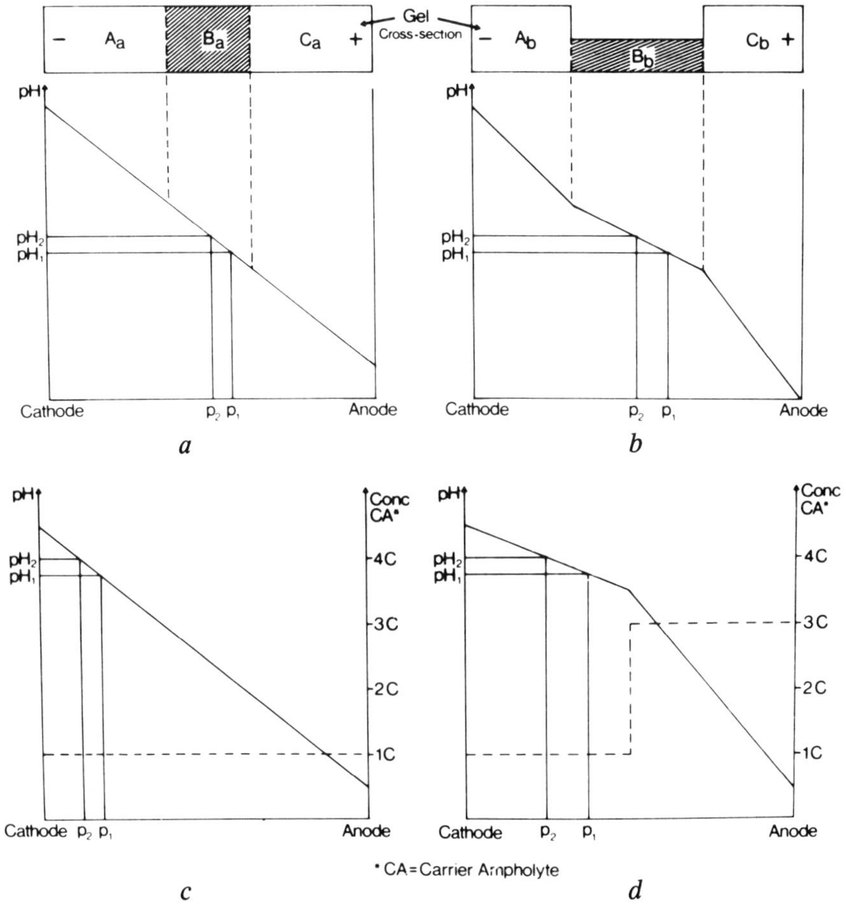

Figure 3.14 (*a*), (*b*) Schematic illustrations of thickness modified pH gradients, (*a*) illustrates the normal experiment with even gel thickness, whereas (*b*) shows the expected pH gradient when the thickness of the indicated part is decreased. The volume within the shaded regions occupies the same percentage of the total gel volume and the same carrier ampholyte species in the two cases. Also indicated are the expected positions of two imaginary proteins p_1 and p_2 with pIs pH_1 and pH_2, respectively. (*c*), (*d*) Schematic illustrations of concentration modified pH gradients. (*c*) pH gradient in the normal case with even concentration of carrier ampholyte over the whole plate. (*d*) Expected pH gradient when half the plate has three times as high a carrier ampholyte concentration as the rest. Also indicated are the positions of two imaginary proteins, p_1 and p_2 with pIs pH_1 and pH_2, respectively (from Läas and Olsson 1981).

3.7 Measurements of pH Gradients

The pI of a protein determined by IEF also represents its isoionic point in the absence of complex-forming ions. By definition, the isoionic point is a measure of the intrinsic acidity of a pure protein, as it is defined as that pH which does not change on addition of small amounts of pure protein. This definition is also applicable to a protein in IEF, as the pH of the iso-electric zone does not alter on addition of more protein.

What are the problems connected with pI estimation in IEF? The major one (often overlooked by scientists in the field) is that pI values are temperature dependent and usually decrease with increasing temperature (Vesterberg and Svensson 1966). Much too often the data in the literature consist of IEF runs at 4°C and pH measurements at 20–25°C (Righetti and Caravaggio 1976, Righetti *et al* 1981c). The difference in pI for the same protein, measured at 25°C and 4°C, could be as high as 0.5 in pH units, the higher value being obtained at the lower temperature. This difference is usually more pronounced in alkaline regions and when a protein has a pI value close to the pK of some of its functional groups. In general, for a strongly acidic protein, dpI/dT should be around -0.005 pH units per degree around 4°C, whereas for a strongly basic protein it should be around -0.03 pH unit per degree. Fredriksson (1978) has published tables which allow the pI, taken by pH measurements at 25°C, of a protein focused at 4°C, to be converted into a true value at 4°C. This has been obtained by measuring the pH shifts which accompany the increase of temperature from 4°C to 25°C of focused Ampholine fractions in sucrose density gradients. According to Gelsema *et al* (1977, 1978) some additives (e.g. glycerol, sorbitol, sucrose but only up to 50%) do not seem to markedly alter the true pI of a protein in IEF (the error could be ± 0.05 pH units in the pH interval $3 < pI < 9$).

In addition to temperature, the other factor that could have a strong influence on pI measurements is the presence of urea. In 6 M urea, Ui (1971) has suggested an overall correction factor of 0.42 pH units, to be subtracted from the apparent pI of the protein in this medium. A correction term ranging from 0.7 to 0.9 pH units has been recommended by Josephson *et al* (1971) in 7 M urea solutions, while Gianazza *et al* (1977) find that the urea effect varies with the various ampholine pH ranges, from 5 to 7.5 hundredths of a pH unit per unit of urea molarity in going from acidic to alkaline pH ranges (i.e. subtract 0.4 pH units at pH 3 and 0.6 pH units at pH 10 to the readings obtained in presence of 8 M urea, the variation between the two extremes being linear).

A third factor which could strongly alter pI measurements above pH 7, is the absorption of CO_2. For a number of alkaline proteins, e.g. Cyt c, chymotrypsinogen, ribonuclease, subtilopeptidase, the pI values measured in thin-layer IEF were consistently lower (by as much as 1 to 2 pH units) than the values determined in density gradient IEF. Delincée and Radola (1978) have found this to be due to CO_2 absorption by the open gel surface, the interference of CO_2 being quite strong above pH 8.2. Thus thin layer IEF above pH 8 should be performed in a CO_2 free atmosphere (e.g. by covering the open gel surface with a plastic foil, by using CO_2-traps (sponges impregnated with NaOH), by flushing the chamber with N_2 or with argon), followed by pH measurements also in a CO_2 free environment.

Table 3.2 List of pI markers for IEF

Protein	*Source material*	*pI(15°C)*	*MW*
Cytochrome c†	Horse heart	10.2	13 000
Myoglobin†	Whale skeletal		
major band	muscle	8.2	17 000
minor band		7.7	17 000
Myoglobin†	Horse skeletal		
major band	muscle	7.4	17 500
minor band		7.0	17 500
Carbonic anhydrase	Bovine erythrocytes	6.1	31 000
β-lactogloblin (A,B)	Bovine milk	5.4, 5.5	35 000
Ovalbumin	Chicken egg	4.8	45 000
Glucose oxidase	*Aspergillus niger*	4.2	186 000
Amyloglucosidase	*Aspergillus niger*	3.6	97 000
Coloured markers†			

† from Marine Colloids, instruction leaflet.

In laboratory practice it is often adequate to map the course of the pH gradient with pI marker proteins. For example, the Isogel pI marker kit (Marine Colloids) contains 8 different proteins with pIs in the 3.6–10.2 pH range (see table 3.2). Because of the presence of minor components, they form 11 easily identifiable spots, sufficient to chart the pH course over the gel length and to interpolate the pIs of the unknowns with accuracy (see figure 3.15) (as the marker pIs are given at 15°C, we suggest you run the IEF experiment at the same temperature). A valid method is always to cut gel segments from anode to cathode (in 5 mm increments), transfer them to a test tube, add 300 μl of 10 mM KCl and immediately read the pH with a combination microelectrode (Righetti and Drysdale 1971). We do not trust pH measurements with surface electrodes, especially the types driven along the gel length by a motor, as the readings are totally unreliable due to lack of equilibration of the glass membrane with the gel surface.

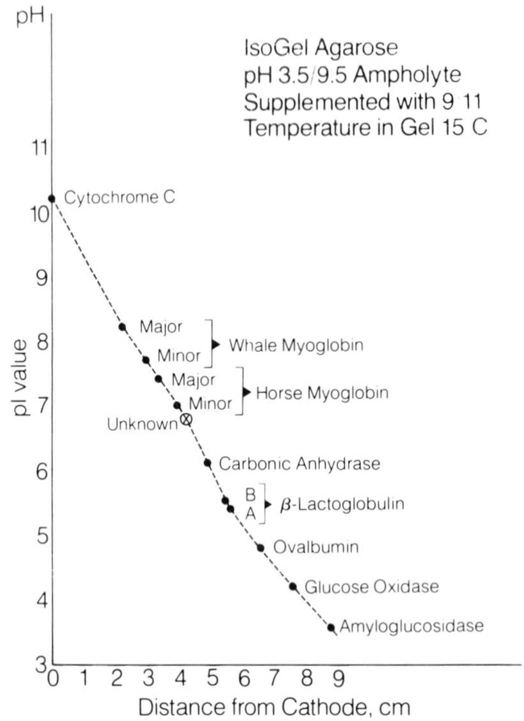

Figure 3.15 Mapping the course of a pH gradient in an IEF experiment with a set of pI marker proteins. The position of each protein (measured as a distance from the cathodic gel end) is plotted against its known pI value and the pI value of any unknown may then be interpolated from the graph. (Courtesy: Marine Colloids 1981 instruction leaflet.)

3.8 Trouble Shooting

It does not pay to have a beautiful, red, slick Ferrari racing car if you do not even know that somewhere in it there is a hood you can raise to check at least some basic functions, like the water and oil levels. So, what do you do when your marvellous machine, your IEF turbo that has made you a winner in so many competitions, abruptly lets you down to hitch-hike your way home? We could try and explain all the methods in about a thousand words but, having found this Marine Colloids' leaflet, we will let it speak on our behalf (see table 3.3). Only two comments are necessary:

(i) this guide has been written for agarose IEF, but most of it will apply to polyacrylamide gels as well;
(ii) please notice point 1. If the 'phenomenon' is optimum performance, the cause has been 'following recommended operating conditions' and the cure will be 'NA' (no one).

Table 3.3 Guide for IsoGel^{TM} agarose isoelectric focusing

Phenomenon	*IEF Band Appearance**	*Cause*	*Cure*
1. Optimum Performance	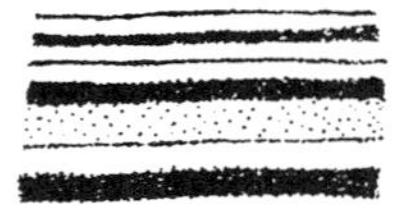	Following recommended operating conditions	N.A.
2. Streaks or gaps perpendicular to bands	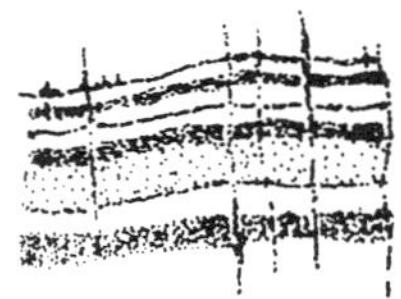	**A.** Particulates in sample **B.** Isoelectric precipitation	**A1.** Prefilter or precentrifuge sample **2.** Remove sample deposit from gel surface with a tissue after removing sample application mask **B1.** Do not prefocus gel **2.** Apply sample in different location
3. Fuzzy bands		**A.** Not focused long enough **B.** Focusing too long	**A.** Increase focusing time or adjust power/voltage **B.** Decrease focusing time or adjust power/voltage

(*Continued*)

Table 3.3 continued

Phenomenon	*IEF Band Appearance**	*Cause*	*Cure*
4. Skewed bands		**A.** Electrode not clean **B.** Uneven electrode contact **C.** Sample applied too close to the edge of the gel **D.** Electrode wicks too short	**A.** Clean electrode **B.** Apply wicks/electrode firmly but gently **C.** Do not apply sample closer than 10 mm from the edge of the gel **D.** Wicks should extend to the edge of the gel
5. Missing or faint bands		**A.** Low M.W. proteins ($\leqslant$10 kD) **B.** Protein not denatured by fixative	**A.** Immunofixation before visualization **B.** Establish appropriate denaturing conditions (increase TCA conc., etc.)
6. Wavy bands		**A.** Salt in sample **B.** Improper anolyte **C.** Old electrolyte solutions **D.** Improper power conditions **E.** Dirty electrodes	**A.** Desalt sample **B.** Replace anolyte **C.** Make fresh **D.** Use proper power **E.** Clean thoroughly before use

7. Arc-shaped bands		**A.** Thin spot in gel	**A1.** Reduce evaporation by keeping gel sealed until used **2.** Improve casting technique especially deaeration **3.** Keep chamber lid closed except when handling gel
8. Burning or sparking		**A.** Dry wicks	**A.** Anodal wick should be wet; cathodal wick should be damp
		B. Conductivity gap in ampholytes	**B.** Supplement deficient region with narrow range ampholytes
		C. Focusing too long	**C.** Reduce focusing time to minimum required to obtain a linear pH gradient
		D. Thin spot in gel	**D.** See #7 above
9. Sample smearing		**A.** Sample overload	**A.** Decrease sample concentration
		B. Sample aggregation	**B.** Solubilize sample with neutral surfactant or in 2% ampholyte

(*Continued*)

Table 3.3 continued

Phenomenon	*IEF Band Appearance**	*Cause*	*Cure*
10. Fluid expression			
A. At cathode only		**A1.** Unblotted wick	**A1.** Blot cathodal wick with filter paper until damp
		2. Focusing too long	**2.** Focus properly
B. Localized elsewhere		**B1.** Salt in sample	**B1.** Desalt sample
		2. Sample overload	**2.** Decrease sample concentration
		3. Uneven electrode contact	**3.** Gently press electrode on wicks
C. In general		**C1.** Use of wrong ampholytes	**C1.** Use IsoGel ampholytes available from Marine Colloids
		2. IsoGel concentration is too low (<1%)	**2.** Increase concentration to 1 or 1.25%
		3. Inadequate curing	**3.** Cure ⩾1 hour under refrigeration
		4. Gel was not adequately blotted	**4.** Use blotting paper
		5. Absence of humectants	**5.** Add up to 10% (W/V) sorbitol and up to 5% (W/V) glycerol to control water activity
11. Condensation inside chamber		**A.** Inefficient cooling	**A.** Adjust cooling water temperature to 10°C
		B. Too high a voltage/power setting	**B.** Adjust 1W for 1st 10 min, 25W limiting

		C. Ampholyte deficiency	**C.** Recheck ampholyte type and formulation
		D. Salt in sample	**D.** Desalt sample
		E. Humid ambient conditions	**E1.** Place humectant cartridge in chamber **2.** Select a chamber with a condensation coil in the lid
12. Incomplete sample uptake		**A.** Gel too thin	**A.** Increase gel thickness
		B. Sample application mask not left on long enough	**B.** Allow at least 10 min. at 1W power setting to ensure complete uptake
		**C.** Gel too wet when sample applied	**C.** Blot gel once just before sample application
		D. Insolubles or aggregates in sample	**D1.** Prefilter or centrifuge sample **2.** Remove sample deposit from gel surface with a tissue after removing sample application mask
13. Gel Thinning **A.** At electrode		**A.** Improper electrolyte	**A.** Decrease sample concentration
B. Near point of application		**B.** Sample overloaded	**B.** Follow storage recommendations
C. Generalized		**C.** Gel stored too long under low relative humidity conditions	

(Continued)

Table 3.3 continued

Phenomenon	*IEF Band Appearance**	*Cause*	*Cure*
14. Uneven pH gradient		**A.** Use of wrong ampholytes	**A.** Use only IsoGel ampholytes
		B. Ampholyte concentration too low	**B.** Ampholyte concentration should be ⩾2.5%
		C. Gel over-focused	**C.** Do not exceed the time required to form a linear pH gradient
		D. Impure distilled water	**D.** Check distilled water for heavy metals and related ionic contaminants
15. Loss of basic region of pH gradient		**A.** Cathodic drift	**A.** Supplement gel with pH9–10
		B. Overfocusing	**B.** Decrease focusing time
16. Current increases with time	N.A.	**A.** Electrode strips applied at the wrong end of the gel	**A.** Apply cathodal (−) electrode to NaOH-soaked wick and anodal (+) electrode to 0.5M HOAc-soaked wick
17. Gel peeling off GelBond backing		**A.** Insufficient gel curing time	**A.** Cure gel for ⩾1hr. at 4–8°C
		B. Gel was cast on the wrong (i.e. hydrophobic) side of GelBond	**B.** Test GelBond surface with water droplet before casting. Droplet spreads on hydrophilic side

18. High background stain	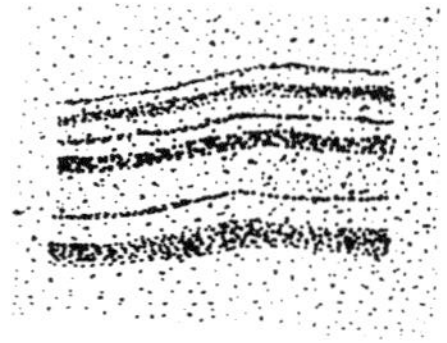	**A.** Residual ampholytes	**A.** Follow recommended press-blotting
		B. Old stain	**B.** Make up fresh stain and make sure all dye is in solution
		C. Gels allowed to settle to bottom of staining dish	**C.** Float plate gel side down in stain for 15–30 minutes
19. Diagonal band migration		**A.** Uneven electrode contact	**A.** Apply electrodes evenly on wicks
		B. Too much pressure/weight on electrodes	**B.** Adjust for light but firm contact
		C. Uneven electrode positioning	**C.** Ensure that electrodes are parallel and perpendicular to the defective sample migration
20. Lateral band migration	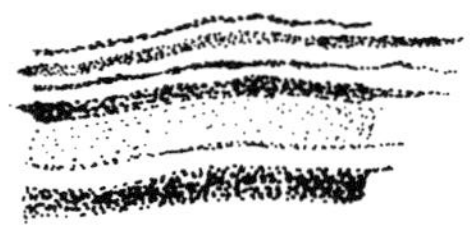	**A.** Too much fluid applied to sample slot	**A.** Reduce sample volume, increase sample concentration
		B. Sample mask left on too long/not long enough	**B.** Remove mask after 10 minutes at 1 watt

* Idealized schematic

3.9 Artefacts: a Unified View

With the widespread use of IEF it has been accepted that microheterogeneity is a phenomenon of wide prevalence, while homogeneity of a protein seems to be an impossible dream. Several explanations can be given for microheterogeneity:

(i) conformers: the components may represent alternative stable conformational substates;
(ii) the separated bands could be caused by possible oligomeric combinations of different subunits;
(iii) in glycoproteins, the isolated IEF zones could correspond to non-uniform distribution of saccharide units among different molecules;
(iv) varying degrees of enzymatic phosphorylation, methylation or acetylation could be responsible for the IEF pattern, as is often the case with nuclear proteins;
(v) there always remains the possibility of proteolytic damage, probably as an artefact of the preparation;
(vi) partial deamidation of asparagine and glutamine residues;
(vii) tightly bound cofactors, or inhibitors, to a protein moiety, further complicated by different charge states of these ligands, such as a redox state.

On the other hand, we have found that there are classes of compounds which display an artefactual heterogeneity upon IEF fractionation, but they belong to limiting structures and sequences which bear no relationship with those commonly found in proteins. We will list these cases below, together with a proper interpretation of these phenomena.

(i) IEF of Nucleic Acids

In 1972 Drysdale and Righetti reported the first fractionation of tRNA species by IEF. Subsequently, several reports appeared on the IEF separation of α- and β-globin mRNAs as well as of ribosomal 18S and 28S RNAs from rabbit reticulocytes (Shafritz and Drysdale 1975, Drysdale 1977). These 'isoelectric fractionations' are unexpected on theoretical grounds, since a method of 'pure charge fractionation', such as IEF, is indifferent to polydispersity in molecular mass where charge-to-mass ratio is constant. Moreover, a true theoretical pI for tRNA (and probably for most nucleic acids) is centred around pH 2.6 (Morris 1978), while all RNAs tested exhibited 'apparent pI values' in the pH range 3.5–5.2. Indeed Galante *et al* (1976), on the basis of the variation of melting points of tRNA in the presence of Ampholine, have demonstrated that the mechanism of IEF fractionation of nucleic acids is the formation of a strong complex between them and the carrier ampholytes used to generate and stabilise the pH gradient in IEF. The complex is strongly pH dependent: it is practically non-existent at neutral pH (pH 7.4), rather weak at pH 5.4 and very strong at pH 4.2 and lower, i.e. in the typical pH range of apparent isoelectric fractionation of tRNAs.

(ii) IEF of Heparin

Nader *et al* (1974) and McDuffie *et al* (1975) reported that heparin could be fractionated by IEF into 21 components, with apparent pI values between pH 3.2 and 4.2. The basis for the fractionation was associated with the polydispersity in M_r the 'low pI' components exhibiting the smallest size (3 000) while the 'high pI' species had the highest M_r (37 000), with graded intervals of around 1 500–2 000 atomic mass units (u).

Heparin fractionation by IEF is even more puzzling than nucleic acid separation, since, while at least the latter are amphoteric, the former is not and is a simple polyanion, with no positive counter-charges. Righetti and Gianazza (1978) and Righetti *et al* (1978a) demonstrated the IEF heparin profile to be artefactually elicited by interaction with carrier ampholytes. The 21 IEF fractions indeed represent 21 different complexes of the same

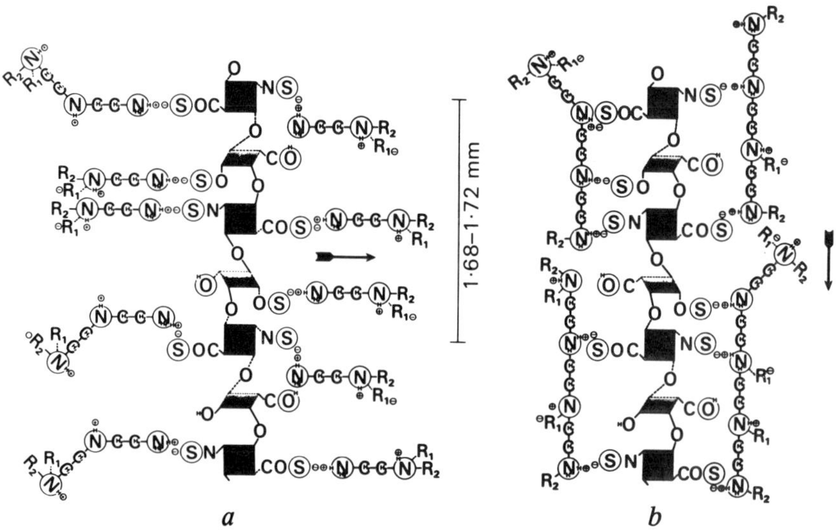

Figure 3.16 An example of a possible artefact in IEF. Model for the interaction between segments of heparin and segments of Ampholine molecules. (*a*) Represents the aggregation of ampholytes on heparin in an excess of Ampholine, e.g. at a ratio of one Ampholine molecule to one heparin charge. (*b*) Shows the complex formed in an excess of heparin, e.g. with one Ampholine species per three charges on heparin at pH = 3.5. Notice how, between the two extremes, the vectorial orientation of the ampholytes varies from perpendicular to parallel to the heparin helix (from Righetti *et al* 1978).

macromolecule with 21 specific Ampholine molecules. The complexes differed markedly in binding strength, the high pI species forming the weakest interactions, which could be extensively split by toluidine blue, in a competitive binding test, while the lowest pI species exhibited the strongest aggregates, which could not be disaggregated by excess basic dye. It was then found that not only the binding strengths, but also the stoichiometries of these complexes could vary considerably, depending upon the relative Ampholine heparin ratios in solution. In an excess of Ampholine, aggregation is perpendicular to the heparin chain, with the end ammonium charge of each Ampholine molecule neutralising one negative charge along the heparin molecule (at this stoichiometry, heparin would bind an amount of Ampholine twice as much its own mass, giving a complex of around 37 000, starting from its original M_r of 12 600). In heparin excess, the bound Ampholine is aligned parallel to the heparin molecule, so that on average one Ampholine species neutralises approximately three negative charges. As shown in figure 3.16, these two models represent two limiting stoichiometries, with any possible ratios in between. In both models unbound, protonated nitrogens in the Ampholine molecules can act as crosslinks or bridges among different heparin helices, thus favouring the formation of huge macromolecular aggregates.

(iii) IEF *of Polyanions*

It is now understood (Gianazza and Righetti 1978) that the anomalous fractionations of nucleic acids and heparin were just particular cases of a more general phenomenon by which all polyanions would elicit complex formation with carrier ampholytes in the pH range 3.2–5.0. As previously reported, the affinity of Ampholine for polyanions is strongly pH dependent, at pH 3 being around 10 000 fold greater than at pH 7. This seems to be caused by the fact that at pH 3 Ampholine species exhibit a stretch of 4 to 5 protonated nitrogens (each two methylene groups apart), while most of its carboxyls bear no negative charge, thus allowing for strong, cooperative binding to polyanions. At pH 7 the net charge will be reversed, and the anionic Ampholine will be repelled by the negatively charged polyanion (Righetti *et al* 1978b). Binding is affected by the type of negative charge, its density and spatial orientation on the polyanion. On the basis of the type of negative charge, the binding strength decreases in the following order

polyphosphate > polysulphate > polycarboxylate.

The minimum length of the polyanion eliciting a measurable binding appears to be of the order of about six negative charges, as demonstrated with hexametaphosphate.

(iv) IEF of Dyes

The same types of phenomena reported with all polyanions were then encountered when focusing acidic dyes (mostly mono- to tetra-sulphonated species) (Righetti *et al* 1977), except that here it is the Ampholine molecule itself which acts as a template to which as many as four to five dye molecules can bind simultaneously. Moreover, the dye can act as a cross-link over several Ampholine species, forming a three-dimensional dye–Ampholine network which aggregates and precipitates in the focusing process. In fact, even in the test tube, simple addition of Ampholine buffered at pH 3 to any polyanions or to most of the dyes tested gives rise to turbidity and eventually to sample precipitation. This 'non-isoelectric precipitation' does not take place if the pH of the polyanion or dye solution is lowered to pH 3 in the absence of CA. These complexes can only be disaggregated by changing the dielectric constant of the solvent, or by high temperatures, or both. On an equimolar basis, the order of dissociating power of the most commonly used disaggregating agents is

$$\text{dimethylformamide} = \text{tetramethyl urea} > \text{dimethylsulphoxide} > \text{urea} > \text{formamide.}$$

A two step binding model was postulated: first a strong, polydentate and undissociated Ampholine–dye salt, followed by a hydrophobic dye–dye interaction.

(v) IEF of Polycations

At the other extreme of the pH scale, when polyArg was run (Gianazza and Righetti 1980) the complex was found to be so strong that the polycation, applied at the anode to a prefocused gel slab, swept away on its wake the focused Ampholine molecules and precipitated as a smear in the 7–8 pH range, where the carrier ampholytes would behave as anions. This 'perturbed Ampholine pattern' is a good indication of complex formation, provided the applied sample is salt-free, since excess salt in the sample could distort the pattern of focused Ampholine. However, this phenomenon seems to be limited to polyArg, since very little, if any, binding occurs with polyLys and no binding at all could be demonstrated with polyHis. Moreover, histones run through the entire gel length, collecting at the cathodic filter paper strip, with no indication of complexes with Ampholine.

In addition to all the classes of artefacts we have listed, there appears to be yet another group of artefacts: when focusing mammalian cells, there is strong evidence that carrier ampholytes can coat the cell surface by binding to sialic acid residues via the protonated nitrogens in the oligoamino backbone (McGuire *et al* 1980). Except for these artefacts generated by peculiar structures, there is practically no evidence that carrier ampholytes will elicit the same multimodal distribution by interacting with proteins. The only real artefact on IEF of a protein has been described by Jonsson *et al* (1978) working with rat incisor phosphoprotein (RIP). RIP binds Ampholine, generating a series of bands with apparent pI values in the pH range 2.5–3.1, whilst its true pI is 1.1. This is not unexpected: RIP is an extremely acidic protein with a charge density similar to that of heparin, so that to all practical effects it behaves as a polyanion. Perhaps lipoprotein lipase also behaves in a similar fashion (Bengtsson and Olivecrona 1977), but in the vast majority of cases the heterogeneity detected in a given sample upon IEF will be inherent in the protein itself, and the scientist will have to find a way to live with it.

Immobilized pH Gradients

It is hard to squeeze into a few pages what is already a fairly large body of data on the new separation technique of immobilised pH gradients (IPG). Luckily, the novelty of the technique and the initial difficulties in grasping the theory underlying it have discouraged many scientists from entering this field, so that we only have to review work coming from a few laboratories. In fact, when we launched this method at the meeting of the International Electrophoresis Society in Athens in the spring of 1982, as a result of an intensive collaborative effort Milano–Bromma–Munich, fellow scientists were so reluctant to pick it up that they left us a two year grace period during which we could make a 360° sweep of the field, extensively developing the technique to the present stage. The following is a list of basic references

(i) the milestone article, laying the foundations of IPGs (Bjellqvist *et al* 1982);

(ii) description of extended pH intervals, spanning 2–6 pH units (computer programs and formulations) (Dossi *et al* 1983, Gianazza *et al* 1983a, 1984a);

(iii) strategy for optimizing preparative runs (manipulations of gel thickness, pH gradient width, buffering power, ionic strength; the discovery of the high loading capacity of 'soft gels'; retrieval of proteins from Immobiline matrices) (Casero *et al* 1985, Ek *et al* 1983, Gelfi and Righetti 1983, 1984);

(iv) co-polymerisation kinetics of Immobilines into polyacrylamide gels and swelling kinetics of dried IPG gels (Righetti *et al* 1984, Gelfi and Righetti 1984);

(v) the first two-dimensional (2D) maps and the use of urea and detergents in IPG systems (Gianazza *et al* 1983b, 1984b, Westermeier *et al* 1983);

(vi) early reviews in the field of IPG and 2D maps (Righetti 1983a, b, 1984, Righetti *et al* 1983b);
(vii) the first (and we hope the last) artefact with the IPG technique (Righetti *et al* 1983a).

3.10 The Chemicals; the Immobiline Matrix

IPGs are based on the principle that the pH gradient, which exists prior to the IEF run itself, is copolymerised, and thus insolubilised, within the fibres of the polyacrylamide matrix. This is achieved by using as buffers a set of seven non-amphoteric, weak acids and bases, having the following general chemical composition: $CH_2{=}CH{-}CO{-}NH{-}R$, where R denotes either one of three different weak carboxyls, with pKs 3.6, 4.4 and 4.6, or one of four tertiary amino groups, with pKs 6.2, 7.0, 8.5 and 9.3 (see table 3.4). During gel polymerisation, these buffering species are efficiently incorporated into the gel (84–86% conversion efficiency at 50°C for 1 h) (Righetti *et al* 1984). Immobiline-based pH gradients can be cast in the same way as conventional polyacrylamide gradient gels by using a density gradient to stabilise the Immobiline concentration gradient, with the aid of a standard, two-vessel gradient mixer. As shown in the formula, these buffers are no longer amphoteric, as in conventional IEF, but are bifunctional: the buffering group is located at one extreme of the molecule and the acrylic double bond, which will be consumed during the grafting process, is at the other extreme. The three carboxyl Immobilines have rather small temperature coefficients (dpK/dT) in the 10–25°C range because of their small standard heats of ionisation (around 1 Kcal/mol) and thus exhibit negligible pK variations in this temperature interval. On the other hand, the four basic Immobilines exhibit rather large pKs (as much as pK = 0.44 for the 8.5 pK species) because of their larger heats of ionisation (6–12 kCal/mol). Therefore, for reproducible runs and pH gradient calculations, all the experimental parameters have been fixed at 10°C. Temperature is not the only variable that will affect Immobiline pKs (and therefore the actual pH gradient generated). Additives in the gel will change the water structure (chaotropic agents, such as urea) or will lower its dielectric constant, and the ionic strength of the solution itself will alter their pK values: for the corrections to be made, see Gianazza *et al* (1983b). The list of seven Immobilines is not exhaustive. For generation of extended pH gradients, with a two-chamber mixer, you will need two additional ones: a strongly acidic ($pK < 1$) and a strongly basic ($pK > 12$) compound with pKs well outside the desired pH range. It is not unreasonable to predict that they will contain a sulphate and a quaternary amino group, respectively.

Figure 3.17 depicts a segment of the hypothetical structure of an Immobiline matrix (*b*) as compared with the partial formula of a carrier ampholyte

Table 3.4 Properties of Immobilines™ Apparent pK values‡ $I = 10^{-2}$ (from Bjellqvist *et al* 1982)

			Polyacrylamide gel‡ T = 5%, C = 3%		*Polyacrylamide gel‡ T = 5%, C = 3% glycerol 25% (w/v)*		*Physical state at room temperature*
	10°C	*25°C*	*10°C*	*25°C*	*10°C*	*25°C*	
Acids with carboxyl as buffering group							
Immobiline pK 3.6	3.57	3.58	—	—	3.68 ± 0.02	3.75 ± 0.02	solid
Immobiline pK 4.4	4.39	4,39	4.30 ± 0.02	4.36 ± 0.02	4.40 ± 0.03	4.47 ± 0.03	solid
Immobiline pK 4.6	4.60	4.61	4.51 ± 0.02	4.61 ± 0.02	4.61 ± 0.02	4.71 ± 0.03	solid
Bases with tertiary amine as buffering group							
Immobiline pK 6.2	6.41	6.23	6.21 ± 0.05	6.15 ± 0.03	6.32 ± 0.08	6.24 ± 0.07	solid
Immobiline pK 7.0	7.12	6.97	7.06 ± 0.07	6.96 ± 0.05	7.08 ± 0.07	6.95 ± 0.06	solid
Immobiline pK 8.5	8.96	8.53	8.50 ± 0.06	8.38 ± 0.06	8.66 ± 0.09	8.45 ± 0.07	liquid
Immobiline pK 9.3	9.64	9.28	9.59 ± 0.08	9.31 ± 0.07	9.57 ± 0.06	9.30 ± 0.05	liquid

† pK values measured with glass surface electrode without any corrections.

‡ Mean values of 10 determinations. Due to the slow response of the electrode the pK values for the amines are uncertain.

molecule (in the former, the long strings represent the polyacrylamide coils, while the short, vertical segments are the cross links). Being copolymerised within the matrix, the Immobiline buffers cannot migrate any longer in the electric field: this means that the pH gradient is stable indefinitely (but it has to pre-exist before the onset of polymerisation) and can only be destroyed if and when the polyacrylamide gel is hydrolysed. At the matrix concentration normally used (5% *T*) and the standard Immobiline levels (around 10 mM buffering ion) there is about one buffering group (or titrant) every 47 acrylamide residues, which means that, statistically, there are about 94 carbon atoms between two adjacent Immobiline species. This is extremely important for the ionic strength of the system, as compared with conventional IEF (Righetti *et al* 1983b).

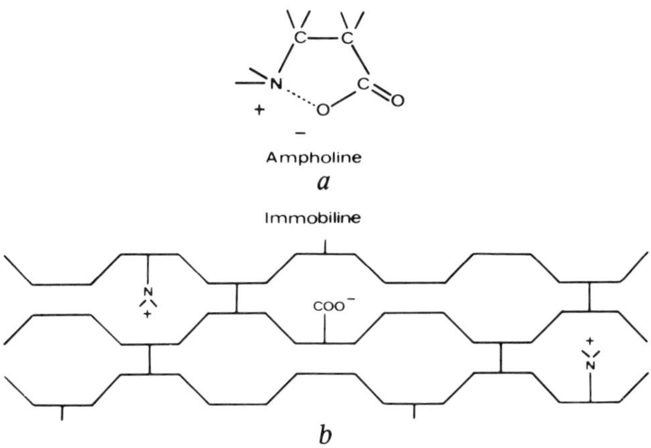

Figure 3.17 Comparison between the ionic strength in Ampholine and Immobiline gels. (*a*) Depicts a segment of an Ampholine molecule and the lower tracing (*b*) shows a portion of an Immobiline gel. An Ampholine molecule, at its pI, is likely to form an inner salt which does not contribute to the ionic strength of the system. The fixed charges in the Immobiline matrix, being spaced about 150 carbon atoms apart, are believed to behave as point charges in the surrounding space, thus effectively contributing to the ionic strength of the mileu (from Righetti *et al* 1983b).

3.11 Narrow and Ultra-narrow pH Gradients

We define, by the above section heading, gradients which span from a minimum of 0.1 to a maximum of 1 pH unit. Within these limits, in general, we work on a 'tandem' principle, i.e. we choose a 'buffering' Immobiline (e.g. a base or an acid), having its pK inside the pH interval we want to generate, and a 'non-buffering' Immobiline (e.g. an acid or a base, respectively), having its pK at least 2 pH units removed from either pH_{min} or pH_{max} of our pH range. This is to provide equivalents of, respectively, the acid or base, in titrating the buffering group but will not itself buffer in the desired pH interval. For these calculations, we used to resort to modified Henderson–Hasselbalch equations (Bjellqvist *et al* 1982) and to rather complex nomograms found in the LKB Application note No 321. Things are much easier now: in Note No 324 you will find no less than 58, 1 pH unit wide gradients, starting with the pH 3.8–4.8 interval and ending with a pH 9.5–10.5 span, spaced apart at 0.1 pH unit increments. In table 3.5 we reproduce the relative recipes, giving the Immobiline volumes needed in the acidic (mixing) chamber to obtain pH_{min} and the corresponding volumes for the basic (reservoir) chamber of the gradient mixer needed to generate pH_{max} of the desired pH interval. It is only between pH 4.6–5.1 (pH_{min}) and 7.2–7.4 (pH_{min}) that we work on the troika principle, as in these regions there are wide gaps in the pKs of neighbouring Immobilines. As an example, let us take the 4.6–5.6 pH interval: there are no available Immobilines with pKs within this pH region, so the nearest species, pKs 4.6 and 6.2 will have to act both as partial buffers and partial titrants and then a third Immobiline will be needed in each vessel, a true titrant, which will bring the pH to the desired value (for pH_{min} pK 3.6 will be used and for pH_{max}, pK 9.3 as titrant, respectively).

If a narrower pH gradient is needed, this can be derived from any of the 58 pH intervals tabulated by a simple linear interpolation of the intermediate Immobiline molarities. Suppose that from a pH 6.8–7.8 range (excellent for most haemoglobin (Hb) analyses) we want to obtain a 7.1 to 7.5 pH gradient (this will resolve neutral mutants which cofocus with HbA). Figure 3.18 shows the graphic method: the limiting molarities of the two Immobilines in the 1 pH unit interval are joined by a straight line (because the pouring of the gradient is done linearly) and then the new pH interval is defined according to experimental needs (in our case, 7.1–7.5 pH). Two lines are drawn from the two new limits of the pH interval, parallel to the ordinates (broken vertical lines). Where they intersect the two sloping lines defining the two Immobiline molarities, four new lines (broken with arrow heads) are drawn parallel to the abscissa and four new molarities of the Immobilines defining the new pH interval are read directly on the ordinates. This process can be repeated for any desired pH interval, down to ranges as narrow as 0.1 pH units.

3.12 Extended pH Gradients

In this case one has to mix several buffering species and the situation becomes considerably more complex. We have solved that with the aid of

Table 3.5 Narrow pH gradients: volumes of Immobiline for 15 ml of each starting solution (2 gels). From LKB Application Note No 324 (1984). The pH range (given in the middle two lanes) is the one existing in the gel during the run at 10°C. For controlling the pH of the starting solutions, the values (control pH) are given at 20°C. LKB Produkter AB, Bromma, Sweden

	Volume (μl) 0.2 M Immobiline pK, Acidic dense solution										Volume (μl) 0.2 M Immobiline pK, Basic light solution						
Control pH at 20°C	*3.6*	*4.4*	*4.6*	*6.2*	*7.0*	*8.5*	*9.3*	*pH range*	*mid point*	*Control pH at 20°C*	*3.6*	*4.4*	*4.6*	*6.2*	*7.0*	*8.5*	*9.3*
3.84 ± 0.03	—	750	—	—	—	—	159	3.8–4.8	4.3	4.95 ± 0.06	—	750	—	—	—	—	591
3.94 ± 0.03	—	710	—	—	—	—	180	3.9–4.9	4.4	5.04 ± 0.07	—	810	—	—	—	—	667
4.03 ± 0.03	—	—	755	—	—	—	157	4.0–5.0	4.5	5.14 ± 0.06	—	—	745	—	—	—	584
4.13 ± 0.03	—	—	713	—	—	—	177	4.1–5.1	4.6	5.23 ± 0.07	—	—	803	—	—	—	659
4.22 ± 0.03	—	—	689	—	—	—	203	4.2–5.2	4.7	5.33 ± 0.08	—	—	884	—	—	—	753
4.32 ± 0.03	—	—	682	—	—	—	235	4.3–5.3	4.8	5.42 ± 0.10	—	—	992	—	—	—	871
4.42 ± 0.03	—	—	691	—	—	—	275	4.4–5.4	4.9	5.52 ± 0.12	—	—	1133	—	—	—	1021
4.51 ± 0.04	—	—	716	—	—	—	325	4.5–5.5	5.0	5.61 ± 0.14	—	—	1314	—	—	—	1208
4.61 ± 0.05	562	—	600	863	—	—	—	4.6–5.6	5.1	5.69 ± 0.04	—	—	863	—	—	105	
4.75 ± 0.05	458	—	675	863	—	—	—	4.7–5.7	5.2	5.79 ± 0.04	—	—	863	863	—	—	150
4.86 ± 0.04	352	—	750	863	—	202	—	4.8–5.8	5.3	5.90 ± 0.04	—	—	863	863	—	—	—
4.96 ± 0.03	218	—	863	863	—	—	—	4.9–5.9	5.4	5.99 ± 0.03	—	—	863	863	—	—	248
5.07 ± 0.03	158	—	863	863	—	—	—	5.0–6.0	5.5	6.09 ± 0.04	—	—	863	803	—	—	338
5.17 ± 0.04	113	—	863	863	—	—	—	5.1–6.1	5.6	6.20 ± 0.04	—	—	863	713	—	—	443
5.24 ± 0.18	1251	—	—	1355	—	—	—	5.2–6.2	5.7	6.34 ± 0.04	337	—	—	724	—	—	—
5.33 ± 0.12	1055	—	—	1165	—	—	—	5.3–6.3	5.8	6.43 ± 0.03	284	—	—	694	—	—	—
5.43 ± 0.12	899	—	—	1017	—	—	—	5.4–6.4	5.9	6.53 ± 0.03	242	—	—	682	—	—	—
5.52 ± 0.09	775	—	—	903	—	—	—	5.5–6.5	6.0	6.63 ± 0.03	209	—	—	686	—	—	—
5.62 ± 0.07	676	—	—	817	—	—	—	5.6–6.6	6.1	6.73 ± 0.03	182	—	—	707	—	—	—
5.71 ± 0.06	598	—	—	755	—	—	—	5.7–6.7	6.2	6.82 ± 0.03	161	—	—	745	—	—	—
5.81 ± 0.06	536	—	—	713	—	—	—	5.8–6.8	6.3	6.92 ± 0.03	144	—	—	803	—	—	—

(Continued)

Table 3.5 continued

Control pH at 20°C	Volume (μl) 0.2 M Immobiline pK, Acidic dense solution: 3.6	4.4	4.6	6.2	7.0	8.5	9.3	pH range	mid point	Control pH at 20°C	Volume (μl) 0.2 M Immobiline pK, Basic light solution: 3.6	4.4	4.6	6.2	7.0	8.5	9.3
5.91 ± 0.05	486	—	—	689	—	—	—	5.9–6.9	6.4	7.02 ± 0.03	131	—	—	884	—	—	—
6.01 ± 0.05	447	—	—	682	—	—	—	6.0–7.0	6.5	7.12 ± 0.03	120	—	—	992	—	—	—
6.10 ± 0.04	416	—	—	691	—	—	—	6.1–7.1	6.6	7.22 ± 0.03	112	—	—	1133	—	—	—
6.11 ± 0.11	972	—	—	—	1086	—	—	6.2–7.2	6.7	7.21 ± 0.03	262	—	—	—	686	—	—
6.21 ± 0.09	833	—	—	—	956	—	—	6.3–7.3	6.8	7.31 ± 0.03	224	—	—	—	682	—	—
6.30 ± 0.08	722	—	—	—	857	—	—	6.4–7.4	6.9	7.41 ± 0.03	195	—	—	—	694	—	—
6.40 ± 0.07	635	—	—	—	783	—	—	6.5–7.5	7.0	7.50 ± 0.03	171	—	—	—	724	—	—
6.49 ± 0.06	565	—	—	—	732	—	—	6.6–7.6	7.1	7.60 ± 0.03	152	—	—	—	771	—	—
6.59 ± 0.05	509	—	—	—	699	—	—	6.7–7.7	7.2	7.70 ± 0.03	137	—	—	—	840	—	—
6.69 ± 0.05	465	—	—	—	683	—	—	6.8–7.8	7.3	7.80 ± 0.03	125	—	—	—	934	—	—
6.78 ± 0.04	430	—	—	—	684	—	—	6.9–7.9	7.4	7.90 ± 0.03	116	—	—	—	1058	—	—
6.88 ± 0.04	403	—	—	—	701	—	—	7.0–8.0	7.5	8.00 ± 0.03	108	—	—	—	1217	—	—
6.98 ± 0.04	381	—	—	—	736	—	—	7.1–8.1	7.6	8.09 ± 0.03	103	—	—	—	1422	—	—
7.21 ± 0.06	1028	—	—	—	750	750	—	7.2–8.2	7.7	8.36 ± 0.05	548	—	—	—	750	750	—
7.31 ± 0.06	983	—	—	—	750	750	—	7.3–8.3	7.8	8.46 ± 0.05	503	—	—	—	750	750	—
7.41 ± 0.05	938	—	—	—	750	750	—	7.4–8.4	7.9	8.56 ± 0.05	458	—	—	—	750	750	—
7.66 ± 0.15	1230	—	—	—	—	1334	—	7.5–8.5	8.0	8.76 ± 0.04	331	—	—	—	—	720	—
7.75 ± 0.12	1037	—	—	—	—	1149	—	7.6–8.6	8.1	8.85 ± 0.03	279	—	—	—	—	692	—
7.85 ± 0.10	885	—	—	—	—	1004	—	7.7–8.7	8.2	8.95 ± 0.03	238	—	—	—	—	682	—
7.94 ± 0.08	764	—	—	—	—	893	—	7.8–8.8	8.3	9.05 ± 0.06	206	—	—	—	—	687	—
8.04 ± 0.07	667	—	—	—	—	810	—	7.9–8.9	8.4	9.14 ± 0.06	180	—	—	—	—	710	—

8.13 ± 0.06	*591*	—	—	—	—	*750*	—	*8.0–9.0*	*8.5*	*9.24 ± 0.06*	*159*	—	—	—	—	*750*	—
8.23 ± 0.06	*530*	—	—	—	—	*710*	—	*8.1–9.1*	*8.6*	*9.34 ± 0.06*	*143*	—	—	—	—	*810*	—
8.33 ± 0.05	*482*	—	—	—	—	*687*	—	*8.2–9.2*	*8.7*	*9.44 ± 0.06*	*130*	—	—	—	—	*893*	—
8.43 ± 0.04	*443*	—	—	—	—	*682*	—	*8.3–9.3*	*8.8*	*9.54 ± 0.06*	*119*	—	—	—	—	*1004*	—
8.52 ± 0.04	*413*	—	—	—	—	*692*	—	*8.4–9.4*	*8.9*	*9.64 ± 0.06*	*111*	—	—	—	—	*1149*	—
8.62 ± 0.04	*389*	—	—	—	—	*720*	—	*8.5–9.5*	*9.0*	*9.74 ± 0.06*	*105*	—	—	—	—	*1334*	—
8.40 ± 0.14	*1208*	—	—	—	—	—	*1314*	*8.6–9.6*	*9.1*	*9.50 ± 0.06*	*325*	—	—	—	—	—	*716*
8.49 ± 0.12	*1021*	—	—	—	—	—	*1133*	*8.7–9.7*	*9.2*	*9.59 ± 0.06*	*275*	—	—	—	—	—	*691*
8.59 ± 0.10	*871*	—	—	—	—	—	*992*	*8.8–9.8*	*9.3*	*9.69 ± 0.06*	*235*	—	—	—	—	—	*682*
8.68 ± 0.08	*753*	—	—	—	—	—	*884*	*8.9–9.9*	*9.4*	*9.79 ± 0.06*	*203*	—	—	—	—	—	*689*
8.78 ± 0.07	*659*	—	—	—	—	—	*803*	*9.0–10.0*	*9.5*	*9.88 ± 0.06*	*177*	—	—	—	—	—	*713*
8.87 ± 0.06	*584*	—	—	—	—	—	*745*	*9.1–10.1*	*9.6*	*9.98 ± 0.06*	*157*	—	—	—	—	—	*755*
8.97 ± 0.05	*525*	—	—	—	—	—	*707*	*9.2–10.2*	*9.7*	*10.08 ± 0.06*	*141*	—	—	—	—	—	*817*
9.07 ± 0.04	*478*	—	—	—	—	—	*686*	*9.3–10.3*	*9.8*	*10.18 ± 0.06*	*129*	—	—	—	—	—	*903*
9.16 ± 0.07	*440*	—	—	—	—	—	*682*	*9.4–10.4*	*9.9*	*10.28 ± 0.06*	*119*	—	—	—	—	—	*1017*
9.26 ± 0.07	*410*	—	—	—	—	—	*694*	*9.5–10.5*	*10.0*	*10.38 ± 0.06*	*111*	—	—	—	—	—	*1163*

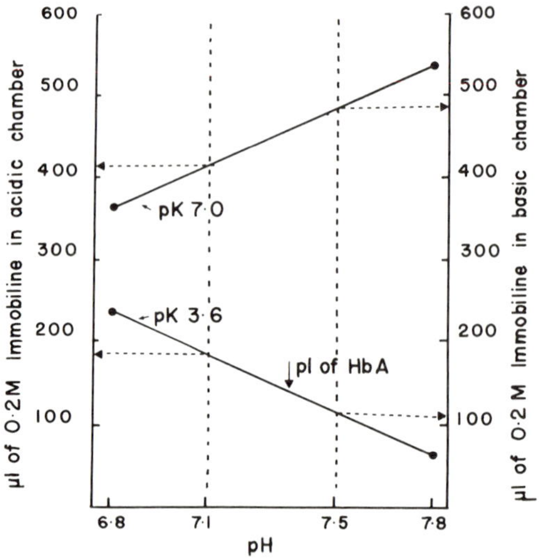

Figure 3.18 Graphic representation of the preparation of narrow (up to 1 pH unit) IPG gradients on the 'tandem' principle. The limiting molarities of 7.0 pK (buffering species) and 3.6 pK (titrant) Immobilines needed to generate a 6.8–7.8 pH interval, as obtained directly from table 3.5 are plotted on a graph. These points are joined by straight lines and the new molarities needed to generate any narrower pH gradients within the stated pH intervals are then obtained by simple linear interpolation (broken vertical and horizontal lines). In this example, a narrow 7.1–7.5 pH gradient is graphically derived (from Rochette *et al* 1984).

computer programs designed specifically for this purpose (Dossi *et al* 1983, Gianazza *et al* 1983b). The basic findings are:

(i) for generating a linear pH gradient the buffering power has to be constant throughout the desired pH interval;
(ii) to avoid deviations from linearity, the titrants should have pKs well outside pH_{min} and pH_{max} of the required pH range (in general, at least 2 pH units removed from the limits of the pH interval).

As a consequence of this, for pH ranges wider than 3 pH units, two additional Immobilines are needed as titrants: one strongly acidic (pK < 1) and one strongly basic (pK > 12). We are routinely using these two species, and we hope they will soon be commercially available.

In table 3.6 we give recipes for ten 2-pH-unit wide gradients, and four

Table 3.6 Broad pH gradients: volumes of Immobiline for 15 ml of each starting solution (23). From LKB Application Note No. 324 (1984). For explanations, see Tables 3.5 and 3.7

Control pH	*Volume (μl) 0.2 M Immobiline pK Acidic dense solution*							*pH*	*mid*	*Control pH*	*Volume (μl) 0.2 M Immobiline pK Basic light solution*						
at 20°C	*3.6*	*4.4*	*4.6*	*6.2*	*7.0*	*8.5*	*9.3*	*range*	*point*	*at 20°C*	*3.6*	*4.4*	*4.6*	*6.2*	*7.0*	*8.5*	*9.3*
3.53 ± 0.06	299	—	223	157	—	—	—	3.5–5.0	4.25	5.06 ± 0.07	212	—	310	465	—	—	—
4.00 ± 0.06	569	—	99	439	—	—	—	4.0–6.0	5.0	6.09 ± 0.14	390	—	521	276	—	—	722
4.54 ± 0.06	415	—	240	499	—	—	—	4.5–6.5	5.5	6.53 ± 0.05	—	—	570	244	235	—	297
5.08 ± 0.03	69	—	428	414	—	—	—	5.0–7.0	6.0	7.01 ± 0.06	—	—	474	270	219	—	320
5.56 ± 0.03	—	—	450	354	113	—	—	5.5–7.5	6.5	7.51 ± 0.09	347	—	—	236	287	284	—
6.06 ± 0.08	435	—	—	323	208	44	—	6.0–8.0	7.0	8.11 ± 0.09	286	—	—	174	325	329	—
6.56 ± 0.13	771	—	—	276	185	538	—	6.5–8.5	7,5	8.66 ± 0.06	192	—	—	153	278	362	—
7.03 ± 0.24	1349	—	—	—	272	372	845	7.0–9.0	8.0	8.94 ± 0.07	484	—	—	—	232	189	546
7.50 ± 0.11	668	—	—	—	445	226	348	7.5–9.5	8.5	9.37 ± 0.06	207	—	—	—	925	139	346
8.10 ± 0.07	399	—	—	—	364	355	94	8.0–10.0	9.0	9.89 ± 0.05	91	—	—	—	329	366	289
4.01 ± 0.05	578	—	110	450	—	—	—	4.0–7.0	5.5	7.02 ± 0.14	302	—	738	151	269	—	876
5.03 ± 0.12	702	—	254	416	133	346	—	5.0–8.0	6.5	8.12 ± 0.07	175	—	123	131	345	346	—
6.04 ± 0.14	779	—	—	402	93	364	80	6.0–9.0	7.5	9.01 ± 0.06	241	—	—	161	449	237	225
6.98 ± 0.07	542	—	—	—	378	351	—	7.0–10.0	8.5	9.88 ± 0.05	90	—	—	—	324	350	280

Table 3.7 4, 5 and 6-pH-unit wide immobilised pH gradients. From Gianazza *et al* (1984a). The 'same concentration' formulations are listed in the left column, the ones with 'different concentrations' on the right. In each mixture the initial and final pHs in the gel phase, at 10°C are recorded; Notes, pH of the limiting solutions (and of the buffer mixture, prior to the addition of titrants) at 20°C (figures before) and at 25°C (figures after the colon); buffer concentrations expressed as $mM\,l^{-1}$ (to convert into $\mu l\,ml^{-1}$, for a 0.2 M Immobiline concentration, multiply by 5)

4–8

Initial pH	: 4.000				
Final pH	: 8.000				
Notes	: 4.068 8.153(7.813): 4.084–8.004(7.689)				
Buffer concentrations					
n. 1 pK	0.80	Cham 1	11.085	Cham 2	0.000
n. 2 pK	3.57	Cham 1	0.864	Cham 2	0.864
n. 3 pK	4.51	Cham 1	5.259	Cham 2	5.259
n. 4 pK	6.21	Cham 1	4.167	Cham 2	4.167
n. 5 pK	7.06	Cham 1	2.379	Cham 2	2.379
n. 6 pK	8.50	Cham 1	6.432	Cham 2	6.432
n. 8 pK	12.00	Cham 1	0.000	Cham 2	0.924

Initial pH	: 4.015				
Final pH	: 8.008				
Notes	: 4.076–8.112 : 4.091–7.986				
Buffer concentrations					
n. 1 pK	0.80	Cham 1	8.643	Cham 2	0.000
n. 2 pK	3.57	Cham 1	1.420	Cham 2	0.000
n. 3 pK	4.51	Cham 1	4.560	Cham 2	6.519
n. 4 pK	6.21	Cham 1	5.222	Cham 2	1.975
n. 5 pK	7.06	Cham 1	2.326	Cham 2	4.586
n. 6 pK	8.50	Cham 1	3.276	Cham 2	3.822
n. 8 pK	12.00	Cham 1	0.000	Cham 2	3.130

5–9

Initial pH	: 5.000				
Final pH	: 9.000				
Notes	: 5.076–9.086(7.544) : 5.085–8.926(7.457)				
Buffer concentrations					
n. 1 pK	0.80	Cham 1	7.527	Cham 2	0.000
n. 3 pK	4.51	Cham 1	7.089	Cham 2	7.089
n. 4 pK	6.21	Cham 1	3.408	Cham 2	3.408
n. 5 pK	7.06	Cham 1	3.180	Cham 2	3.180
n. 6 pK	8.50	Cham 1	4.683	Cham 2	4.683
n. 7 pK	9.59	Cham 1	1.839	Cham 2	1.839
n. 8 pK	12.00	Cham 1	0.000	Cham 2	4.464

Initial pH	: 5.022				
Final pH	: 9.037				
Notes	: 5.100–9.080 : 5.110–8.921				
Buffer concentrations					
n. 1 pK	0.80	Cham 1	11.404	Cham 2	0.000
n. 3 pK	4.51	Cham 1	7.989	Cham 2	3.498
n. 4 pK	6.21	Cham 1	3.102	Cham 2	3.493
n. 5 pK	7.06	Cham 1	1.908	Cham 2	2.933
n. 6 pK	8.50	Cham 1	11.020	Cham 2	4.091
n. 7 pK	9.59	Cham 1	1.694	Cham 2	2.692
n. 8 pK	12.00	Cham 1	0.000	Cham 2	0.438

6–10

Initial pH : 6.000
Final pH : 10.000
Notes : 6.039–9.886 : 5.988–9.733

Buffer concentrations

n. 1 pK	0.80	Cham 1	12.060	Cham 2	0.000
n. 1 pK	0.80	Cham 1	1.323	Cham 2	1.323
n. 4 pK	6.21	Cham 1	3.351	Cham 2	3.351
n. 5 pK	7.06	Cham 1	3.603	Cham 2	3.603
n. 6 pK	8.50	Cham 1	3.711	Cham 2	3.711
n. 7 pK	9.59	Cham 1	4.299	Cham 2	4.299
n. 8 pK	12.00	Cham 1	0.000	Cham 2	0.000

Initial pH : 5.994
Final pH : 9.996
Notes : 6.035–9.876 : 5.986–9.723

Buffer concentrations

n. 2 pK	3.57	Cham 1	12.919	Cham 2	1.370
n. 4 pK	6.21	Cham 1	3.748	Cham 2	4.583
n. 5 pK	7.06	Cham 1	3.339	Cham 2	4.962
n. 6 pK	8.50	Cham 1	3.585	Cham 2	3.283
n. 7 pK	9.59	Cham 1	3.891	Cham 2	4.475

4–9

Initial pH : 4.000
Final pH : 9.000
Notes : 4.071–9.086(8.007) : 4.088–8.925(7.876)

Buffer concentrations

n. 1 pK	0.80	Cham 1	11.649	Cham 2	0.000
n. 2 pK	3.57	Cham 1	0.657	Cham 2	0.657
n. 3 pK	4.51	Cham 1	5.394	Cham 2	5.394
n. 4 pK	6.21	Cham 1	3.888	Cham 2	3.888
n. 5 pK	7.06	Cham 1	2.943	Cham 2	2.943
n. 6 pK	8.50	Cham 1	4.734	Cham 2	4.734
n. 7 pK	9.59	Cham 1	1.857	Cham 2	1.857
n. 8 pK	12.00	Cham 1	0.000	Cham 2	3.399

Initial pH : 4.019
Final pH : 8.996
Notes : 4.067–8.980 : 4.081–8.825

Buffer concentrations

n. 1 pK	0.80	Cham 1	11.793	Cham 2	0.000
n. 2 pK	3.57	Cham 1	2.832	Cham 2	4.130
n. 3 pK	4.51	Cham 1	4.253	Cham 2	7.780
n. 4 pK	6.21	Cham 1	5.353	Cham 2	2.635
n. 5 pK	7.06	Cham 1	0.585	Cham 2	4.252
n. 6 pK	8.50	Cham 1	4.814	Cham 2	3.056
n. 7 pK	9.59	Cham 1	4.196	Cham 2	4.654
n. 8 pK	12.00	Cham 1	0.000	Cham 2	7.415

(*Continued*)

Table 3.7 Continued

5–10

Initial pH	: 5.000				
Final pH	: 10.000				
Notes	: 5.076–9.892(8.171) : 5.086–9.745(8.034)				
Buffer concentrations					
n. 1 pK	0.80	Cham 1	9.102	Cham 2	0.000
n. 3 pK	4.51	Cham 1	7.179	Cham 2	7.179
n. 4 pK	6.21	Cham 1	3.210	Cham 2	3.210
n. 5 pK	7.06	Cham 1	3.531	Cham 2	3.531
n. 6 pK	8.50	Cham 1	3.729	Cham 2	3.729
n. 7 pK	9.59	Cham 1	4.275	Cham 2	4.275
n. 8 pK	12.00	Cham 1	0.000	Cham 2	5.922

Initial pH	: 4.993				
Final pH	: 10.034				
Notes	: 5.068–9.933 : 5.075–9.779				
Buffer concentrations					
n. 1 pK	0.80	Cham 1	7.605	Cham 2	0.295
n. 3 pK	4.51	Cham 1	6.260	Cham 2	0.805
n. 4 pK	6.21	Cham 1	4.044	Cham 2	0.454
n. 5 pK	7.06	Cham 1	3.717	Cham 2	5.677
n. 6 pK	8.50	Cham 1	3.088	Cham 2	4.197
n. 7 pK	9.59	Cham 1	1.730	Cham 2	3.687

4–10

Initial pH	: 4.000				
Final pH	: 10.000				
Notes	: 4.079–9.891(8.658) : 4.097–9.742(8.502)				
Buffer concentrations					
n. 1 pK	0.80	Cham 1	13.656	Cham 2	0.000
n. 3 pK	4.51	Cham 1	5.712	Cham 2	5.712
n. 4 pK	6.21	Cham 1	3.609	Cham 2	3.609
n. 5 pK	7.06	Cham 1	3.342	Cham 2	3.342
n. 6 pK	8.50	Cham 1	3.765	Cham 2	3.765
n. 7 pK	9.59	Cham 1	4.305	Cham 2	4.305
n. 8 pK	2.00	Cham 1	0.000	Cham 2	4.431

Initial pH	: 4.026				
Final pH	: 9.968				
Notes	: 4.105–9.826 : 4.121–9.675				
Buffer concentrations					
n. 1 pK	0.80	Cham 1	15.849	Cham 2	0.000
n. 3 pK	4.51	Cham 1	4.740	Cham 2	2.446
n. 4 pK	6.21	Cham 1	6.645	Cham 2	0.000
n. 5 pK	7.06	Cham 1	0.000	Cham 2	6.799
n. 6 pK	8.50	Cham 1	6.633	Cham 2	1.977
n. 7 pK	9.59	Cham 1	3.776	Cham 2	5.641
n. 8 pK	12.00	Cham 1	0.000	Cham 2	0.715

3-pH-unit spans. There are still a few missing, namely three of 4 pH units, two of 5 pH units and one spanning the widest possible pH interval (pH 4–10). These have just been computed and can be found in table 3.7. Two formulations are available: on the left side, identical molarities of the buffering Immobilines are used in the two chambers of the gradient mixer, and then titrated with strong acid and bases to the two extremes; on the right, the buffering species are present in unequal amounts in the two vessels (designed as Cham 1 and Cham 2) so that a gradient of buffering species will also be generated (for explanations, see Gianazza *et al* 1984a). With all these recipes, preparation of any Immobiline gel will be just as easy as cooking spaghetti (we hope you realise that, besides spaghetti 'alla carbonara' and 'al colpo di lupara' (mafiosi, of course), there are at least another 80 recipes, just like with your Immobilines).

3.13 On Buffering Capacity and Ionic Strength

It should be appreciated that, in contrast to conventional IEF, where the ionic strength (I) prevailing during the separation is anybody's guess (in 2% focused Ampholine it has been estimated to be around 1 mequiv L^{-1}) (Righetti 1983a), in IPGs the I value becomes a true physico-chemical parameter, under the full control of the scientist performing the experiment. For the buffering power (β) the discrepancies between the two systems are less pronounced, as the β value can be determined with accuracy and is comparable in both systems. Yet, in conventional IEF, I and β values are in any event subject to strong variations in terms of peaks and troughs within adjacent, focused Ampholine zones, which means in practice, within distances of the order of 10 to 50 μm along the separation track. This is of no minor consequence, since, through its influence on activity factors, the ionic strength will affect the pK values of ionisable groups in proteins. In IPGs, given a known molarity of grafted buffers and titrants, the ionic strength in the gel can simply be calculated by the following relationship

$$I = \sum C(\mathrm{A}_i)\,\frac{10^{[\mathrm{pH}-\mathrm{pK}(\mathrm{A}_i)]}}{10^{[\mathrm{pH}-\mathrm{pK}(\mathrm{A}_i)]}+1} = \sum C(\mathrm{B}_i)\,\frac{1}{10^{[\mathrm{pH}-\mathrm{pK}(\mathrm{B}_i)]}+1} \qquad (3.1)$$

where $C(\mathrm{A}_i)$ is the concentration of acidic Immobiline with $\mathrm{pK} = \mathrm{pK(A)}$, and $C(\mathrm{B}_i)$ is the concentration of basic Immobiline with $\mathrm{pK} = \mathrm{pK}(\mathrm{B}_i)$. It should be noted that the pK values of the Immobilines also vary with the ionic strength. From the Debye–Hückel (1924) law, the variation for Immobiline pK can be given approximately by

$$\mathrm{pK} = \mathrm{pK}_0 - 0.5Z^2\sqrt{I} \qquad (3.2)$$

where pK_0 is the pK value at an I of zero and Z is equal to -1 for acids and $+1$ for bases; thus, the pK increases with I for acids, and decreases for bases. When Immobilines are used according to recommendations, these pK variations, which are less than 0.03 pH units, can normally be neglected; however, they should be kept in mind when using extremely narrow pH gradients (approximately 0.01 pH unit/cm) or when using high Immobiline molarities (e.g. 30 mM), as in these instances the band positions might be influenced.

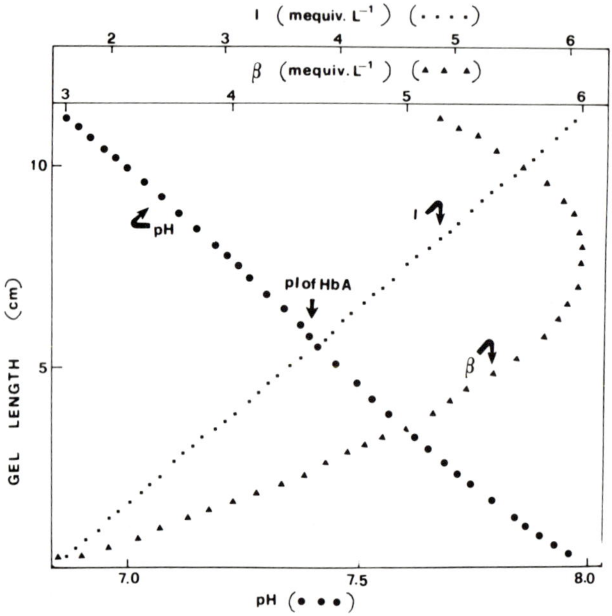

Figure 3.19 Computer simulation of pH, β and I courses in the pH 6.8–7.8 Immobiline gradient for haemoglobin analysis, having the limiting composition as plotted in figure 3.18. The same data could be calculated with the aid of equations (3.1) and (3.3) and, for the pH gradient, by eluting fractions from the gradient mixer and reading the pH in solution of 0.5 ml eluates. The pI position of HbA is marked by a vertical arrow (from Gelfi and Righetti 1983).

The β power is another quantitiy that is well defined in an Immobiline pH gradient. As for any solution containing weak monofunctional acids and bases, the buffering capacity is given by the equation

$$\beta = 2.3 \sum_{i=1}^{l+m} C_i \frac{K_i[\mathrm{H}^+]}{K_i + [\mathrm{H}^+])^2} \tag{3.3}$$

where C_i is the molar concentration of the ith Immobiline having a dissociation constant of K_i. Equations (3.1) and (3.3) are part of our computer program for extended pH gradients (Dossi *et al* 1983) which, given any mixture of Immobilines in any pH range, will automatically simulate and optimise the generated pH gradient and calculate the accompanying I and β values. With the aid of equations (3.1) and (3.3), the I and β power courses can also be calculated manually, with any pocket calculator, usually at 0.1 pH unit increments.

The buffering capacity must be high enough to make the pH gradient insensitive to impurities (e.g. acrylic acid from the acrylamide and Bis monomers) and should also be even, in order to minimise the effects of small disturbances in forming the gradient when casting the gel. For analytical purposes, a value of 3–5 mequiv l^{-1} pH^{-1} will give well functioning pH gradients. A higher buffering capacity will give more sharply focused bands (Gelfi and Righetti 1983); however, such gels will start to swell during the staining and destaining steps if the total molar concentration of Immobilines exceeds 30 mM. As an example, in figure 3.19 you can find a computer-derived graph on the courses of pH gradient, ionic strength (I) and buffering power (β) of the pH 6.8–7.9 immobilised pH gradient we routinely use for Hb separation (the molarities of buffering ion and titrant are plotted in figure 3.19. In a way, this is a primitive graph: with the more sophisticated program we have recently developed, we can arrange for constant I values, if needed, and for a smoother β course (i.e. decreased deviation from the linearity of the pH gradient course). What we want to emphasise here is that, with a bit of patience and wisdom, you can also calculate the experimental parameters of your separation (I, β) (equations (3.1) and (3.3)) once the working pH interval and the molarities of buffer and titrant are known (see tables 3.5–3.7) and, by educated guesses, try to improve them if needed.

3.14 On Electro-endosmosis

Electro-endosmosis is not normally a problem in Immobiline pH gradients, as the gel will not have any net charge after traces of catalysts and non-incorporated Immobilines have been electrophoretically transported away. Generally at low and high pH values the presence of H^+ and OH^- means that the matrix adopts a net charge which will result in water transport towards the cathode at low pH and towards the anode at high pH. A trough could form close to the electrodes at the extreme pH ranges, and eventually the gel could dry out and burn. This phenomenon will in general not occur within the pH ranges for which Immobiline is recommended, but it would be wise to include glycerol (20%) in the washing step in gels below pH 5 and above pH 9.

Carbon dioxide from the air is also expected to result in electro-endosmosis. Delincee and Radola (1978) were the first to describe the effect of CO_2. Gaseous CO_2 dissolves in the gel, especially at pH > 6.3 (the solubility of carbon dioxide increases with pH), forming HCO_3^- ions as follows:

$$CO_2 + H_2O \rightarrow H^+ + HCO^-.$$

While this acidification in conventional IEF causes the part of the pH gradient above pH 6.3 to drift towards the cathode, by charging and electrophoretically mobilising the focused carrier ampholytes (possibly by salt formation), it cannot act on IPGs by the same mechanism, but it will certainly alter the slope of the theoretical pH gradient depending on the local ratio HCO_3^-: Immobiline (it is in fact like introducing a new buffer with pK 6.3 into the immobilised pH gradient). The HCO_3^- ion migrates electrophoretically (figure 3.20) from the cathodic side towards the anode. At pH 6.3, carbon dioxide gas starts to form, and is liberated from the gel. At the same time OH^- is formed according to the equation:

$$HCO_3^- \rightarrow OH^- + CO_2 \text{ (gas)}. \qquad (3.4)$$

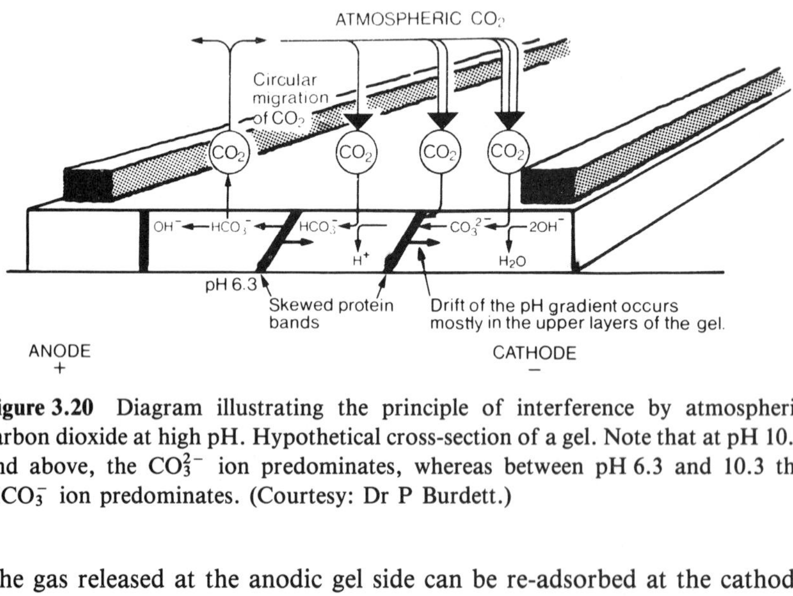

Figure 3.20 Diagram illustrating the principle of interference by atmospheric carbon dioxide at high pH. Hypothetical cross-section of a gel. Note that at pH 10.3 and above, the CO_3^{2-} ion predominates, whereas between pH 6.3 and 10.3 the HCO_3^- ion predominates. (Courtesy: Dr P Burdett.)

The gas released at the anodic gel side can be re-adsorbed at the cathode and thus recirculated through the system as depicted in figure 3.20. In analogy with this, volatile amines will cause exactly the same disturbances but in the opposite direction (it is not recommended, therefore, to use ethanolamine, ethylenediamine and the like as catholytes). Even in IPG a very marked effect on band sharpness is observed if carbon dioxide is excluded from the system in alkaline pH ranges. For this, the IEF cell should be airtight, flushed with inert gas (denser than air, such as argon, nitrogen is lighter and would float!) and/or covered on the gel-free space with sponges impregnated with NaOH or $Ca(OH)_2$. A note of caution should be given to scientists working with 2D techniques: the 2-mercaptoethanol added to the sample will behave in a very similar way to CO_2: it is in fact a buffer with pK 9.5. This compound ionises at the basic gel end and is electrophoretically driven along the pH gradient. The remedy is to apply the sample at the anodic gel end: below pH 7 the -SH group will not buffer or be ionised (Righetti *et al* 1982).

As a final remark, we have also found a strong electro-endosmotic flow when trying to cast Immobiline gels on porous membranes for direct protein blots after the IEF step without removal of the gel from the supporting foil. We have tried several potential candidates: denitrated cellulose nitrate, cellulose acetate (including the best IEF brands), cellophane, porous polyethylene, Zeta Probe, Pall Biodyne: you name it, the results have been disastrous. We do not understand why IPGs should be so sensitive to porous membranes, and what exactly the mechanism underlying this electro-endosmotic flow is, but at present IPGs and porous supports seem to be incompatible. For protein blots, there is a simple solution (Bos *et al* 1984): instead of casting the gel onto the hydrophilic side of Gel bond PAG, to which it adheres rather strongly, cast it onto the hydrophobic side, from which it can be peeled away easily at the end of the IPG run.

3.15 Polymerisation Kinetics

In copolymerisation chemistry, it is often reported that the composition of the copolymer formed differs from the initial input composition because the monomers differ in reactivity towards free radical addition. Thus, with less than 100% incorporation of monomers into the polymer, there is a possibility that the concentration ratios between the Immobilines built into the gel will differ from the ratios in the starting solution; this could have serious effects on the pH gradient generated, e.g. by changing its slope and the theoretically computed pH interval. To minimise this effect, all Immobilines are acrylamide derivatives, but even with this precaution it cannot be excluded that the resulting pH values depend to a certain extent on the polymerisation efficiency. Using techniques described by Gelfi and Righetti (1981a, b) we have studied the effects of the following parameters on Immobiline gels:

(i) level of persulphate, from 0.015 to 0.058%;
(ii) level of TEMED, from 0.024 to 0.096%;
(iii) temperature range, from 20 to 60°C.

The optimum polymerisation efficiency (in the range 84–8% incorporation for the seven Immobilines) was found at 0.047% TEMED, 0.033% persulphate, 50°C and pH > 7. Figure 3.21 gives an example of the effect of temperature on the extent and rate of reaction of Immobilines: as the temperature is lowered, the reactivity rate diverges greatly for the different Immobiline chemicals, with a consequent lowering of the incorporation levels in the gel matrix. Curiously, at 60°C, the incorporation efficiency is lowered slightly for some species (the alkaline ones). Polymerisation for 1 h at 50°C, as previously suggested, appears to be just the right solution: all Immobilines seem to come to a confluence point at this temperature, by exhibiting very similar reactivity ratios and incorporation efficiencies (Righetti *et al* 1984). Another important lesson has been learned from these experiments: when casting extended pH gradients (e.g. pH, 3.5–9.5, the widest possible with Immobiline chemicals) it is imperative that the acidic end of the pH gradient be titrated (with sodium hydroxide solution) to at least around pH 8, so as to ensure a uniform reactivity ratio between the two pH extremes.

When Immobiline gradients are cast by standard techniques with a gradient mixer equilibrated in air, an 84–8% incorporation efficiency is the

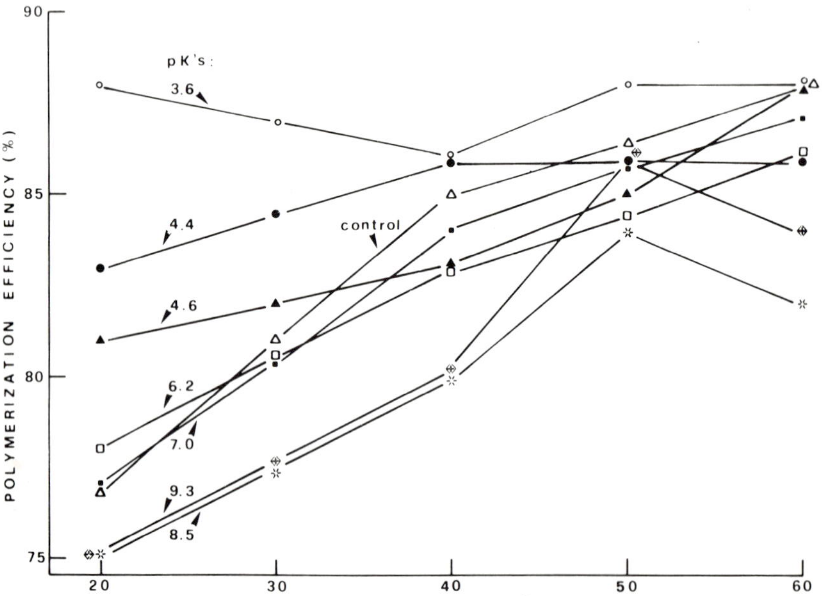

Figure 3.21 Polymerisation efficiency of the seven Immobiline species as a function of temperature. The percentage incorporation into the matrix is putative, as it is based on the ratio between initial and final absorbancies at 285 nm (disappearance of double bonds). The best convergence (similar reactivity ratios) is only obtained at 50°C. Note that the critical parameter is not so much a 100% incorporation efficiency, but the achievement of a ratio of incorporation between the buffering ion and the titrant as close as possible to unity (from Righetti *et al* 1984).

best that can be achieved, and this for most practical applications will be adequate. However, if better than 96% incorporation is required for some special experiments, this can be achieved by completely excluding oxygen from the polymerisation mixture. We have achieved this by building a gradient mixer that provides for anaerobic conditions, but the experimental manipulations become more complex.

A note of caution: this is already a complex chemistry, since we are trying to copolymerise, possibly with the same efficiency, as many as nine different chemicals at one time (in extended pH gradients, seven Immobilines and acrylamide and Bis), so we would not advise to be smarter and try and beat the system. Any time we tried that, we were, in fact, thoroughly beaten. For, example, when trying to make highly cross-linked (i.e. highly porous) gels by using 30% or more Bis we had the porous structure, but not the Immobilines grafted to it! Even trying to change the crosslinker is often troublesome (acrylaide, for instance, gave us blurred and highly distorted patterns). Composite agarose (the best IEF brand)–acrylamide gels are also besieged with difficulties and gave us extremely poor results, possibly because the incorporation of Immobilines into the polyacrylamide is impeded by the onset of agarose gelation.

3.16 Methodology: Casting an Immobiline Gel

For the cassette, the basic assembly is just as depicted in figure 3.7 (as Immobiline gels have, in general, to be washed, it does not pay to cast them in glass cylinders). We suggest binding to the cover glass pocket-forming strips (up to 20 may be required), cut out of adhesive embossing tape (e.g. Dymo), so as to apply the sample in a liquid free layer. We suggest gluing the intact length of tape and then cutting out and removing 3 mm wide segments, perpendicular to it, so as to leave separate rectangles of tape glued to the glass. The depressions formed by the Dymo tape in the gel layer are about 200 μm deep and can usually accommodate volumes up to 10–12 μl. A double volume can be arranged for by gluing two embossing strips, one on top of the other. It is always best, however, to make sure that, after opening the cassette, the pockets are sealed at the bottom by a thin film of polyacrylamide gel. The final assembly is shown in figure 3.22.

Notice that the two upper clamps on the top of the cassette have been removed and that two paper clips inserted instead. This greatly facilitates the insertion of the plastic tubing conveying the solution from the mixer to the chamber by forcing the top rims of the glass slabs to diverge and thus widen the 0.5 mm gap. The gradient mixer is positioned about 5–8 cm from the chamber top (the liquid will flow down by gravity) and is filled with the acidic and basic solutions. One should remember to fill only one chamber first, and then to remove the air bubble from the channel connecting the two chambers by gently opening the central valve. The gel chamber is standing vertically on a levelling table and the capillary tubing is inserted in the middle. At this point the stirring is started (500 rpm), the catalysts added (TEMED and persulphate, in this order), both valves are opened and the density (and pH) gradient is allowed to flow into the gel cassette. It takes about 10 min before the onset of polymerisation occurs at 20°C, but in a hot summer (and in 8 M urea) things will move faster (2–3 min > 30°C). Once the whole gradient has been poured, the clips are quickly removed and the two upper clamps fastened in their position.

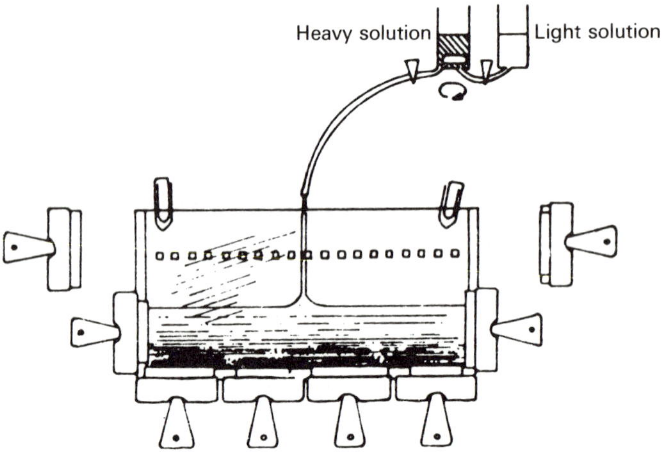

Figure 3.22 Gel cassette and gradient mixer for casting an Immobiline gel. Note that the two upper clamps are removed and two paper clips inserted instead, to widen the gap (0.5 mm) between the two glass slabs. After the gradient mixer has been emptied, the two clips are quickly removed and the clamps placed back *in situ* (the liquid level will thus rise to the proper height). (Courtesy: Dr A Gorg.)

After standard polymerisation (1 h, 50°C) the gel cassette is removed from the oven, the eight clamps unfastened and the supporting glass plate gently prised open with the tip of a spatula. The Gel Bond PAG foil, with its bound polyacrylamide gel layer, can now be lifted from one corner and gently peeled off from the other glass plate.

The first operation to perform at this point is a weighing step (after blotting any traces of liquid around the ridges of the foil, if necessary), as the gel has to be washed and will swell in water during this procedure. It is a good idea to mark the weight of the gel on the plastic backing. The gel is now washed in 1 l of distilled water for 30–40 min for 0.5 mm gels (use double this time for 1 mm gels). The washing step is essential: TEMED, persulphate and 12–16% unpolymerised Immobilines have to be removed, otherwise huge plateaus of free acid and free base will form at the anode and cathode, respectively, and will prevent the protein from focusing. After washing, the gel should be blotted with soft tissue and then, with the aid of a fan, reduced to its original weight. This step is essential, as gels containing too much water will 'sweat' during the IEF run and droplets of water will form on the surface. It is at this point that (if you have not quit in despair owing to the many manipulations involved) you can apply your sample.

3.17 Artefacts

When describing a new technique it is important to define the limits of its validity within which the method behaves as predicted. Outside these borders the method could give erroneous results likely to lead to misinterpretation of experimental data. We have already described all the possible artefacts of conventional IEF (*see* pp 29–35). The IPG technique is no exception to this rule, although artefacts will occur by a different mechanism: as the charge density of the macromolecule to be separated approaches the charge density of the matrix, a strong interaction will occur, which will result either in total sample precipitation at the application point, or in extended smears covering a wide gel surface.

We have found that IPG matrices interact strongly with at least two classes of proteins, histones and the histone-like 'high-mobility group' (HMG) chromatin proteins, forming insoluble complexes. The nature of these interactions has been demonstrated to be purely ionic: the complexes are split by high ionic strength (0.5 M sodium chloride) and/or by altering the pH (full disaggregation being obtained at pH 5.5 and 11.5). By preparing soluble homo-Immobiline polymers (polymerised in the absence of a crosslinker) formed either from a pure carboxyl or a pure amino surface, we have demonstrated (Righetti *et al* 1983a) that histones and HMGs bind preferentially with carboxyl Immobiline polymers (see figure 3.23). Thus, one should not ask the impossible from the IPG technique: nucleic acids, heparin and polyanions will also not be amenable to fractionation in IPG matrices, and will produce a curtain of molecules smeared over the gel surface or simply precipitated at the application point. Except for these limitations, we have found that IPGs perform normally with all the proteins we have tried (having pIs in the 3.5–10 pH range) except one: human serum albumin (HSA). As seen in the separations of Görg *et al* (1983) and Cleve *et al* (1982), HSA produces long smears between 4.7 and 5.2 pH instead of focusing regularly. We believe that HSA recognises pK 4.4 and 4.6 Immobilines as ligands. But these immobilines are unfortunately needed as buffers in the pH region in which HSA is isoelectric. These complexes are, however, sensitive to 8 M urea, so that practically normal patterns are obtained when running 2D maps by the O'Farrell technique (Gianazza *et al* 1984).

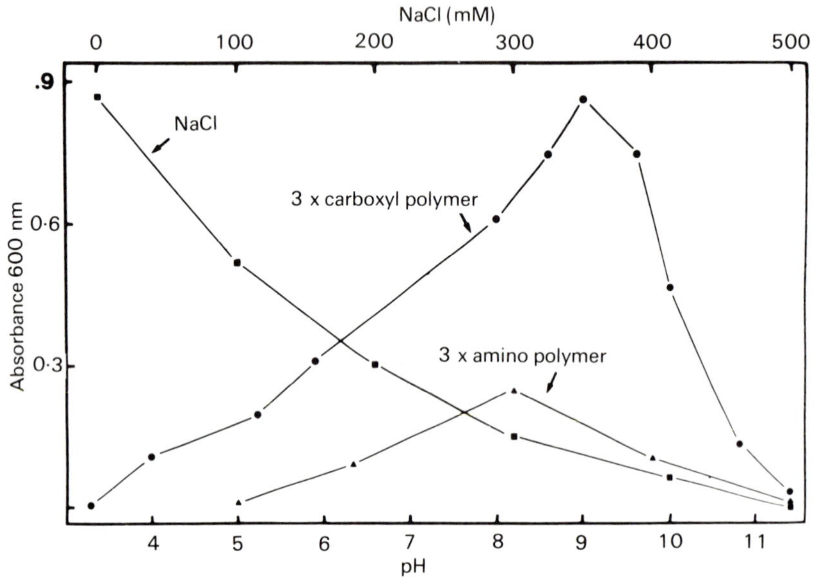

Figure 3.23 Formation and disaggregation of complexes between soluble, carboxyl and amino Immobilines and histone-like (HMG) proteins. Polymers with pure carboxyl and amino surfaces, having a three-fold higher concentration of Immobilines than standard gels (about 30 mM), were used in this experiment, and the stability of their complexes with HMG proteins was studied as a function of pH. For the carboxyl polymer, the disaggregation of its complex with the HGM protein by increasing sodium chloride molarities is also plotted (from Righetti *et al* 1983a).

3.18 Conclusions

We believe that conventional IEF has been fully developed and has now definitely reached a plateau. IPGs, though in their infancy, have been already extensively developed, both in their analytical as well as preparative aspects. Highly desirable would be the introduction of highly porous matrices (but that will be quite a tough project) as well as of ready-made plates with selected narrow ranges for clinical screening (this is foreseeable in the near future). We expect that there will be an explosive growth in the use of IPGs, as the technique possess all the desirable features of conventional IEF, while being completely free of its drawbacks.

Acknowledgments

Support from a five year grant from Consiglio Nazionale delle Ricerche (CNR, Roma) Progetto Finalizzato 'Salute dell'Uomo', Sottoprogetto 'Meccanismi di Invecchiamento'.

References

Almgren M 1971 *Chem. Scripta* **1** 69–75

Altland K and Kaempfer M 1980 *Electrophoresis* **1** 57–62

Ambler J and Walker G 1979 *Clin. Chem.* **25** 1320–2

Arnott S, Fulmer A, Scott W E, Dea I C M, Moorhouse R and Rees D A 1974 *J. Mol. Biol.* **90** 269–84

Beccaria L, Chiumello G, Gianazza E, Luppis B and Righetti P G 1978 *Am. J. Haematol.* **4** 367–74

Bengtsson G and Olivecrona T 1977 *Isoelectric Focusing and Isotachophoresis* ed. B J Radola and D Graesslin (Berlin: De Gruyter) pp 189–95

Bianchi Bosisio A, Loehrlein C, Snyder R S and Righetti P G 1980 *J. Chromatogr.* **189**, 317–30

Bjellqvist B, Ek K, Righetti P G, Gianaza E, Görg A and Postel W 1982 *J. Biochem. Biophys. Methods* **6** 317–39

Bos E S, Van Der Doelen A A, Van Der Struik E, Bergink E W and Schuurs A H W M 1984 *Electrophoresis '84* ed. V. Neuhoff (Weinheim: Verlag-Chemie) p 93

Brown R K, Caspers M L, Lull J M, Vinogradov S N, Felgenhauer K and Nekic M 1977 *J. Chromatogr.* **131** 223–32

Casero P, Gelfi C and Righetti P G 1985 *Electrophoresis* **6** 59–69

Caspers M L, Posey Y and Brown R K 1977 *Anal. Biochem.* **79** 166–80

Charlionet R, Martin J P, Sesboue' R, Madec P J and Lefebvre F 1979 *J. Chromatogr.* **176** 89–101

Chrambach A and Hjelmeland L M 1984 *Electrophoresis '83* ed. H Hirai (Berlin: de Gruyter) pp 81–97

Cleve H, Patutschnick W, Postel W, Weser J and Görg A 1982 *Electrophoresis* **3** 342–45

Constans J, Viau M, Gouaillard C, Bouisson C and Clerc A 1980 *Electrophoresis '79* ed. B J Radola (Berlin: de Gruyter) pp 701–10

Cossu G, Manca M, Gavina P M, Bullitta R, Bianchi Bosisio A, Gianazza E and Righetti P G 1982 *Am. J. Haematol.* **13** 149–57

Creighton T E 1979 *J. Mol. Biol.* **129** 235–64

Crowle A J and Cline L J 1977 *J. Immunol. Methods* **17** 379–385

Cuono C B and Chapo G A 1982 *Electrophoresis* **3** 65–75

Debye P and Huckel E 1924 *Physik. Z.* **24** 305–30

Delincee H and Radola B J 1978 *Anal. Biochem.* **90** 609–23

Dossi G, Celentano F, Gianazza E and Righetti P G 1983 *J. Biochem. Biophys. Methods* **7** 123–42

Drysdale J W 1977 *Electrofocusing and Isotachophoresis* ed. B J Radola and D Graesslin (Berlin: De Gruyter) pp 241–52

Drysdale J W and Righetti P G 1972 *Biochemistry* **11** 4044–52
Ek K, Bjellqvist B and Righetti P G 1983 *J. Biochem. Biophys. Methods* **8** 134–55
Ek K, Gianazza E and Righetti P G 1980 *Biochim. Biophys. Acta* **626** 356–65
Ek K and Righetti P G 1980 *Electrophoresis* **1** 137–40
Frants R R, Noordhock G T and Eriksson A W 1978 *Scand. J. Lab. Clin. Invest.* **38** 457–62
Fredriksson S 1978 *J. Chromatogr.* **151** 347–55
Galante E, Caravejgio T and Richel P G 1976 *Biochim. Biophys. Acta* **442** 309–315
Garoff H and Ansorge W 1981 *Anal. Biochem.* **115** 450–7
Gelfi C and Righetti P G 1981a *Electrophoresis* **2** 213–219
—— 1981b *Electrophoresis* **2** 220–228
—— 1983 *J. Biochem. Biophys. Methods* **8** 156–71
—— *1984 Electrophoresis* **5** 253–66
Gelsema J W, De Ligny C L and Van Der Veen N G 1977 *J. Chromatogr.* **140** 149–55
—— 1978 *J. Chromatogr.* **154** 161–74
Gianazza E, Artoni F and Righetti P G 1983a *Electrophoresis* **4** 321–6
Gianazza E, Celentano F, Dossi G, Bjellqvist B and Righetti P G 1984a *Electrophoresis* **5** 88–97
Gianazza E, Dossi G, Celentano F and Righetti P G 1983b *J. Biochem. Biophys. Methods* **8** 109–33
Gianazza E, Frigerio A, Tagliabue A and Righetti P G 1984b *Electrophoresis* **5** 209–216
Gianazza E, Gelfi C and Righetti P G 1980 *J. Biochem. Biophys. Methods* **3** 65–75
Gianazza E, Pagani M, Luzzana M and Righetti P G 1975 *J. Chromatogr.* **109** 357–64
Gianazza E and Righetti P G 1978 *Biochim. Biophys. Acta* **540** 357–64
—— 1980 *Electrophoresis '79* ed. B J Radola (Berlin: de Gruyter) pp 129–40
Gianazza E, Righetti P G, Bordi S and Papeschi G 1977 *Electrofocusing and Isotachophoresis* ed. B J Radola and D Graesslin (Berlin: De Gruyter) pp 173–9
Görg A, Postel W, Weser J, Weidinger S, Patutschnich S and Cleve H 1983 *Electrophoresis* **4** 153–7
Görg A, Postel W and Westermeier R 1978 *Anal. Biochem.* **89** 60–70
Grubhofer N and Borja C 1977 *Electrofocusing and Isotachophoresis* ed. B J Radola and D Graesslin (Berlin: de Gruyter) pp 111–20
Haff L A, Lasky M and Manrique A 1979 *J. Biochem. Biophys. Methods* **1** 275–286
Harada S 1975 *Clin. Chim. Acta* **85** 275–83
Harada S, Agarwal D P and Goedde H W 1980 *Electrophoresis '79* ed. B J Radola (Berlin: de Gruyter) pp 687–93
Horejši V 1979 *J. Chromatogr.* **178** 1–13
Janik B and Dane R G 1981 *Electrophoresis '81* ed. R C Allen and P Arnaud (Berlin: de Gruyter) pp 77–82
Jonsson M, Fredriksson S, Jontell M and Linde A 1978 *J. Chromatogr.* **157** 235–42
Josephson R V, Maheswaran S K, Morr C V, Jenness R and Lindorfer R K 1971 *Anal. Biochem.* **40** 476–82
Just W W 1980 *Anal. Biochem.* **102** 134–44
Krishnamoorthy R, Bianchi Bosisio A, Labie D and Righetti P G 1978 *FEBS Lett.* **94** 319–323

Läas T and Olsson I 1981 *Anal. Biochem.* **114** 167–168
Lostanlen D, Gacon G and Kaplan J C 1980 *Eur. J. Biochem.* **112** 179–183
McDuffie N M, Dietrich C P and Nader H B 1975 *Biopolymers* **14** 1473–86
McGuire J K, Miller T Y, Tipps R W, Snyder R S and Righetti P G 1980 *J. Chromatogr.* **194** 323–33.
Marine Colloids 1981 *Instruction leaflet* (FMC Corp. Rockland MN)
Morris C J O R 1978 *J. Chromatogr.* **159** 33–46
Nader H B, McDuffie N M and Dietrich C P 1974 *Biochem. Biophys. Res. Commun.* **57** 488–493
Olsson I and Läas T 1981 *J. Chromatogr.* **215** 373–8
Perrella M, Cremonesi L, Vannini Parenti I, Benazzi L and Rossi Bernardi L 1980 *Anal. Biochem.* **104** 126–132
Perrella M, Heyda A, Mosca A and Rossi Bernardi L 1978 *Anal. Biochem.* **88** 212–24
Perrella M, Samaja M and Rossi Bernardi L 1979 *J. Biol. Chem.* **254** 8748–50
Qureshi R A and Punnett H H 1981 *Electrophoresis '81* ed. R C Allen and P Arnaud (Berlin: de Gruyter) pp 83–7
Radola B J 1973 *Ann. NY Acad. Sci.* **209** 127–43
Radola B J 1980 *Electrophoresis* **1** 43–56
Righetti P G 1983a *Isoelectric Focusing: Theory, Methodology and Applications* (Amsterdam: Elsevier)
—— 1983b *Trends Anal. Chem.* **2** 193–6
—— 1984 *J. Chromatogr* **300** 165–23
Righetti P G, Brost B C W and Snyder R S 1981a *J. Biochem. Biophys. Methods* **4** 347–63
Righetti P G, Brown R P and Stone A L 1978a *Biochim. Biophys. Acta* **542** 232–44
Righetti P G and Caravaggio T 1976 *J. Chromatogr.* **127** 1–28
Righetti P G, Delpech M, Moisand F, Kruh J and Labie D 1983a *Electrophoresis* **4** 393–8
Righetti P G, Ek K and Bjellqvist B 1984 *J. Chromatogr.* **291** 31–42
Righetti P G, Gacon G, Gianazza E, Lostanlen D and Kaplan J C 1978b *Biochem. Biophys. Res. Commun.* **85** 1575–81
Righetti P G, Gelfi C and Bianchi Bosisio A 1981b *Electrophoresis* **2** 291–5
Righetti P G and Gianazza E 1978 *Biochim. Biophys. Acta* **532** 137–46
—— 1979 *Protides Biol. Fluids* **27** 711–4
—— 1981 *Electrophoresis '81* ed. R C Allen and P Arnaud (Berlin: de Gruyter) pp 655–65
Righetti P G, Gianazza E and Bjellqvist B 1983b *J. Biochem. Biophys. Methods* **8** 89–108
Righetti P G, Gianazza E, Brenna O and Galante E 1977 *J. Chromotogr.* **137** 171–81
Righetti P G and Hjerten S 1981 *J. Biochem. Biophys. Methods* **5** 259–72
Righetti P G, Krishnamoorthy R, Gianazza E and Labie D 1978c *J. Chromatogr.* **166** 455–60
Righetti P G, Menozzi M, Gianazza E and Valentini L 1979 *FEBS-Letters* **101** 51–5
Righetti P G, Pagani M and Gianazza E 1975 *J. Chromatogr.* **109** 341–56
Righetti P G, Tudor G and Ek K 1981c *J. Chromatogr.* **220** 115–194

Righetti P G, Tudor G and Gianazza E 1982 *J. Biochem. Biophys. Methods* **6** 219–27

Rochette J, Righetti P G, Bianchi Bosisio A, Vertogen F, Schneck G, Boissel J P, Labie D and Wajcman H 1984 *J. Chromatogr.* **285** 143–52

Rosen A, Ek K and Aman P 1979 *J. Immunol. Methods* **28** 1–11

Saravis C A, Cantaroa W, Marasco P V, Burke B and Zamcheck N 1980 *Electrophoresis* **1** 191–3

Savaris C A and Cook 1979 *Marine Colloid's Instruction Leaflet* (Marine Colloid: Rockland, Mn)

Saravis C A, O'Brien M and Zamcheck N 1979 *J. Immunol. Methods* **29** 97–100

Serwer P 1980 *Biochemistry* **19** 3001–5

—— 1983 *Electrophoresis* **4** 375–82

Shafritz D A and Drysdale J W 1975 *Biochemistry* **14** 61–8

Studier F V 1973 *J. Mol. Biol.* **79** 237–42

Svensson H 1961 *Acta Chem. Scand.* **15** 325–41

—— 1962a *Acta Chem. Scand.* **16** 456–66

—— 1962b *Arch. Biochem. Biophys. Suppl.* **1** 132–40

Ui N 1971 *Biochim. Biophys. Acta* **229** 567–81

Valentini L, Gianazza E and Righetti P G 1980 *J. Biochem. Biophys. Methods* **3** 323–38

Valerio F, Bottino P and Cimberle R 1983 *Chromatography and Mass Spectrometry in Biomedical Sciences* ed. P Frigerio (Amsterdam: Elsevier) pp 221–9

Vesterberg O 1969 *Acta Chem. Scand.* **23** 2653–66

—— 1980 *Electrophoresis '79* ed. B J Radola (Berlin: de Gruyter) pp 95–104

Vesterberg O and Hansen L 1978 *Biochim. Biophys. Acta* **534** 369–73

Vesterberg O and Svensson H 1966 *Acta Chem. Scand.* **20** 820–34

Vinogradov S N, Lowenkron S, Andonian H R, Bagshaw J, Felgenhauer K and Pak S J 1973 *Biochem. Biophys. Res. commun.* **54** 501–6

Westermeier R, Postel W, Weser J and Görg A 1983 *J. Biochem. Biophys. Methods* **8** 321–30

Williams K W and Soderberg L 1979 *Int. Lab.* Jan/Feb. 45–53

Chapter 4

High Resolution Two-dimensional Polyacrylamide-gel Electrophoresis

by
M J Dunn and A H M Burghes

4.1 Introduction

The inherently high resolving power of electrophoretic techniques makes them ideal tools for the separation and analysis of complex protein mixtures. However, traditional methods of gel electrophoresis are generally not able to resolve all the components of such complex mixtures. By combining two different techniques to produce a two-dimensional (2D) separation procedure, resolving power can be increased by several orders of magnitude. To maximise resolution in such a system, the components of a complex mixture should be spread uniformly across the second-dimension gel, rather than being aligned along a diagonal. The aim of a 2D separation should, therefore, be to use a combination of techniques which separate proteins on the basis of independent parameters in each dimension.

The history of 2D electrophoresis dates back over 25 years to the experiments of Smithies and Poulik (1956) which showed that a 2D combination of paper and starch gel electrophoresis substantially improved the separation of serum proteins compared with either method on its own. A substantial increase in the power of electrophoretic separations was achieved by the advent of polyacrylamide gel electrophoresis (PAGE) (Raymond and Weintraub 1959). It was soon realised that gels of different concentrations were able to resolve preferentially different components of complex mixtures and the advantages of this approach for 2D separations of serum proteins were clearly demonstrated (Raymond and Nakamichi 1964). Band sharpness and resolution can, therefore, be improved by the use of gradient polyacrylamide gels in which the proteins being separated are migrated through pores of progressively decreasing size (Slater 1965; Margolis and Kenrick 1967). This approach was further improved by the use of a low concentration, discontinuous gradient polyacrylamide gel system, separating predominantly by mobility in the first dimension followed by second-dimension convex gradient polyacrylamide gels, separating according to size (Margolis and Kenrick 1969). This system thus approaches the ideal of the proteins being separated in the two dimensions according to separate properties.

Methods designed specifically for the separation of ribosomal proteins were developed using gels of different pH in the two dimensions (Kaltschmidt and Wittmann 1970a, b). This system with modifications has remained the method of choice for the analysis of ribosomal proteins and will be discussed in more detail later.

A major innovation in 2D methodology was the adoption of isoelectric focusing (IEF) techniques. This approach made it possible for the first-dimension separation to be based solely on charge, and could be used in combination with either single concentration (Dale and Latner 1969, Macko and Stegemann 1969) or gradient (Emes *et al* 1975, Latner and Emes 1975) polyacrylamide gels. The combination of IEF with a second-dimension electrophoresis in polyacrylamide gels containing the anionic detergent sodium dodecyl sulphate (SDS), separating proteins on the basis of polypeptide molecular weights, then provided a means for separating proteins according to two truly independent parameters, i.e. charge and size (Stegemann 1970).

Application of IEF techniques to samples with a wide range of solubility properties required the development of modified IEF procedures. The inclusion of urea in IEF gels improved resolution of EDTA-extractable erythrocyte membrane proteins (Bhakdi *et al* 1974) and non-histone nuclear proteins (Barret and Gould 1973, MacGillivray and Rickwood 1974, Suria and Liew 1974). The addition of non-ionic detergents, such as Triton X-100, to IEF gels containing 8 M urea was found to improve 2D separations of membrane proteins (Bhakdi *et al* 1975a, Chignell and Wingfield 1974).

The foregoing developments made it possible to apply a 2D PAGE technique combining IEF and SDS-PAGE to the high-resolution analysis of proteins of various cells and tissues (Iborra and Buhler 1976, Klose 1975a, b,). These developments culminated in the publication of a paper by O'Farrell (1975) describing a 2D PAGE procedure optimised for the separation of *Escherichia coli* (*E. coli*) proteins. This technique involves a first-dimension separation in cylindrical $T = 4\%$, $C = 5\%$ IEF gels containing 9 M urea and 2% (w/v) NP-40 with the discontinuous SDS gradient PAGE slab system of Laemmli (1970). Upon this procedure are based most current methods of high-resolution 2D PAGE.

In this review we will restrict our discussion to 2D-techniques based either

(i) on a combination of two electrophoretic procedures as used, for example, in the analysis of ribosomal proteins or
(ii) on the high-resolution 2D PAGE method of O'Farrell (1975).

Combination of electrophoresis, IEF or isotachophoresis (ITP) with immuno-electrophoresis in the second dimension produces a very powerful tool for the characterisation of complex protein mixtures. This approach is described in detail in this volume by Heegard and Bøg-Hansen (see chapter 5). The elegant methods for titration curve analysis of protein mixtures provided by electrophoresis in stationary pH gradients, formed by a first dimension electrophoresis in the absence of the sample, are fully discussed in the article by Righetti *et al* (see chapter 3).

4.2 Two-dimensional Electrophoresis under Non-denaturing Conditions

Techniques of 2D electrophoresis under non-denaturing conditions predated the development of 2D PAGE methods under denaturing conditions based on the technique of O'Farrell (1975). The inherently low resolution capacity and limited applicability to different types of samples of the former procedures have resulted in their general replacement by high-resolution 2D PAGE under denaturing conditions. However, 2D electrophoresis under non-denaturing conditions is still of importance in studies of soluble proteins where their native physiochemical properties and biological activities are to be maintained.

The technique most often used is based on the procedure developed by Latner and his co-workers (Dale and Latner 1969, Emes *et al* 1975) and modified by Manabe *et al* (1979, 1982b). In this technique proteins are separated in the first dimension by rod-gel IEF in the absence of urea and detergents. The first-dimension gels, without prior equilibration, are applied in the second dimension to 4 to 21% linear gradient polyacrylamide slab gels in Tris buffer. An example separation using this system is shown in figure 4.1.

This system has been converted to a microscale version by decreasing gel size in both dimensions and adapting it so that 16 first- and second-dimension gels can be run simultaneously (Manabe *et al* 1982a). In another recent modification (Gahne and Juneja 1984), first-dimension, flat-bed IEF gels were run using immobilised pH gradients (pH 4–6). Gels of 18 samples were run on plastic sheets (Serva) which were simply cut up into strips after

IEF for application to the second-dimension gels which contained 'step' layers of 8%, 4% and 12% polyacrylamide.

Some examples of the use of this technique are the analysis of serum (Manabe *et al* 1982, 1983a, Manabe and Okuyama 1981), cerebrospinal fluid (Manabe *et al* 1982a) and salivary (Manabe *et al* 1982a, Marshall 1984) proteins. As this technique is carried out in the absence of denaturing agents it is possible to demonstrate enzyme activities directly in the gels after 2D separation. For example, Kadofuku *et al* (1983) have investigated α-amylase iso-enzymes present in human body fluids using this procedure. In addition, as protein–protein interactions are not disrupted in this method it is possible to study protein–protein complexes. For example, Manabe *et al* (1983a) have applied this technique to the analysis of three phenotypes of haptoglobin and their haemoglobin complexes in human sera. Heterogeneity in the isoelectric point (pI) of the haptoglobin–haemoglobin complexes was observed, suggesting heterogeneous binding of an $\alpha\beta$-half molecule of haemoglobin to a β-chain of haptoglobin.

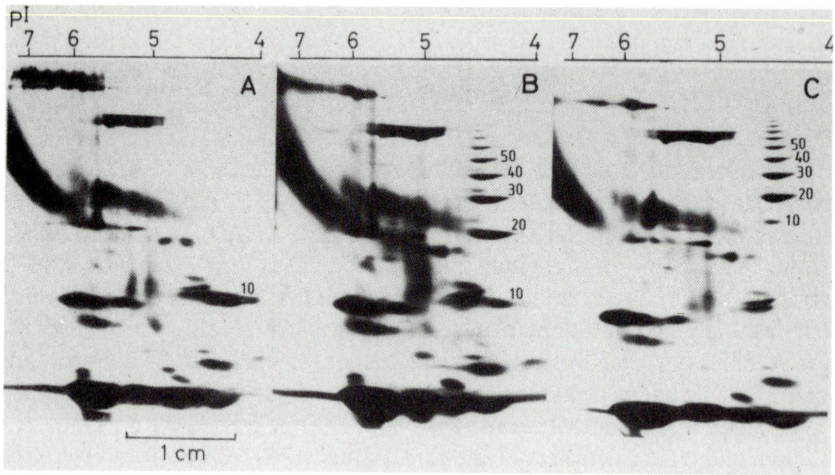

Figure 4.1 Sections of micro-2D-PAGE patterns of human sera run under non-denaturing conditions. (A), (B) and (C) represent the sera of three haptoglobin phenotypes, Hp 1–1, Hp 2–1, and Hp 2–2 respectively. Taken with permission from Manabe *et al* (1983a).

4.3 Denaturing 2D Systems using Electrophoresis in both Dimensions

Although techniques using electrophoresis under denaturing conditions in both dimensions were of importance in the development of 2D PAGE technology, the almost universal adoption of IEF for the first dimension has largely rendered these procedures obsolete. However, they are still important in the 2D separation of proteins from particular sources.

4.3.1 Ribosomal Proteins

The 2D method routinely used to separate eukaryotic and prokaryotic ribosomal proteins was first described by Kaltschmidt and Wittman (1970a, b). A discontinuous buffer system is used in the first-dimension rod gels, followed by a continuous one in the second-dimension slab gels, urea being present at a high concentration (6 M) in both dimensions. A large pore acrylamide sample gel, polymerised in the middle of the resolving gel, was used in the original method (Kaltschmidt and Wittman 1970a). In this way proteins migrating to the anode and cathode were resolved simultaneously. However, it was found that substantial loss of sample occurred due to protein immobilisation in the sample gel (Kaltschmidt and Wittman 1970b). To overcome this problem an agarose sample gel was substituted for the polyacrylamide sample gel (Howard and Traut 1973). The first-dimension gel is of a low concentration (4%) to minimise molecular sieving, while the second-dimension gel is of a high concentration (18%). A different pH is also used in each dimension (pH 8.6 followed by pH 4.5) so that in addition to size the proteins are also separated according to different charged groups in each dimension. In a modification of this system, separation in both the first and the second dimension is carried out at acidic pHs thus allowing all ribosomal proteins to migrate cationically (Knopf *et al* 1975).

The main variations on this 2D procedure for ribosomal proteins have been developed as a consequence of the advantages of SDS-PAGE for the separation of proteins according to their polypeptide molecular weight (Shapiro *et al* 1967; Weber and Osborn 1969). A low percentage gel containing urea is usually run at low pH in the first dimension followed by second-dimension SDS-PAGE in the presence (Martini and Gould 1971) or absence (Mets and Bogorad 1974) of urea. An additional advantage of using SDS-PAGE in the second dimension is the greatly reduced time required for the separation. In a further modification of this procedure (Hoffman and Dowben 1978), the non-ionic detergent, Triton X-100, was included in the first-dimension separation so that proteins were separated by their ability to bind detergent as well as by their mobility.

Difficulties can arise in relating the protein spots obtained by one 2D-separation procedure with those obtained by a different procedure. For example, in the case of ribosomal proteins a general nomenclature for the proteins was derived from the coordinates of each spot defined by the Kaltschmidt and Wittman (1970b) procedure. As new and improved methods requiring less sample and time, and giving improved resolution were developed (see above), there was a need to correlate the spots observed with the protein species classified according to the original nomenclature. This necessitated co-electrophoresing each purified protein together with the total proteins and determining its specific position by the new methods.

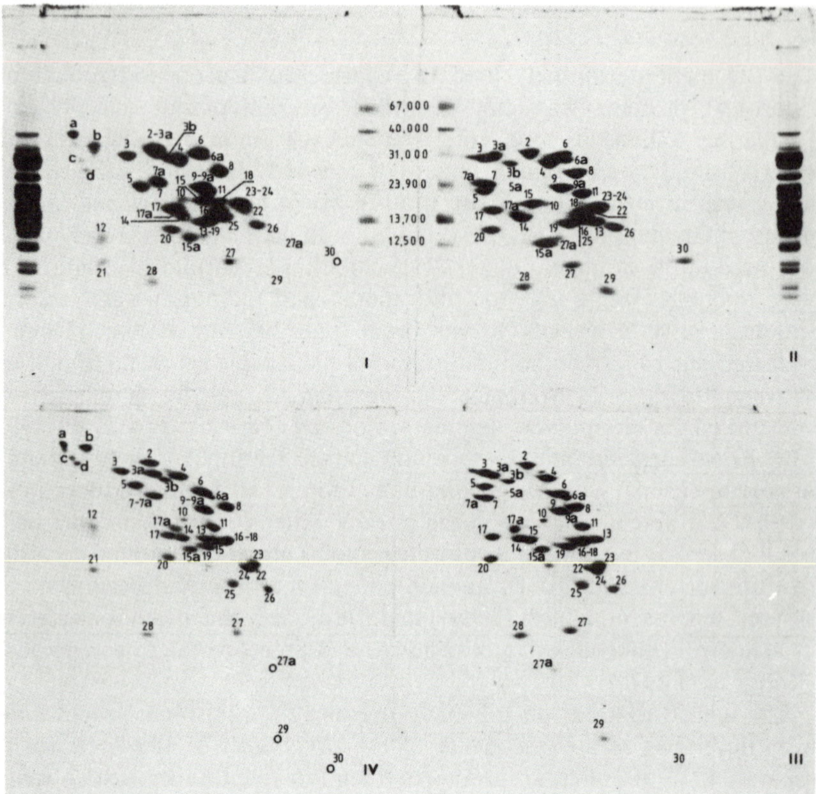

Figure 4.2 Analysis of proteins from unwashed 40s ribosomal subunits of rat liver by four methods of 2D PAGE. The first dimension was run either at acidic (I, IV) or basic (II, III) pH. Separation in the second dimension was carried out either at acidic pH (III, IV) or in the presence of SDS (I, II). One-dimensional SDS-PAGE separation of 40s ribosomal subunits and reference molecular weights are presented on the edges. Taken with permission from Madjar *et al* (1979a).

To overcome this problem a method which avoids the necessity for purification of individual components has been devised for ribosomal proteins in which the same protein preparation is analysed in four different two-dimensional gel systems (Madjar *et al* 1979b). The first dimension separation is run either at acidic or basic pH and the second dimension is carried out either at acidic pH or in the presence of SDS. An example of this 'method of four corners' as applied to ribosomal proteins of rat liver is shown in figure 4.2.

4.3.2 Histones

The highly basic character and similarity of molecular weights of the histones, the major protein components of chromosomes, make them very difficult to separate using techniques based on the O'Farrell (1975) 2D PAGE procedure. Histones can be separated, essentially on the basis of their charge, using the acetic acid–urea gel system developed by Panyim and Chalkley (1969), in which the proteins are separated on 15% acrylamide gels containing 2.5 M urea and 0.9 M acetic acid (pH 2.7). Several successful 2D separation procedures have been described using this procedure in either the first or the second dimension.

In one approach, separation of histone proteins by charge using the acetic acid–urea system in the first dimension is combined with a second-dimension separation on the basis of size by SDS-PAGE (Davie and Candido 1978, Sinclair and Rickwood 1981). It should be noted, however, that there is evidence for anomalous behaviour of histones in SDS-PAGE. Another problem with this procedure is that lateral migration of proteins results in 'tailing' of the protein pattern with consequent loss of resolution. As a solution to this problem, a technique using SDS-PAGE in the first dimension followed by an acetic acid–urea gel in the second dimension has been proposed (Davie 1982). Before application to the second dimension, the first dimension gels were equilibrated in a buffer containing 1% protamine sulphate and 6 M urea. It is claimed that the combined action of urea and protamine resulted in the splitting of the SDS-protein complexes into an insoluble protamine dodecyl sulphate complex and free protein (Davie 1982).

Alternative 2D separation systems for histone proteins are based on the use of acetic acid–urea gels containing non-ionic detergents such as Triton X-100, Triton DF-16 or Lubrol-WX (Franklin and Zweidler 1977). The gels containing detergent are usually used in a 2D combination with detergent-free acetic acid–urea gels (Bhatnagar and Bellve 1978, Bonner *et al* 1980). This procedure produces an essentially diagonal distribution of spots. However, certain histone species and variants bind non-ionic detergents and are consequently differentially retarded in their electrophoretic mobility. These retarded species are then found as spots which are not aligned with the diagonal. A potential problem in this technique is that oxidation, especially of methionine residues, can result in altered mobilities on acetic acid–urea–detergent gels (Zweidler 1978) leading to artefactual heterogeneity. Triton has reduced affinity for oxidised histones and it is therefore essential to maintain the proteins in a reducing environment, for example by the addition of an agent such as dithiothreitol to the system. Davie (1982) has recommended that the stacking gel for the acetic acid–urea–detergent system should be polymerised by riboflavin catalysis rather than with ammonium persulphate which could oxidise the proteins during electrophoresis. Charge separation by acetic acid–urea–detergent gels can also be used as the second dimension following first-dimension size separation of histone proteins by SDS-PAGE (Davie 1982).

4.3.3 Non-histone Nuclear Proteins

Non-histone nuclear proteins are usually analysed by 2D PAGE procedures based on the method of O'Farrell (1975). However, as a consequence of solubility problems, alternative procedures, such as a combination of acetic acid–urea gels and SDS-PAGE (Orrick *et al* 1973), are sometimes used. Nucleosomes can initially be separated using low percentage (5%) polyacrylamide gels containing EDTA and triethanolamine, followed by separation of their constituent proteins by second-dimension acetic acid–urea gels (Goodwin *et al* 1981).

4.3.4 Membrane Proteins

Several methods based on the 2D PAGE method of O'Farrell (1975) have been described for the analysis of membrane proteins. However, sample solubility prior to and during the first-dimension IEF run is critical and protein components of low solubility produce ill-defined spots or are absent from the final 2D pattern. Combinations of two electrophoretic techniques containing detergents and other protein solubilising agents have been used in an attempt to overcome this problem. Examples of such 2D procedures for the analysis of membrane proteins are acetic acid–urea–Triton X-100 gels against SDS-PAGE (Fernandes *et al* 1978), a combination of SDS-PAGE and SDS–Triton CF10–urea gels (Imada *et al* 1978, Imada and Sueoka 1980) and gels containing chloral hydrate against gels containing organic-base dodecyl sulphates (Booth 1977). It should be noted that chloral hydrate must be used under acidic conditions to avoid its reactivity with proteins at alkaline pH, and it is therefore unsuitable for use with IEF systems. Unfortunately, these procedures generally result in rather poor resolution with the proteins typically being distributed along a diagonal, due to the same separation criterion (presumably size) predominating in both dimensions. This limits the applicability of these techniques but they can yield useful information under some circumstances. For example, using a combination of acetic acid–urea–Triton X-100 gels and a neutral SDS gel system, Ames and Nikaido (1981) were able to investigate labile phosphate-containing proteins of *Salmonella typhimurium* and *E. coli* which would be lost in systems employing basic conditions.

4.4 The O'Farrell System

O'Farrell (1975) introduced several innovations designed to improve resolution in 2D PAGE and his method is the foundation of modern 2D PAGE technology. The mixture of 9 M urea and non-ionic detergent, NP-40, used for sample separation was found to solubilise the majority of *E. coli* proteins. A combination of IEF in rod gels containing urea and NP-40 with the gradient SDS-PAGE system of Laemmli (1970) was used as these methods are high-resolution techniques and separate proteins according to independent parameters, i.e. charge and molecular weight. Gradients were engineered to obtain maximum separation of *E. coli* proteins and high sensitivity of detection was achieved using autoradiography. Features of the O'Farrell system are listed in table 4.1 and compared with those features which we believe to be desirable for optimal resolution in 2D PAGE. Some of these may be difficult to achieve in practice, but many papers have been published attempting to resolve some of these problems as applied to a wide variety of protein samples. For example, in some instances the order of dimensions has been reversed with SDS-PAGE being followed by flat bed IEF in the second dimension (Shackelford *et al* 1981, Shackelford and Strominger 1980, Siemankowski *et al* 1978, Singer *et al* 1978, Tuszyncki *et al* 1979). This approach was adopted to overcome problems of sample solubility as it is assumed that after SDS-PAGE, the proteins do not reaggregate or become insoluble during subsequent steps in the procedure. However, a major disadvantage of these systems is that they result in more lateral diffusion, and therefore larger spots, due to the less restrictive nature and greater porosity of the second-dimension IEF gels (Dunn and Burghes 1983a).

Table 4.1 Features and disadvantages of the O'Farrell (1975) 2D PAGE procedure compared with features required for optimal resolution. Taken with permission from Dunn and Burghes (1983a)

Features of the O'Farrell system	*Optimal features*	*Disadvantages of the O'Farrell system*
Separation by charge and molecular weight	Separation by two different physiochemical parameters	Hydrophobic properties not analysed
Samples solubilised using 9.5 M urea and 2% NP-40	Complete solubilisation and disruption of all non-covalent interactions. Complete entry of sample into both dimensions. Solubilising agents should be stable and have low viscosity.	Incomplete solubilisation and disaggregation. Incomplete entry of sample into gels. Urea is viscous and problems with carbamylation.
Polyacrylamide support medium	Stable and inert support medium over a wide pH range. No sieving during charge separation. Good sieving properties for molecular weight separation.	Deamidation at extreme alkaline pH. Sieving of high molecular weight proteins in IEF.

(*Continued*)

Table 4.1 Continued

Features of the O'Farrell system	*Optimal features*	*Disadvantages of the O'Farrell system*
IEF run for 30 to 59.2 Vh cm^2. Long focusing time to get sharper spots.	Stable pH gradients. True equilibrium and maximum resolution. Separate systems to detect mobility variants *i.e.* amino acid substitutions involving $pK_a = pI$.	Cathodic drift. Final pH gradient does not extend above pH 7. Equilibrium not tested.
Low field strength used.	High field strengths to minimise focusing time and to sharpen bands.	High field strengths cannot be used due to heating and gradient drift.
Final gradient pH 4.5 to pH 7.0. Ampholine used.	Optimisation of pH gradient to fit distribution of protein pI values. Maximise number of ampholyte species.	Gradient cannot be extended due to excess drift and loss of resolution. Only one type of ampholyte used.
Equilibration of first dimension gel to allow reaction of proteins with SDS.	Minimisation of diffusion and protein loss with complete elution of proteins and no streaking.	Diffusion and loss of proteins giving reduced resolution.
Laemmli stacking gel.	Rapid elution with stacking into second dimension.	Slow and incomplete elution.
Exponential gradient polyacrylamide SDS gels for second dimension.	Good resolution of all proteins in second dimension.	High molecular weight proteins poorly resolved. Low molecular weight proteins form diffuse spots.
	Use of large, thin gels to maximise area of gel used for separation.	
Good reproducibility under standard conditions.	Good reproducibility to facilitate inter-gel comparisons.	Pattern matching can be difficult.
	Simultaneous processing of multiple gels.	Single gel procedure.
High sensitivity using radiolabelled proteins.	High sensitivity of detection with minimum spreading of spots. High dynamic range.	Spreading of spots. Limited dynamic range.
	Quantitative evaluation of gels. Identification and characterisation of separated polypeptides.	Limited qualitative and quantitative analysis.

4.5 Sample Preparation and Solubilisation

The diverse nature of samples which are subjected to 2D PAGE means that no one particular method of sample preparation is suitable for all types of sample. Many samples, such as those from body fluids, often require concentration before 2D PAGE analysis. Special precautions must be exercised with many samples to prevent protein modification during handling which can result in artefacts in the final 2D map. Some samples must be specially treated to remove non-protein components, e.g. phenolic pigments must be removed from preparations of leaf proteins (Hari 1981).

4.5.1 Concentration and Removal of Salt

Samples of body fluids are amenable to analysis by 2D PAGE, but many of these are very dilute and contain high levels of salts which would interfere with the IEF dimension of 2D PAGE. Materials such as urine, cerebrospinal fluid (CSF), saliva, aqueous humour from the eye and blister fluid fall into this category. Before analysis by 2D PAGE is attempted, these samples must be dialysed to remove salt and then concentrated by lyophilisation or precipitation.

Very sensitive detection methods such as silver staining are very advantageous for use with body fluids such as CSF as in this way excessive concentration and the application of large amounts of material to the gel which can interfere with resolution can be avoided (Merril *et al* 1981). For CSF samples, the proteins were concentrated four-fold by dialysis against 10% polyethylene glycol and 15 μl of the concentrated samples were applied to the gels (Merril *et al* 1981). Silver staining and other sensitive detection methods are discussed in detail by Merril *et al* in chapter 7 of this book.

Urinary proteins have been analysed extensively by 2D PAGE. Various methods have been used to remove the salts, such as ion exchange chromatography, ultrafiltration, dialysis and lyophilisation. Anderson and his coworkers (Anderson *et al* 1979, Edwards *et al* 1982a, b) have developed methods for concentration of urine samples for analysis by 2D PAGE. The two methods that they found most successful were

(i) dialysis followed by lyophilisation, and
(ii) gel exclusion chromatography.

Flow charts showing the steps in these two procedures are shown in figure 4.3. Marshall *et al* (1985) have reported obtaining satisfactory 2D patterns from unconcentrated or dialysed urine using a silver staining technique. However, they commented that the separation obtained was not as good if salts were not removed from the sample. An important factor to remember when using lyophilisation is that protein modification, such as deamidation of asparagine and glutamine residues or oxidation or cysteine residues, can occur (O'Farrell 1975). In the case of body fluids like urine it is quite likely that considerable protein modification occurs *in vivo*. For other body fluids similar methods can be used, except that such an extensive dialysis step is unnecessary. For example, salivary proteins have been successfully analysed by 2D PAGE following dialysis and concentration using a centrifugal concentrator (Rubin and Penny 1983).

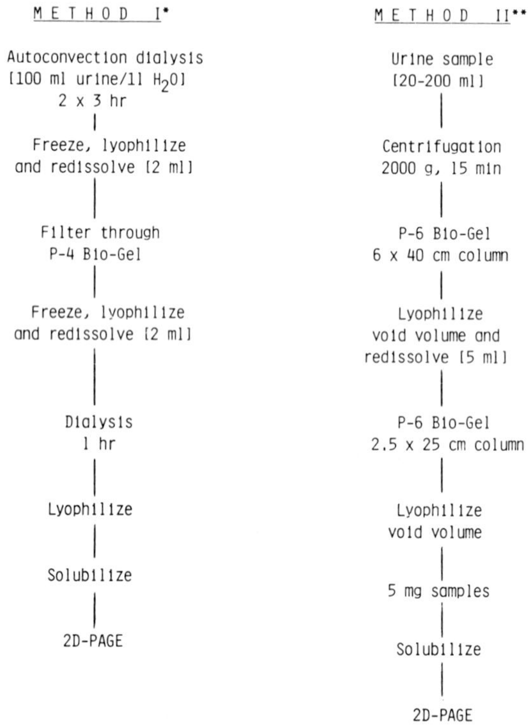

Figure 4.3 Methods for concentration and removal of salts from urine samples for analysis by 2D PAGE (*Anderson *et al* 1979, **Edwards *et al* 1982b).

A novel method of sample concentration for IEF has been described by Vesterberg and Hansen (1978). This involves applying the sample to a hydrophobic chromatography matrix. Under high salt conditions, the proteins bind but the salt in the sample elutes from the column. The proteins can then be eluted by low salt or a detergent–urea mixture. However, this method has not found widespread application, and appears not to have been used for 2D PAGE.

Some types of fluid sample require only concentration prior to 2D PAGE. This can be achieved conveniently using apparatus such as the Minicon concentrator (Amicon Ltd) (Marshall *et al* 1985). Another convenient method is precipitation with acetone. Although it is possible to get loss of sample using this procedure, we found that precipitation of 100 μl of sample with 1 ml cold acetone for 1 h, followed by centrifugation and storage at −20°C results in no qualitative loss of bands (Dunn and Burghes 1983c). However, the possibility of quantitative losses has not been excluded. A 2D gel of proteins secreted into the medium by skin fibroblasts and concentrated using this method is shown in figure 4.4. Acids such as trichloroacetic accid (TCA) can be used but there is a risk of protein modification and the precipitates can be difficult to redissolve. Immunoprecipitates can be similarly analysed, although the current popularity of protein blotting techniques has diminished the requirement for such analysis.

Figure 4.4 A 2D PAGE gel of ^{35}S-methionine-labelled proteins secreted into the medium by skin fibroblasts in culture. The sample was concentrated by precipitation with cold acetone. The first dimension was a flat-bed IEF gel (anode at left). The second dimension was a 7–20% gradient SDS-PAGE gel.

4.5.2 Preparation of Tissue Samples

Most procedures use homogenisation in solubilisation buffer to disrupt the tissue. Large pieces of tissue should first be chopped into smaller pieces using a scalpel or with a suitable electric blender. Devices such as the Ultra-Turrax or Polytron are ideal for the subsequent homogenisation of tissue fragments. Care should be exercised to minimise heating and foaming during homogenisation. It is possible to add detergent after homogenisation or to include a suitable antifoam reagent in the solubilisation mixture.

With the advent of modern sensitive visualisation techniques, such as silver staining, it has become possible to run 2D gels of cryostat sections of tissue. This approach has many advantages

(i) less tissue is used.
(ii) pathological changes can be monitored by histological analysis of serial sections, and
(iii) a much wider range of tissue sample should be available for analysis.

Cryostat sections can either be collected on a suitable probe and placed in solubilising buffer or be applied to the hydrophilic surface of squares of Gel Bond and scraped off into solubilisation buffer (Thompson *et al* 1981). In addition to using sensitive staining methods it is also possible to radio-label cryostat tissue sections, for example using ^{14}C-iodoacetamide (Giometti and Anderson 1981, Giometti *et al* 1983) or reductive methylation using ^{3}H-sodium borohydride or ^{14}C-formaldehyde (Kuhn and Wilt 1980). The latter technique has been shown not to result in charge heterogeneity of the proteins (Means and Feeney 1968) but it should be remembered that this is a potential hazard with such post-synthetic labelling procedures (Dunn and Burghes 1983b).

4.5.3 Cultured Cells

Cells grown in tissue culture are most conveniently prepared for 2D PAGE by direct lysis in the culture dish or well, using a small amount of solubilisation buffer. Samples should be stored at −70°C. Nucleic acids present in the sample in small amounts do not interfere with the first dimension, but larger samples should first be treated with DNase and RNase (Bravo 1984, Garrels 1979).

4.5.4 Plant Proteins

Plant seed (Görg *et al* 1981, Westermeier *et al* 1981) and membrane (Booz and Travis 1981) preparations can be treated in the same way as animal tissue. However, leaf proteins must be first extracted with acetone to remove phenolic pigments (Hari 1981).

4.5.5 General Precautions for Sample Preparation

It is essential to minimise the effects of protein modifications and degradation during sample preparation for 2D PAGE. It should be noted that certain types of sample can be rich in proteases. The best method is to handle the samples rapidly and to keep them cold at all times. Protease inhibitors can be added but certain of these can result in modification of charged groups of proteins and are therefore unsuitable for use with 2D PAGE. Samples containing urea should not be heated otherwise charge heterogeneity will be introduced due to carbamylation of the proteins by isocyanate formed by decomposition of urea.

4.5.6 Sample Solubilisation

We have recently discussed in detail solubilisation procedures for 2D PAGE (Dunn and Burghes 1983a) and it is our intention here only to comment briefly on possible procedures. The method most commonly used is that described by O'Farrell (1975) and uses a combination of the non-ionic detergent, NP-40, and 8 M urea. This combination of reagents is satisfactory for the majority of samples, but certain proteins such as histones, ribosomes and membrane proteins are more resistant to disaggregation and solubilisation.

O'Farrell (1975) found that SDS could be included in the solubilisation procedure for 2D PAGE without adverse effects. This procedure was further developed by Ames and Nikaido (1976) for membrane proteins. Samples were initially solubilised in SDS and NP-40 was subsequently added. The ratios of SDS to protein (1 : 3) and SDS to NP-40 (1 : 8) were carefully controlled to achieve effective solubilisation while minimising the effects of high levels of SDS on the IEF dimension. Although there have been reports of difficulties with this method, we find that it generally works well, with some minor loss of resolution in the acidic region, provided that the amount of SDS loaded is not too high. It was originally recommended to heat the samples in SDS at 70°C (Ames and Nikaido 1976), but we believe that it is better either to heat them at 100°C or to not heat them but include 8 M urea. The advantages of this method can be illustrated by reference to studies of the erythrocyte membrane. Band 3 protein and spectrin are not solubilised by a urea/NP-40 mixture (Rosenblum *et al* 1982, Rubin and Milikowski 1978), whereas these components are present in 2D maps if an SDS solubilisation method is used. A caveat to the use of SDS solubilisation for 2D PAGE is that it seems quite possible that some proteins might not remain in a soluble state when the SDS concentration is lowered by the addition of a non-ionic detergent resulting in artefacts in the IEF dimension.

Alternative solubilisation procedures have been described and found to be preferable to SDS for some samples (Dunn and Burghes 1983a). For example, a urea/NP-40 mixture used under alkaline conditions (pH 10.3) was very effective for plant seed and membrane proteins (Horst *et al* 1980, Basha 1979). Microsomal proteins appear to be particularly resistant to solubilisation and a selective procedure using deoxycholate has been used to analyse separately those proteins not solubilised by SDS (Kaderbhai and Freedman 1980). Certain proteins, for example cytochrome P450, tend to precipitate during IEF (Kaderbhai and Freedman 1980, Vlasuk and Walz 1980), which can be overcome by application of the sample at acid pH.

The non-ionic detergents, Triton X-100 and NP-40, have polyoxyethylene head groups, the hydrophobic character of which may favour protein solubility. The zwitterionic, sulphobetaine (SB)-type detergents are effective solubilising agents (Gonenne and Ernst 1978, Hjelmeland *et al* 1979) and are compatible with IEF as they behave as 'poor carrier ampholytes' (Hjelmeland *et al* 1978, 1979). We have found these detergents to be generally unsatisfactory for 2D PAGE (Burghes *et al* 1982b) as they precipitate in the presence of 8 M urea and used alone are less effective than a urea/NP-40 mixture. However, an alternative zwitterionic detergent, CHAPS (3-[(cholamidopropyl)dimethylammonio]-1-propane sulphonate), has been used recently for the 2D separation of microsomal proteins in the presence of 8 M urea (Perdew *et al* 1983). We now routinely use a urea/CHAPS mixture for solubilisation of samples prior to 2D PAGE (Dunn *et al* 1984, 1985), but it should be pointed out that using this method some insoluble material still remains associated with the sample application point.

Garrels has described a method (Bravo 1984, Garrels 1979) which involves initial extraction of samples in SDS, digestion of nucleic acids with RNase and DNase, followed by lyophilisation and resolubilisation in a urea/NP-40 mixture. This procedure is particularly useful for samples containing substantial amounts of nucleic acids, but care should be exercised that proteolysis does not occur during the RNase/DNase digestion step. The following recommended procedure (Bravo 1984) should be carried out in the cold to minimise protein degradation:

Cells scraped in 20 mM Tris (pH 8.8), 2 mM $CaCl_2$ and dispersed through a narrow gauge needle are treated with 0.3% SDS to solubilise the proteins. A solution containing nucleases (10 × nuclease solution : 1 $\mu g\ ml^{-1}$ DNAase I, 500 $\mu g\ ml^{-1}$ RNAase A, 0.5 M Tris (pH 7), 50 mM $MgCl_2$) is then added and the sample is left in the cold unit until it is no longer viscous. Following freeze-drying the sample is resuspended in solubilisation buffer and stored at $-70^\circ C$.

At present, then, there are three principal methods of sample solubilisation for 2D PAGE.

(i) urea/non-ionic detergent (Triton X-100, NP-40).
(ii) urea/CHAPS, and
(iii) SDS treatment followd by the addition of urea/NP-40.

However, an improvement in the effectiveness of these techniques is still required to achieve entry of all sample proteins into first-dimension IEF gels (Dunn and Burghes 1983a). Surprisingly, there has been little investigation of alternative denaturants to urea, as increasing the denaturing potential should disrupt more protein interactions. In addition, an alternative denaturant could overcome problems associated with the decomposition of urea resulting in possible sample protein charge modification.

4.6 IEF Dimension

4.6.1 Gel Composition

High-resolution 2D PAGE is carried out under denaturing conditions, with both urea and NP-40 present in the first-dimension IEF gels (O'Farrell 1975, Anderson and Anderson 1978a). IEF gels containing 8 M urea require longer focusing times to attain equilibrium due both to an increase in viscosity of the system (Dunn and Burghes 1983a) and to an increase in Stokes radius of proteins following their denaturation by urea (Creighton 1979, Olsson and Läas 1981).

First dimension IEF is usually done using a polyacrylamide gel matrix. Rather weak gels (3% to 5% *T*) must be used to minimise the effects of molecular sieving, but it is important to remember that the migration of high molecular weight proteins can still be impeded. Polyacrylamide gels of very high porosity can be prepared using high levels of *N*, *N'*-methylenebis-acrylamide (Bis) crosslinker (50% to 60% *C*), but such gels are not suitable for 2D PAGE as they are mechanically unstable and hydrophobic in character (Bianchi Bosisio *et al* 1980, Righetti 1983). A variety of alternative crosslinking agents to Bis are available (see table 3.1 in Righetti 1983), but certain of these such as *N,N'*-diallyltartardiamide (DATD) and *N,N'*,*N'''*-triallylcitrictriamide (TACT) should be avoided as they are inhibitors of gel polymerisation (Gelfi and Righetti 1981, Righetti 1983).

Highly purified (Rosen *et al* 1979, Saravis and Zamchek 1979) and charge-balanced agarose preparations are now available for IEF, and these have the potential to overcome the restriction of focusing high molecular proteins. Little notice has been taken so far of the possible advantages of agarose as a medium for first-dimension IEF gels in 2D PAGE. This is probably due to two factors

(i) there are difficulties associated with the use of high levels of urea in agarose gels (Dunn and Burghes 1983a, Olsson and Läas 1981), and
(ii) focusing times are restricted due to pH gradient instability associated with residual electro-endosmotic properties of the medium (Serwer and Hayes 1982, Thompson *et al* 1982).

Hirabayashi (1981) has described a 2D system using agarose IEF in the presence of 6 M urea in the first dimension, but the results obtained were not entirely satisfactory and the method has not gained general acceptance.

Non-ionic detergent (Triton X-100, NP-40) is usually included in IEF gels for 2D PAGE at a level of 2%, although this can be reduced to 0.5% without any adverse effect on sample solubility or resolution of the proteins (Burghes *et al* 1982a). Recently, the zwitterionic detergent CHAPS has been found to give improved sample solubility for IEF (Perdew *et al* 1983) and this detergent can be incorporated at 0.5% in first-dimension IEF gels containing 8 M urea (Dunn *et al* 1984, 1985).

The final, and most important, components of the IEF gel on which the quality of the first-dimension separation depends are the carrier ampholytes which form the pH gradient. Proteins can vary by fractional charge and to achieve good resolution it is, therefore, essential to maximise the number of carrier ampholyte species present in the gel (Charlionet *et al* 1979, Svensson 1967). We have recently discussed the methods used for the synthesis of different types of carrier ampholytes and the interested reader is referred to this and other recent reviews on this topic (Allen *et al* 1984, Dunn and Burghes 1983a, Righetti 1983). The most important point here is that the different commercial ampholyte preparations (Ampholine from LKB, Pharmalyte from Pharmacia, Servalyte from Serva) are synthesised by different procedures and, therefore, contain different ampholyte species. From this consideration it is, perhaps, not surprising that the various commercial ampholytes give better resolution in different pH regions of IEF gels (Burghes *et al* 1982b). Applying this strategy to 2D PAGE, we have shown that the use of carrier ampholyte mixtures for the IEF dimension can enhance resolution in the final 2D maps (Burghes *et al* 1982a, 1983). Moreover, by manipulation of these ampholyte mixtures using wide and narrow pH range preparations it is possible to engineer the final shape of the pH gradient to match the distribution of protein components within the gel (Burghes *et al* 1982a).

4.6.2 Rod IEF Gels

In the original 2D PAGE procedure (O'Farrell 1975) first-dimension IEF was carried out using a rod-gel system and this continues to be the most popular system for 2D PAGE. Gel tubes with an internal diameter of 1.2 mm to 3.0 mm are used and are cut to an appropriate length (12 to 20 cm). The simplest way to prepare rod IEF gels is to seal the bottom of the tubes with several layers of Parafilm or a small plug of concentrated polyacrylamide. The gel solution should be loaded into the tubes using a long, narrow-gauge, blunt-ended needle. The needle should initially be inserted into the bottom of the tube and slowly withdrawn to avoid trapping air bubbles. The tubes should all be filled to the same level so that all gels are of the same length for the sake of inter-gel reproducibility. The gel mixture is then overlaid with water and allowed to polymerise for at least 2 h. Although this method is simple, it suffers from the disadvantage that variability between gels is difficult to control, making comparisons of the final 2D maps difficult. This problem can be partially overcome by casting IEF gels simultaneously in batches and a relatively simple device to do this has been described recently (Bravo 1984, Garrels 1983). A more sophisticated system to cast large batches of IEF gels has been described by Anderson and Anderson (1978a).

In the standard O'Farrell (1975) method the upper electrolyte tank contains 20 mM NaOH and the lower electrolyte is 10 mM H_3PO_4. After the gels have been prefocused, the samples are applied, using a suitable microsyringe or dispenser, to the top of the IEF gels and underneath a layer of buffer containing urea, detergent and ampholytes which protects them from the very basic electrolyte solution. The gels are then run for the appropriate time. After the IEF run is complete, the gels are removed from the tubes by injecting water around the gels or by application of air pressure, but care must be taken not to break the rather delicate gels. Anderson and Anderson (1978a) have described a system for ejecting 20 gels at a time from their tubes. Recently, Ramasamy *et al* (1985) have described an apparatus which automatically ejects batches of 20 IEF gels and loads them directly onto the second-dimension SDS slab gels.

The standard rod-gel IEF system does produce good high-resolution protein separation patterns, but there are certain problems associated with its use. The most severe problem is the inherent pH gradient instability associated with the cathode region of these gels (see §4.6.7) so that the final pH gradients rarely extend above pH 7 with the consequent loss of basic proteins from the 2D maps. Treatment of the glass walls of the tubes with methylcellulose (Vanderhoff *et al* 1977) can help to reduce electroendosmosis during IEF (Burghes *et al* 1982a). Gradient stability can also be improved by modification of the electrolytes so that both stronger catholyte (Duncan and Hershey 1984, Tracy *et al* 1982a) and anolyte solutions (Duncan and Hershey 1984) are used.

IEF in gel rods is often performed in apparatus designed for analytical gel electrophoresis. Such apparatus is not ideal for IEF as it is important to use small electrolyte volumes and to position the electrodes close to the ends of the gels in order to minimise gradient drift and loss of proteins and ampholytes from the gels (Righetti 1983). Moreover, it is advantageous if such apparatus can be cooled by a thermostatically controlled recirculation system so that higher field strengths with reduced run times can be used.

A 2D pattern of fibroblast proteins obtained using a modified rod-gel IEF system is shown in figure 4.5. The gel tubes were treated with methylcellulose and anolyte pH was adjusted (Burghes *et al* 1982a). The pH gradient extended to pH 10 using this system so that cathodic proteins are now represented in the 2D map, but these spots are still streaked even if very long IEF runs (20 000 V hr) are used.

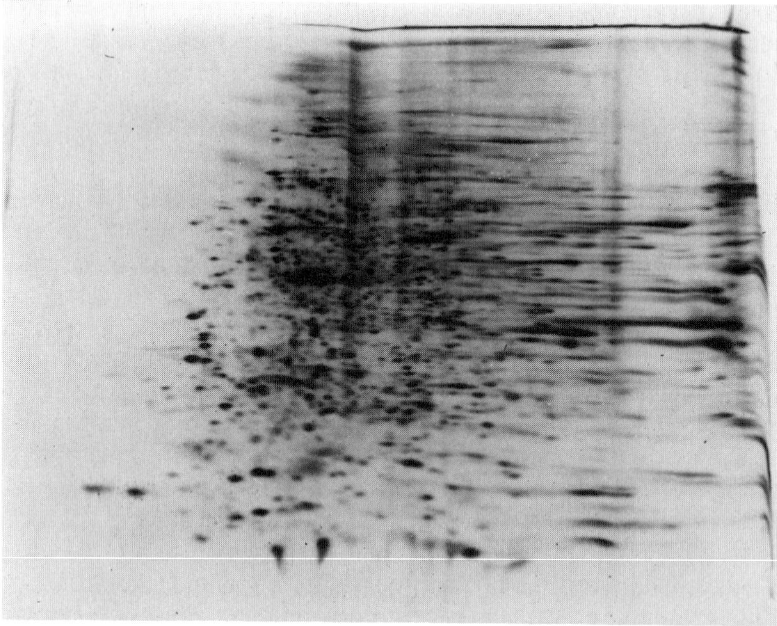

Figure 4.5 A 2D PAGE gel of ^{35}S-methionine-labelled skin fibroblast proteins using a rod IEF gel in the first dimension. The glass tubes were treated with methylcellulose to reduce electro-endosmosis and the IEF gels run for 10 000 V h. The second dimension was 7–20% linear gradient SDS-PAGE. Anode of first dimension at left. Taken with permission from Burghes *et al* (1982a).

4.6.3 Flat-bed IEF Gels

Slab gels for the first dimension have the advantage that a series of samples can be applied to the same gel. This should ensure greater reproducibility of the separation between samples and thereby overcome some of the problems associated with inter-gel comparisons. Some systems have been described using vertical slabs (Ames and Nikaido 1976, Ferreira and Eichinger 1981, Piperno *et al* 1977), but these are not satisfactory unless specially modified apparatus is available. One-dimensional IEF is usually carried out on horizontal flat-bed systems and several investigators have attempted to devise 2D PAGE systems based on this approach. Rather thick gels were used in initial procedures (Goldsmith *et al* 1979, Iborra and Buhler 1976, Rangel-Aldao *et al* 1979) but more recently advantage has been taken of the improvements possible using thin (Burghes *et al* 1982a) and ultrathin (Görg *et al* 1980, 1981) gels. In the method we have developed, 0.5 mm thick first-dimension IEF gel slabs are cast on a plastic support using a cassette with a simple, U-shaped gasket (Burghes *et al* 1982a). The reverse side of the plastic gel support can be annotated with marker pen to delineate the sample tracks and provide sample identification numbers.

The success of the flat-bed IEF technique for 2D PAGE depends on reliable binding of IEF gel to the plastic sheet, as after the first dimension the individual sample tracks must be cut out and transferred, still on their supports, to the second dimension. In our original method (Burghes *et al* 1982a) such binding was achieved by treating the plastic with Dow Corning Prime Coat 1200, followed by reaction with silane A174. However, gel adherence was somewhat unreliable if the acrylamide–urea mixture was deionised prior to use. At that time we found that commercially available plastic supports did not bind polyacrylamide reliably. However, more recently we have found that by increasing the catalyst concentration two-to-three-fold over that used in the O'Farrell (1975) method, reliable binding of polyacrylamide gels to Gel Bond PAG sheets was obtained in the presence of 8 M urea and 0.5% NP-40 or CHAPS (Dunn *et al* 1984). Moreover, this binding was more reliable than that to the specially derivitised plastic described originally.

After polymerisation is complete the IEF gel is placed on the flat-bed apparatus. We use a thin layer of Dow Corning silicone fluid 200/105 (BDH Chemicals), which is a low viscosity, non-conducting and non-volatile fluid with excellent thermal transfer properties, to ensure uniform contact between the gel and the cooling plate. In our original method (Burghes *et al* 1982a) the gel was run with its upper surface exposed to the atmosphere. In this case the atmosphere must be humid and NaOH pellets or an N_2-atmosphere should be used to minimise the deleterious effects of CO_2. Recently, we have covered the upper surface of the gel with a plastic sheet in which sample wells have been punched (Dunn *et al* 1984). The sheet is sealed in place with paraffin wax and weighted with glass plates. In this way desiccation and the effects of the atmosphere can be avoided.

In flat-bed IEF, advantage can be taken of the use of small electrolyte volumes, but the electrolyte solutions must be stronger than those used for rod IEF. We use four layers of glass fibre paper strips (Cuono and Chapo 1982) as electrode wicks. Our recommended electrolytes are 1 M oxalic acid as the anolyte and 1 M NaOH containing 0.25 M arginine and 0.1 M lysine as the catholyte. The gels should be run at 15°C, lower temperatures not being practicable because of problems with the crystallisation of urea. A typical 2D gel obtained using our flat-bed procedure is shown in figure 4.6.

There is obviously great flexibility in the choice of sample application site using a flat-bed system. Thus, problems associated with the application of samples at the acid and alkaline extremities of the gels can be avoided. Various methods can be used to apply the samples and these are discussed in §4.6.4. Running conditions for first-dimension gels are considered in §4.6.6.

After IEF is complete, the cover-sheet (if used) is removed and the gel covered with Saranwrap food wrapping film. For non-radioactive samples, the individual sample tracks must be cut from the gel using the lane markings made at the stage of gel casting. The strips can then be frozen and stored at $-70^\circ C$. For radiolabelled samples, the gel, in contact with a sheet of X-ray film, can be sandwiched between two glass plates and placed at $-70^\circ C$. After overnight exposure, the film is then developed and the resulting auto-radiograph used as a guide to cut out the sample tracks.

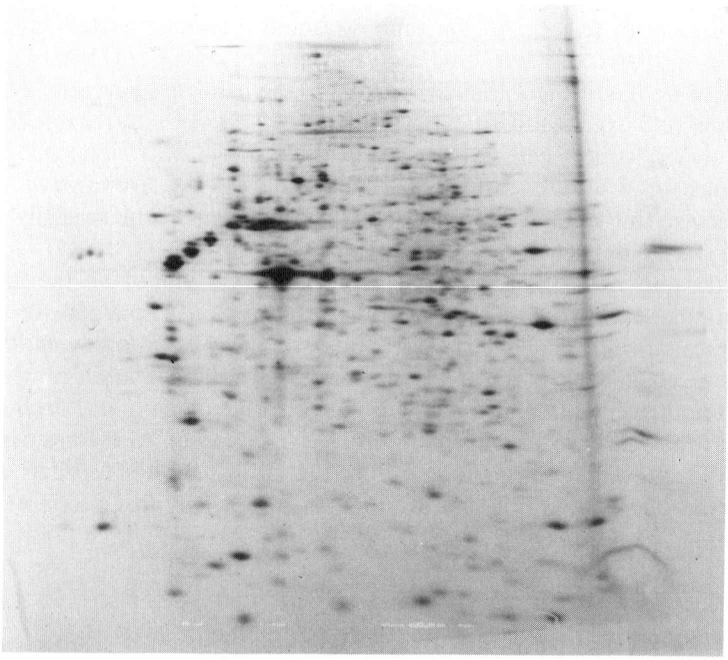

Figure 4.6 Separation of ^{35}S-methionine-labelled fibroblast proteins using a flat-bed 2D PAGE system. The first dimension gel contained Pharmalyte 3–10 (0.92% w/v), Servalyte 2–11 (0.92% w/v), Ampholine 3–10 (0.92% w/v), Servalyte 4–6 (0.12% w/v), Pharmalyte 5–8 (0.24% w/v) and CHAPS (0.5% w/v). The inter-electrode distance was 13.5 cm and the gel was focused for 88 $V\,h/l^2$. The second dimension was an 8–20% gradient SDS-PAGE gel of the shape shown in figure 4.12. Taken with permission from Burghes *et al* (1985).

It is often argued by proponents of the rod IEF gel system that flat-bed systems are more prone to skewed bands. However, it is interesting to note that it is very rare for separation patterns obtained using rod IEF gels to be illustrated in reports of 2D PAGE (Allen *et al* 1984, Dunn and Burghes 1983a). Indeed, when such gels are shown slanted bands are often observed (e.g. figure 6 of Burdett *et al* 1981). In fact, distorted zones can readily occur in both rod and flat-bed IEF gels (Finlayson and Chrambach 1971) and skewing depends on running conditions rather than the nature of the gel system *per se* (Jonssen 1980). In rod IEF, the gels are more mechanically unstable than in a flat-bed system, which can result in compression of the gel, or in the worst case its simply sliding out of the tube! Distorted bands can result from such physical disturbance. The cathode end of rod IEF gels is prone to collapse (Kuhn and Wilt 1980) which can lead to uneven penetration of catholyte into the gel and result in skewed bands (Jonssen 1980). This should not occur in horizontal flat-bed IEF gels. However, in flat-bed IEF the electrode wicks should be parallel to each other and good electrode contact is essential. The electrode design of the apparatus is crucial here so that heavy gauge wire or ribbon electrodes should be used. It should be noted that some commercial flat-bed IEF apparatus is somewhat deficient in this respect.

Rod IEF gels have a larger sample capacity than the flat-bed systems and they are more tolerant to salts. In flat-bed IEF, high protein or salt concentrations in the sample can readily result in distorted protein zones (Jonssen 1980), so that more control over sample constitution is necessary. Rod IEF gels are probably easier to apply to the second-dimension SDS slab gels, but as they are not bound to a solid support they are more prone to stretching which can distort the final 2D maps, thereby complicating inter-gel comparisons. Lastly, it should be noted that some of the recent developments for the first dimension of 2D PAGE, such as rehydratable IEF gels and immobilised pH gradients (see §4.7), would be very difficult to apply using a rod IEF system.

4.6.4 IEF Sample Application

Using a rod-gel system, samples must be applied directly to the top of the gel. Although it is possible to incorporate the sample into the gel solution before polymerisation, this procedure is not recommended because of the adverse effects of the gel constituents (e.g. catalysts, urea) and the heat of polymerisation on the proteins to be separated. For top loading, the sample can be readily layered onto the top of the gel, after prefocusing, with a suitable micropipette or microsyringe. It is usual to overlay the sample with a layer of buffer to protect the proteins from the basic or acidic environment of the electrolyte at the gel extremities.

There is much more flexibility of sample application when a flat-bed IEF system is used, both with respect to the position on the gel at which the sample is loaded and to the method of its application. It is possible, although not recommended, to apply small sample volumes as droplets directly onto the gel surface, but alternative procedures must be used for larger samples. In this context, it should be noted that large sample volumes are easier to apply using a rod-gel system. It is possible to form sample wells as indentations in the gel during casting, but this method is not recommended as it results in disturbance of the electrical field during the IEF run, often leading to distortion of the protein bands. It is possible to absorb the protein sample onto a suitable substrate such as filter paper, cellulose acetate, or dessicated polyacrylamide. However, strips of different materials adsorb uncontrolled amounts of sample and it is possible that particular proteins are not readily eluted from these materials. Fully swollen Sephadex or polyacylamide gel might work well if used in this manner. It is much better to use application strips made of silicone rubber and containing slots to accommodate the sample. We have recently found (Dunn *et al* 1984) that it is beneficial to cover the IEF gel with a plastic sheet during the IEF run as this minimises effects of the atmosphere, dessication and crystallisation of urea. Slots are punched in this cover sheet to form sample application wells.

4.6.5 Time of IEF Run

IEF is an equilibrium technique as proteins have no net charge at their isoelectric point (pI) (Svensson 1961, 1962). The time required for a protein to focus will vary depending on the particular protein under investigation and on the running conditions, such as field strength and gel length (Delincee and Radola 1978). In 2D PAGE procedures, focusing times are usually reported in terms of volt hours (V h) (Anderson and Anderson 1978a, O'Farrell 1975). However, if this parameter is to be used as a standard for reproducibility of the IEF dimension, it is essential to understand that these values are only meaningful if the system being used is totally defined. For example, the length of the IEF gels used in 2D PAGE varies widely between laboratories and it is surprising how frequently gel length is not documented in published procedures. It is important to realise that the number of V h required to achieve equilibrium varies with the square of the length of the gel (Voris and Young 1980), as for a longer gel the protein has both a longer distance to migrate and, at the same applied voltage, experiences a lower field strength. This has led us to suggest that $V\,h\,l^{-2}$ (where l = separation length of the IEF gel in cm) might be a better unit to describe the IEF dimension of 2D PAGE (Dunn and Burghes 1983a). The use of $V\,h\,l^{-2}$ does not, however, compensate for the resistance of the gel. The parameter watt hours (W h) can be used here and we have suggested (Dunn and Burghes 1983a) that by using W h/vol, allowance can be made for changes in gel volume so that the term can be applied to gels of any size.

We have recently calculated the IEF conditions used in a variety of published procedures (see table 2 in Dunn and Burghes 1983a). Surprisingly, conditions used range from as many as $80\,V\,h\,l^{-2}$ to as few as $12.9\,V\,h/l^{-2}$. Using one-dimensional IEF gels containing urea and non-ionic detergent, we found that as many as $87\,V\,h\,l^{-2}$ were required for coincidence of proteins migrated from the anode and cathode (Burghes *et al* 1982b), and we are now routinely using 88 $V\,h/l^{-2}$ for the IEF dimension of 2D PAGE .

Very small charge differences between proteins will not be resolved until equilibrium is attained and we have observed the number of bands to increase for complex protein mixtures as equilibrium is approached (Burghes *et al* 1982b). A variety of methods can be used to determine equilibrium, such as coincidence of protein bands when samples are co-migrated from the anode and the cathode, and constancy of pattern over long focusing times (Delincee and Radola 1978, Finlayson and Chrambach 1971). As we have discussed above, generally rather extended focusing times are required to obtain equilibrium in IEF gels containing urea and detergent. It is difficult to ensure that all proteins, especially those of high molecular weight, are at equilibrium even if high values of $V\,h\,l^{-2}$ are used. However, it should be remembered that in 2D PAGE it is the protein separation that is of interest rather than equilibrium *per se*. Thus, useful 2D maps can often be obtained using non-equilibrium IEF techniques (see §4.6.8 on NEPHGE), but such gels may not be as reproducible and small charge differences between proteins may not be resolved.

4.6.6 Monitoring pH Gradients

The polypeptide spots in the first dimension of 2D PAGE can be characterised in terms of their position, that is their isoelectric point (pI), in the IEF gel. To achieve this the pH gradient must be calibrated with some accuracy. Often, pH gradients are determined simply by slicing the IEF gel into small segments and eluting in water or 8 M urea. KCl (10 mM) can be added to increase the conductivity of the solution (Ames and Nikaido 1976). It should be noted that if the gel is eluted into water, the concentration of urea in the final solution is unknown. In any case, the presence of high concentrations of urea make the measurement of pH difficult as urea decreases the activity coefficient of H^+ ions (Bull *et al* 1964, Ui 1973). In addition, urea affects the proteolytic group because the solvent is charged (Gelsema *et al* 1978, 1979). Some authors have suggested the use of a correction factor to compensate for the effects of urea on measured pH (Ui 1973). However, the use of a single correction factor is incorrect as the magnitude of the urea effect is dependent on pH, the correction factor becoming greater with increasing pH (Gelsema *et al* 1978, 1979). With rod IEF gels the situation is even more complicated as measurements must be made on a separate gel leading to inaccuracies because of variability and stretching.

On IEF gel slabs, it is possible to measure pH directly on the gel surface with a flat membrane electrode. We have found that the electrode assembly supplied by Desaga is very satisfactory for these measurements. Such measurements are difficult on a gel rod, although a suitable automated apparatus has been described (Chidakel *et al* 1977). However, it must be remembered that these measurements are also being made in the presence of urea. An additional complication is that the temperature at which the IEF run and pH measurement are made should be the same.

The alternative approach to direct measurement of pH is to use a series of marker proteins to calibrate the pH gradient. It is possible for many types of sample to use certain identified polypeptide components of the sample as internal markers (e.g. actin, tubulin, tropomyosin, vimentin and lactic dehydrogenase) and to refer to the *x*-axis position of a protein in relation to one of these selected standards (Bravo 1984, Bravo *et al* 1981, Bravo and Celis 1982). Alternatively it is possible to add purified proteins of known pI to the sample being analysed (Peters and Comings 1980). Several commercial pI marker kits for IEF are available. However, it should be remembered that these are not usually suitable for use with 2D PAGE under denaturing conditions, as the pI values are those established for the native proteins, so that the consequences of dissociation of the proteins into subunits and the effects of urea cannot be corrected. An exception here is a set of acetylated cytochrome C markers available from Calbiochem which have the additional advantage of being coloured, so that the pH gradient can be monitored while the IEF run is in progress.

It is possible to generate a series of pI markers by heating a suitable protein in the presence of urea for varying lengths of time. Steinberg *et al* (1977) used mild carbamylation to calibrate charge shifts on 2D gels. This method was further developed by Anderson and his coworkers (Anderson and Hickman 1979, Hickman *et al* 1980, Tollaksen *et al* 1981) and at present is the best method for calibrating pH gradients. The loss of a free amino group below pH 8.5 results in a unit change in the charge of a molecule, so that the products appear in the final 2D map as a horizontal row or 'train' of spots of constant mass and spaced apart by about 0.1 pH units. The length of the chain depends on the amino acid composition of the protein, so that human haemoglobin β-chain gives a row of 12 spots, while bovine erythrocyte carbonic anhydrase produces 20 spots, and rabbit muscle creatine phosphokinase forms a series of 30 components (Righetti 1983). Proteins with amino acid substitutions will have fractional charge changes (Righetti 1979) and, in addition, various side chains can influence the pK_a of the group resulting in an increased variety of charge changes. Furthermore, the precise amino acid composition of a protein can place it almost any fractional charge distance from another protein. Thus, proteins with fractional charge differences can be located between adjacent carbamylated standard spots. In non-equilibrium pH gradient electrophoresis (NEPHGE) gels, such standards are useful for calibrating spot positions, but they do not indicate the true pH at the alkaline end of the gels (Hickman *et al* 1979).

In a similar way, ampholytes are composed of varying amino and carboxylic groups and thus can produce a varied distribution. Hence, groups which occur between carbamylation standards do not necessarily indicate a lack of carrier ampholyte species in these regions, but rather may indicate that a large number of carrier ampholytes exist between the two carbamylation standards. The resulting gap can be described as a local pH gradient flattening (Dunn and Burghes 1983a). Thus carbamylated standards are unsuitable for pH gradient optimisation and the protein sample under investigation must be used to establish the exact ampholyte composition required to distribute the spots over the whole gel.

4.6.7 pH Gradient Instability

A major disadvantage of the 2D PAGE system of O'Farrell (1975) is that the IEF dimension is subject to severe cathodic drift, resulting in loss of basic proteins from the final 2D map (O'Farrell *et al* 1977, O'Farrell and O'Farrell 1977). This pH gradient instability results in gradients extending only between pH 4 and pH 7 at the completion of the IEF dimension. This problem of resolution at basic pH resulted directly in the development of transient state non-equilibrium procedures (see §4.6.8 on NEPHGE).

It should be remembered that acrylamide itself contains various agents that can contribute to cathodic drift

(i) trace impurities of acrylic acid in the gel,
(ii) covalent incorporation of polymerisation catalysts as terminal groups in polyacrylamide chains, and
(iii) hydrolysis of amide groups to acrylic acid above pH 10.

It is, therefore, essential to prepare freshly and deionise acrylamide solutions prior to IEF gel preparation (Burghes *et al* 1982c). To overcome the other problems, Righetti and Macelloni (1982) suggested the use of 'balanced matrices' obtained by the covalent incorporation into acrylamide matrices of tertiary or quarternary amines in stoichiometric amounts as compared with the gel negative charges. Dimethylaminopropylmethacrylamide and methacrylamidepropyltrimethylammonium chloride have been used and were found to result in IEF gels showing little cathodic drift after 5000 V h (Righetti and Macelloni 1982).

Attempts have been made to extend pH gradients but proteins which focused in the region above pH 7 still tend to be streaked and poorly resolved (Piperno *et al* 1977). Gradient stability is also improved by incorporation of arginine and lysine (Breithaupt *et al* 1978, Tracy *et al* 1982a) and alkaline carrier ampholytes (Tracy *et al* 1982a) into the gel, and by the use of strong catholyte solutions (Tracy *et al* 1982a). In our laboratory, we (Burghes *et al* 1981, 1982a) have attempted to improve focusing of cathodic proteins by treating gel tubes with methylcellulose to reduce electroendosmosis (Vanderhoff *et al* 1977) and by the manipulation of anolyte pH as suggested by Chrambach and his coworkers (Nguyen and Chramback 1977, An der Laan and Chrambach 1980). Although the pH gradient, as measured in the presence of 8 M urea, extended to pH 10.5, cathodic proteins were still poorly resolved and streaked (Burghes *et al* 1981, 1982a).

The cathodic ends of IEF gels often shrink in rod systems (Kuhn and Wilt 1980) and become thin in flat-bed systems (Burghes *et al* 1981, 1982a). We have previously suggested (Burghes *et al* 1982a) that proteins, when applied at the cathode, experience difficulty in traversing this region, resulting in streaking of the basic proteins. This effect may be due to a poor distribution of ampholytes in this region (Burghes *et al* 1982a, 1983a). Alkaline hydrolysis of the polyacrylamide matrix will also exacerbate drift which could act to retard protein migration in this region. In a rod-gel system, the sample must perforce be applied at one extremity of the IEF gel. Anode application may make it possible to obtain good cathodic resolution, but new problems may occur due to the proteins being exposed to an acidic environment (Dunn and Burghes 1983). In contrast, using a flat-bed IEF system such as we have described (Burghes *et al* 1982b), it is possible to apply the samples at any selected point on the gel surface. Thus, proteins do not have to migrate through the cathodic region, and using this system both cathodic and anodic proteins are resolved into discrete spots, with the pH gradient extending to pH 10. An additional advantage of the flat-bed IEF system is that small electrolyte volumes can be used which ameliorates gradient drift (Righetti and Drysdale 1976).

4.6.8 Non-equilibrium Systems (NEPHGE)

As we have discussed in section 4.6 the original 2D PAGE system is subject to severe cathodic drift resulting in the loss of basic proteins (O'Farrell *et al* 1977). Although gradient stability can be improved in the rod IEF gel system (Burghes *et al* 1982a), protein spots are still streaked in the cathodic region of 2D maps. In order to overcome these problems of resolution at basic pH a non-equilibrium system, termed non-equilibrium pH gradient electrophoresis (NEPHGE), was developed (O'Farrell *et al* 1977). In this technique, samples are loaded at the acidic end of the gels which are electrophoresed for a relatively short time. Run times which have been used vary from 800 to 3000 V h (5.1 to 21 V h l^{-2}) (Dunn and Burghes 1983a). Proteins are separated by this procedure in the presence of a rapidly forming pH gradient according to a number of parameters

(i) by pH stacking or isotachophoresis, i.e. by their mobility,
(ii) the mobility of a protein will progressively decrease due to titration of its ionisable groups, and
(iii) some acidic proteins will focus (O'Farrell *et al* 1977).

It should be noted that the position of a protein under non-equilibrium conditions in a polyacrylamide gel will be a function of both its charge and its size and thus the separation parameters used for the two dimensions of 2D PAGE are not independent. Reproducibility of transient state focusing is difficult to control and is sensitive to experimental conditions, carrier ampholytes, focusing time, gel length and sample composition (Dunn and Burghes 1983a). It is therefore essential in comparative studies to run samples under identical conditions. In addition, acidic proteins are often not well resolved using the NEPHGE system and some minor charge differences cannot be distinguished (O'Farrell *et al* 1977).

Although basic proteins can be separated using non-equilibrium gels, this approach suffers from the major disadvantage that two different gels are necessary for the analysis of a given sample and some types of sample separate poorly in the NEPHGE system (Tracy *et al* 1982b). However, the horizontal slab gel IEF method we have developed as the first-dimension in 2D PAGE (see §4.6.3) is capable of resolving in an equilibrium system those basic proteins which are usually separated using the NEPHGE procedure (Burghes *et al* 1982a). In this way only a single gel is required for the complete analysis of a given sample, which reduces the number of 2D gels which have to be run and eases the task of inter-gel comparisons.

It is important to realise that in theory, non-equilibrium and equilibrium methods are able to resolve different types of charge mutation (Righetti 1979). Neutral amino acid substitutions involving an amino acid with a side chain group with a pK_a equal to the pI of the protein will not affect the equilibrium position of the protein but may alter its mobility in a non-equilibrium system (Righetti 1979). Thus, it is possible that the flat-bed equilibrium IEF procedure and the NEPHGE system can yield complementary information on certain types of sample.

4.7 New Alternatives for the First Dimension of 2D PAGE

4.7.1 Isotachophoresis

It is in theory possible to use isotachophoresis rather than IEF to separate proteins according to charge in the first dimension of 2D PAGE. This would have advantages under conditions where the particular proteins to be separated require a high ionic concentration for their solubility. However, as far as we are aware, no successful system using isotachophoresis as the first dimension in 2D PAGE has been described.

4.7.2 Buffer Focusing

It is possible to generate pH gradients for IEF using simple, even non-amphoteric, buffers (Nguyen and Chrambach 1976, 1980). Recently an improved buffer focusing system consisting of 47 components (Polybuffer

47) has been described by Cuono and Chapo (1982). This mixture generated a broad pH gradient with reasonably even conductivity and stability. The mixture is claimed to have the advantages of

(i) increased reproducibility as the constituents are defined, and
(ii) elimination of artefactual interactions of proteins with ampholytes (Cuono and Chapo 1982, Cuono et al 1983).

The attributes of buffer focusing—reproducibility, narrow bands, absence of protein interactions and ease of manipulation of pH gradients—appear ideal for 2D PAGE and we have recently examined the suitability of Polybuffer 47 for first-dimension IEF gels (Burghes *et al* 1985). A typical 2D PAGE separation of ^{35}S-methionine-labelled fibroblast proteins using 11% (w/v) Polybuffer 47 in IEF gels containing 8 M urea and 0.5% (w/v) CHAPS is shown in figure 4.7. It can be seen that the proteins appear to be concentrated into approximately four bands along the first dimension with some separation into discrete bands at the acidic end of the gel. This pattern should be compared with a conventional 2D PAGE pattern (figure 4.6), where the protein spots from this complex mixture are distributed over the entire gel.

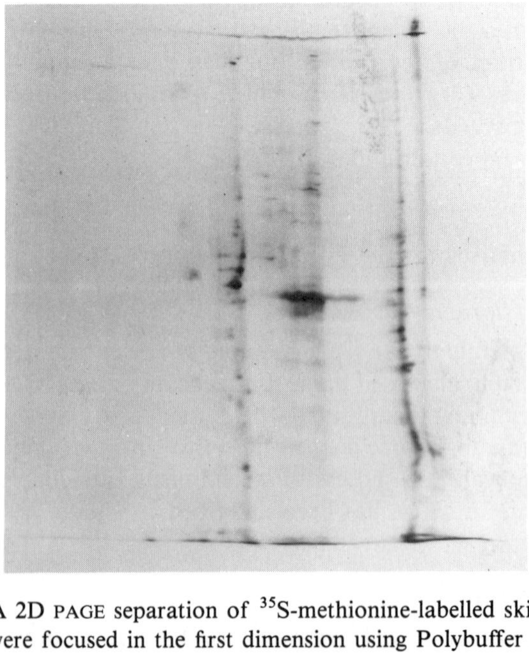

Figure 4.7 A 2D PAGE separation of ^{35}S-methionine-labelled skin fibroblast proteins which were focused in the first dimension using Polybuffer 47 for 100 Vh/l^2 with a 10 cm interelectrode distance, The second-dimension was a 10% SDS-PAGE gel. Taken with permission from Burghes *et al* (1985).

The apparently anomalous distribution of proteins by buffer focusing was further examined using IEF gels containing mixtures of carrier ampholytes and buffer focusing components and diagonal gel techniques (Burghes *et al* 1985). These experiments suggested that in buffer focusing, proteins are confined to certain regions of the gels and appear to be concentrated at interfaces. We are forced to conclude that buffer focusing using Polybuffer 47 alone is not suitable for the first dimension of 2D PAGE. However, mixtures of carrier ampholytes with low amounts (2%) of Polybuffer 47 produce good 2D separations with pH gradients extended into the very alkaline and acidic regions (figure 4.8).

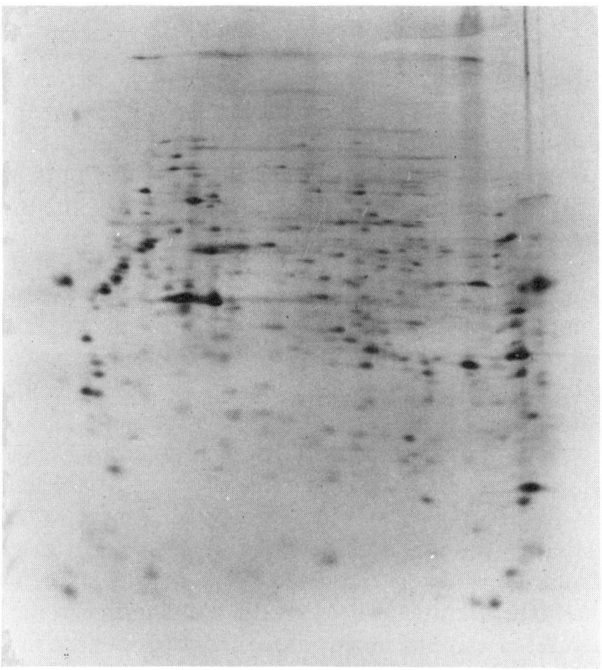

Figure 4.8 A 2D PAGE separation of ^{35}S-methionine-labelled fibroblast proteins using a mixture of Polybuffer 47 and ampholytes. The IEF first-dimension gel contained 2% w/v Polybuffer 47, 0.92% w/v Pharmalyte 3–10, 0.92% w/v Servalyte 2–11, 0.92% Ampholine 3.5–10, 0.12% Servalyte 4–6, 0.24% w/v Pharmalyte 5–8 and was focused for 88 Vh/l^2. Taken with permission from Burghes *et al* (1985).

4.7.3 Rehydratable Gels

Radola and his co-workers (Frey *et al* 1984) have recently described the preparation of rehydratable ultrathin gels for IEF which in dry form may be stored for extended periods and which prior to use can be rehydrated with solutions of any desired composition. Such a system has a number of advantages.

(i) It is easy to prepare batches of gels under defined conditions which can be stored, thereby improving inter-gel reproducibility.
(ii) Reagents such as unreacted catalysts, unpolymerised monomers, linear polymers and salts which can interfere with IEF can be washed out the gels after polymerisation.
(iii) Some additives can interfere with gel polymerisation so that these can be soaked into the gels prior to IEF.
(iv) Rehydratable gels are the only practical solution to the use of immobilised pH gradients in the presence of 8 M urea (Altland *et al* 1984, Gelfi and Righetti 1984, Gianazza *et al* 1984a).

Rehydratable gels are cast bound to a suitable support in the normal manner, washed to remove contaminants which might interfere with IEF and dried. Frey *et al* (1984) state that gels should be impregnated with 10% (v/v) glycerol prior to drying. In addition, the gels should not be heated at elevated temperatures during the drying process. Three methods of rehydration have been described.

(i) A 'rolling technique' suitable for use with ultrathin gels (Frey *et al* 1984). The film with the dry gel is rolled onto a calculated amount of the required solution, care being taken to spread the solution uniformly over the entire gel surface. After contact for 10–15 s, the gel is lifted, turned through 90° and rolled again over the residual solution. This technique is economical as the solution is quantitatively absorbed by the gel layer.
(ii) Righetti and his co-workers use measurement of solvent regain to monitor rehydration (Gelfi and Righetti 1984, Gianazza *et al* 1983a, 1984a, b). After polymerisation, the gel is weighed prior to washing and drying and from this value the combined weights of the plastic backing and of the gel monomers are subtracted to give a tare-free value for 100% solvent regain. After washing and drying, the gels are soaked in a bath containing an excess of the required solution for reconstitution. The gel weights are taken at various time intervals after blotting excess liquid from the gel surface and support. In the case of solutions containing 8 M urea, solvent regain must be corrected for the density increment.
(iii) Volume, rather than weight, is used to control gel rehydration (Altland *et al* 1984) and this is the method we consider to be the most reliable. After polymerisation, the wet gel is weighed. The thickness of the dried gel is then measured, so that the dry gel volume can be calculated and the displacement of this volume accounted for when using solutions such as 8 M urea for reswelling. The dry gel is reassembled into a cassette of the same dimensions as that in which it was cast and reswollen in a controlled amount of the required solution.

It has been reported that rehydration of dried polyacrylamide gels in solutions containing both 8 M urea and 2% NP-40 is difficult. Using excess solution in an open bath system, equilibrium swelling was only obtained after 24 h (Gelfi and Righetti 1984). Using a cassette system, we have found (Dunn *et al* 1986) that gels never reswell fully in the presence of both 8 M urea and 0.5% detergent (Triton X-100 or CHAPS). However, by substituting dimethylacrylamide for acrylamide it was possible to achieve complete rehydration in 1 h using either an open bath or cassette for reswelling (figure 4.9). It appears, therefore, that dimethylacrylamide may be the polymer of choice when hydrophobic gels are to be used. It is of interest to note that ethylene urea was also compatible with this system and complete reswelling of dimethylacrylamide gels in this reagent was possible.

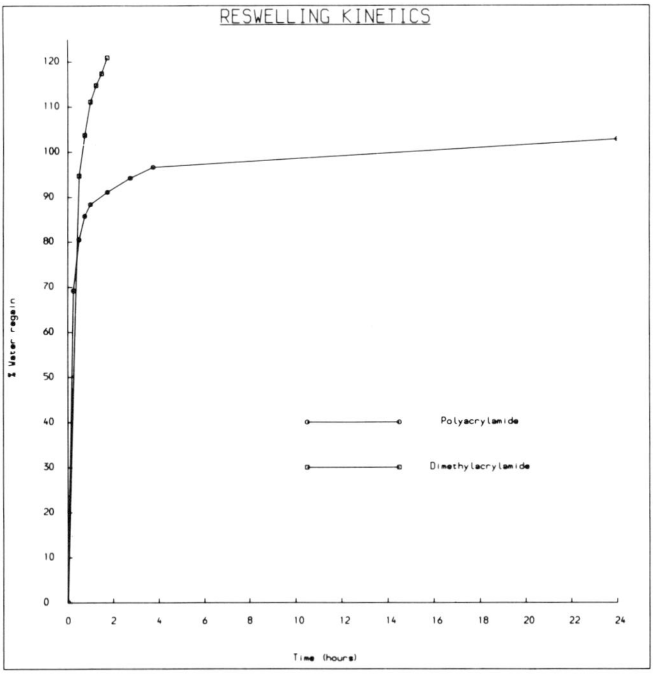

Figure 4.9 Reswelling kinetics of dried gels prepared using acrylamide and dimethylacrylamide in the presence of both 8 M urea and 0.5% Triton X-100 using an open bath system. The dimethylacrylamide gels rapidly reswell and in an open system become over-rehydrated. In contrast, polyacrylamide gels are only fully rehydrated after 24 h under the same conditions.

4.7.4 Immobilised pH Gradients

An interesting and exciting recent innovation in IEF has been the development of 'Immobiline' reagents (LKB) for the production of immobilised pH gradients. This technological advance is discussed in detail in chapter 3 by Righetti *et al* and only those aspects relevant to 2D PAGE will be discussed here. The Immobilines are a series of acrylamide derivatives in which the buffering groups forming the pH gradient are covalently attached to the polyacrylamide backbone. Tertiary amines are used to produce basic Immobilines and carboxylic groups for the acidic species (Righetti *et al* 1983). The IEF gel is made by casting a gradient using the appropriate Immobiline solutions as described in detail by Righetti in chapter 3. The Immobiline reagents polymerise with acrylamide via vinyl bonds and in this way a defined pH gradient can be obtained.

Initially, only narrow range pH gradients could be produced, which limited the application of this technology in 2D PAGE separations. However, recently methods have been developed for the generation of wide-range pH gradients (Gianazza *et al* 1983a, b, 1984c) and even for the formation of non-linear pH gradients where the shape of the gradient can be matched to the statistical distribution of protein isoelectric points (Gianazza *et al* 1985a). It should be pointed out that very wide (pH 3–10) immobilised pH gradients can only be generated using two non-buffering species, very acidic ($pK < 1$) and alkaline ($pK = 9.95$) Immobilines. These are unfortunately not available commercially at present but it is likely that they will eventually be obtainable from LKB. The advantages of immobilised pH gradients for 2D PAGE are improved reproducibility, straight protein bands and the ability to withstand high voltages and long focusing times without gradient drift. Righetti and his co-workers have published a series of papers on the use of immobilised pH gradients for the first, IEF dimension of 2D PAGE (Gianazza *et al* 1984a, b, 1985a, b). Satisfactory 2D maps of serum proteins were obtained by their procedure, but it should be pointed out that the gels contained 8 M urea but did *not* contain any detergent. We believe that the presence of non-ionic or zwitterionic detergent in addition to 8 M urea is essential for the solubility during IEF of many types of protein samples examined by 2D PAGE.

We have recently started to examine immobilised pH gradient IEF gels for the first dimension of 2D PAGE. Our data are somewhat preliminary at present but we feel they warrant inclusion here as this methodology appears to have great potential for 2D separations. Much of this methodology is based on that of Altland and Rossman (1985) and we are grateful to Dr Altland (University of Giessen) for advice on the use of Immobiline gels. The immobilised gels are cast on Gel Bond PAGE using either the two-chamber method (Gianazza *et al* 1984b, c) or the sophisticated computer-controlled stepmotor-driven high-precision burette apparatus described by Altland (Altland and Altland 1984a, b). A commercial version of the latter equipment is available from Desaga and has great flexibility for casting non-linear pH gradients of any desired shape. The gel contains only the appropriate Immobilines and freshly deionised acrylamide. We are using dimethylacrylamide instead of acrylamide for gels which will be run in the presence of both 8 M urea and 0.5% non-ionic or zwitterionic detergent as we have found that such gels can be more readily rehydrated (see §4.4.3). The gel is allowed to set at room temperature for 2 h, with the catalyst concentration adjusted to give the required polymerisation kinetics. It should be noted that Altland (personal communication) recommends titrating stock Immobiline solutions to pH 6.7 and storing them frozen in aliquots, but finds that the alkaline species (pK 8.5 and pH 9.3) still have problems of instability. After polymerisation, the gel is removed from the cassette and washed for 1 h in 1% (w/v) glycerol. The gel is dried under a fan at room temperature, its thickness measured and is stored, sealed in a plastic bag, in a refrigerator.

For rehydration in the appropriate solution the dry gel is reassembled into a cassette of the same dimensions as that in which it was originally cast (see §4.4.3). Altland and Rossman (1985) and ourselves (Dunn *et al* 1986) have independently found it beneficial to include low levels (0.5% (v/v)) of carrier ampholytes of the appropriate pH range in immobilised pH gradient gels as this improves the final IEF separation, especially when low protein loads are used. It also appears to minimise sideways spreading of samples from their application wells and distortion of adjacent sample tracks. We believe that in Immobiline gels it is difficult for current to pass through the gels as there are no free ions in solution to carry the current, but only rigid, regularly spaced Immobiline molecules. In conditions of high sample loadings, such as is the case with serum, the proteins within the sample can carry the current, but severe problems will exist in the case of low sample loads. By including carrier ampholytes in the gel, these will act as free ions to carry current but the Immobiline components will still predominate to determine the pH gradient.

After rehydration, the gel (10 cm inter-electrode distance) is prefocused for 1 h at 3000 V, 4 mA, 12 watts limiting using 0.5% (v/v) carrier ampholyte solution as electrolytes to remove salts (Altland and Rossman 1985). The electrolytes are then changed to 0.01 M NaOH and 0.5 M phosphoric acid, the samples applied and the run continued under the same electrical conditions for a total of 9000 V h. Examples of 2D PAGE patterns of serum proteins and ^{35}S-methionine-labelled fibroblast proteins obtained using pH 4 to 7 immobilised pH gradients are shown in figures 4.10 and 4.11 respectively.

It is our belief that once problems associated with the stability of the alkaline Immobilines have been resolved, this methodology combined with the use of a large (20 × 20 cm^2) gel format (see §4.9.4) with the gels bound to plastic supports in both dimensions, will form the basis of a highly reproducible high-resolution 2D PAGE system.

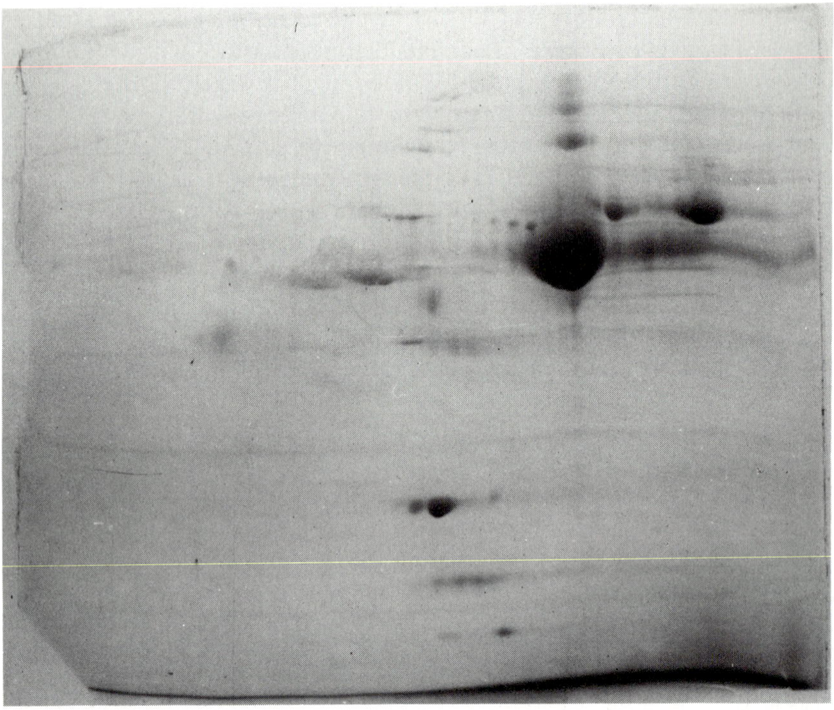

Figure 4.10 A 2D PAGE separation of human serum proteins using the Immobiline system for the first dimension. A pH 4–7 immobilised pH gradient IEF gel was cast on Gel Bond PAG, washed and dried. The gel was rehydrated in 8 M urea and run as described in the text. The second dimension is a 8–25% gradient SDS-PAGE gel of the shape shown in figure 4.12. The anode of the first dimension gel is at the left.

4.8 Equilibration

After IEF and prior to application to the second dimension gel, the first-dimension gels are usually equilibrated in a buffer containing SDS under reducing conditions. Gel rods are normally equilibrated for 30–40 minutes (O'Farrell 1975), while thin-layer slab gels require only 5–10 minutes (Burghes *et al* 1982a) and ultrathin gels only 1–2 minutes (Görg *et al* 1980). The reduced equilibration times that can be used with thin slab IEF gels are advantageous as diffusion of the protein bands should be minimised. O'Farrell (1975) found, using a 30 min equilibration time, that bands could broaden by 40% resulting in a serious loss of resolution. In addition, protein is leached out of the first-dimension gel during equilibration and loss of as much as 5–50% of the original total sample has been reported with extended equilibration times (Bravo 1984, Burghes *et al* 1982a, O'Farrell 1975, Peterson and McConkey 1976, Tracy *et al* 1982b). The problems

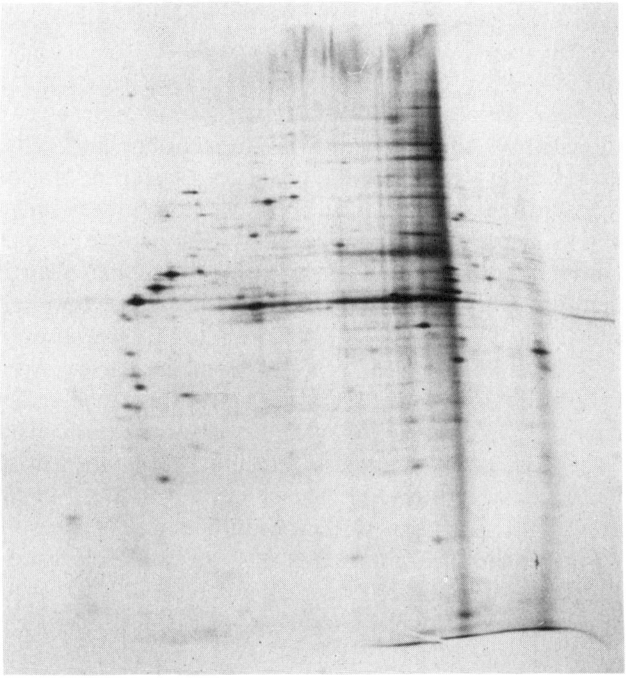

Figure 4.11 A 2D PAGE separation of ^{35}S-methionine-labelled skin fibroblast proteins using the Immobiline system for the first dimension. The gel was prepared and run as described in the caption to figure 4.10. The anode of the first dimension gel is at the left.

associated with equilibration have prompted some investigators to omit this step and apply the IEF gel directly to the second dimension slab (Klose and Feller, 1981, Rosenmann *et al* 1982). However, omission of the equilibration step can result in streaking, especially of high molecular weight proteins, in the second dimension (O'Farrell 1975). Therefore, short equilibration times, while reducing diffusion and loss of sample, may result in decreased efficiency of transfer of proteins between dimensions. For these reasons we prefer to equilibrate IEF gels before application to the second dimension.

There may be particular problems associated with the equilibration of Immobiline immobilised pH gradient gels. Görg *et al* (1984) have reported horizontal streaking of spots in the second dimension which they attribute to electro-endosmotic effects of the immobilised pH gradient interfering with migration of proteins from the first to second dimensions. To overcome this problem they have suggested adding Immobiline reagents to the equilibration buffer and/or to a 'starting gel' which overlays the second-dimension SDS gel.

In an alternative approach, it has been recommended that, rather than use an equilibration step, first-dimension IEF gels should be fixed prior to application to the second dimension. Hirabayashi (1981) used fixation with 10% TCA and 5% sulphosalicylic acid for agarose IEF gels, which were then washed for 2 h before application to the second-dimension SDS gels. Although this method was found to result in less sample loss than normal equilibration (Hirabayashi 1981), the efficiency of elution of sample from the first dimension was not investigated. Jackle (1979) recommended fixing IEF gels in 50% methanol, 10% acetic acid and then staining them with Coomassie brilliant blue R-250 before equilibration and transfer to the second dimension. Görg *et al* (1981) reported that it was preferable to use Coomassie brilliant blue G-250 as R-250 resulted in streaking in the SDS dimension. Complete removal of proteins from the IEF gel strip into the SDS gel was claimed (Görg *et al* 1981) since no protein could be detected by restaining the IEF strip on completion of SDS electrophoresis. This method apparently does not work satisfactorily using IEF gel rods as significant amounts of protein were found to remain in the gel (Görg *et al* 1981). A possible caveat with regard to this procedure is that Radola (1980) has shown that acid alcohol fixation of thin IEF gels can leach proteins out of the gel. In addition, Coomassie dyes may compete with the same binding sites as SDS and therefore inhibit interaction of the detergent with the protein (Dunn and Burghes 1983a). However, as Coomassie dyes complex with proteins in a manner which may be analogous to the binding of SDS, these dyes may actually act in concert with SDS in a mixed micelle state.

4.9 SDS Dimension

4.9.1 Preparation of SDS Gels

During development of his 2D PAGE procedure, O'Farrell (1975) found that the distribution of proteins in the 2D gel could significantly affect resolution. If proteins are clustered into small regions of the gel, much of the area available for protein separation is wasted and resolution will consequently be reduced. Comparison of subunit molecular weights for a large series of proteins from a variety of sources show that they tend to be clustered around a mean value (Gianazza and Righetti 1980). Thus, for most samples an SDS gel of a single polyacrylamide concentration will not produce the uniform distribution of spots necessary for optimal resolution in 2D PAGE.

One approach to this problem is to run each sample on a series of SDS gels of different polyacrylamide concentrations (Garrels 1979). In this way a set of 2D maps is obtained, each one resolving different regions of the molecular weight spectrum. A series of five different concentrations is usually recommended (7.5, 9, 12, 15 and 18% polyacrylamide), with a 15%

gel being claimed to be the best compromise for routine analysis of total eukaryotic cellular extracts (Bravo 1984).

This approach has the advantage of simplicity and reproducibility of gel casting, but is rather laborious and complicates analysis as for each sample several 2D patterns must be concatenated to produce a single 2D map representative of the total components of that particular sample, However, it should be pointed out that gels of a single polyacrylamide concentration are very useful for analysing particular regions of interest and to increase resolution in that area.

The alternative approach is to use SDS gradient polyacrylamide gels which have the capacity to resolve proteins over a wide range of molecular weights. O'Farrell (1975) found that an exponential 5 to 22.5% polyacrylamide gradient gel gave good resolution for *E. coli* proteins. Since then most investigators have followed this recipe without consideration of whether this gradient results in optimal distribution of the protein components in the particular sample under investigation. Only if this is done will the maximum area of the gel be used for effective separation of the proteins, thereby maximising resolution for that particular sample. For most proteins, it is advantageous to flatten the mid-region of the polyacrylamide gradient to maximise separation in the region where the majority of proteins are found, while using steeper gradients in the high and low molecular weight regions where fewer components usually occur (Burghes *et al* 1982b). The polyacrylamide gradient shape which we currently find to be optimal for the resolution of human skin fibroblast proteins is shown in figure 4.12.

The appropriate gradient for optimal resolution of any sample can be established empirically for the particular type of sample under investigation. However, it may be better to use the approach recently described by Gianazza *et al* (1985) and applied to the construction of immobilised pH gradients to optimise separation in IEF. This involves statistical analysis of the distribution of protein molecular weights (or pI's) and using weighting factors to derive the appropriately shaped gradient to give the optimal resolution of these components. This approach can be applied to maximise resolution in both dimensions. It may also be necessary to take into account the abundance of each species as this has a significant effect on spot size and thus influences resolution.

Various procedures of varying complexity have been described for casting SDS gradient polyacrylamide gels. The simplest approach is to use a two-chamber gradient mixer of fixed shape (Görg *et al* 1981, Margolis and Kenrick 1967, Rothe and Purkhanbaba 1982). Such devices are commercially available and should produce reproducible gradients, but their lack of flexibility limits their usefulness for casting gradients of variable shape. A two-chamber system of variable complementary shape is used in the Isodalt system of Anderson and Anderson (1978b, 1979) which is capable

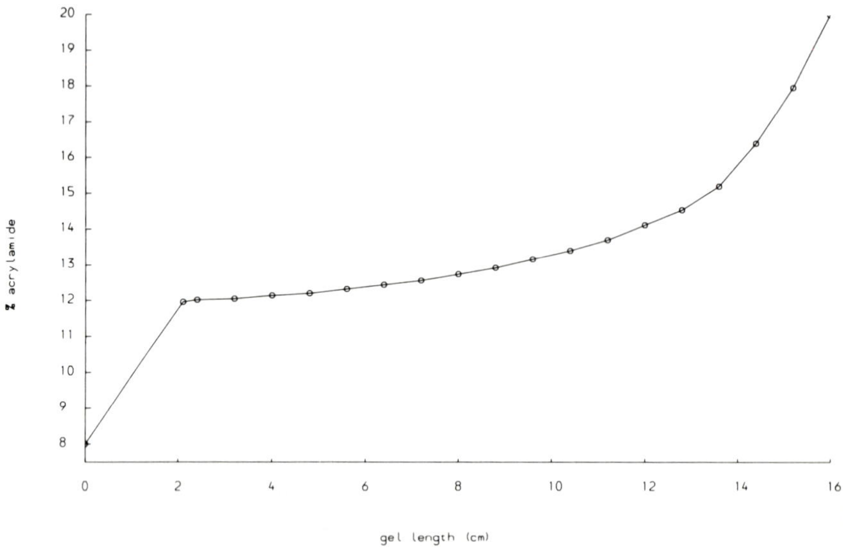

Figure 4.12 Modified polyacrylamide gradient shape used for second-dimension SDS-PAGE gels. Taken with permission from Burghes *et al* (1985).

of producing gradients of user-defined shape. However, the reproducibility with which gradients can be formed with this device depends on the accuracy with which the moveable baffle between the chambers can be positioned. In addition, it is difficult, using this apparatus to reproduce identical conditions in different laboratories. Electronic gradient forming apparatus such as the Ultrograd (LKB) can be used to cast gradient gels (Sanders *et al* 1980, Groppi and Browning 1980) in a flexible and reproducible way (Poduslo 1981). We have used an Ultrograd to cast gels for 2D PAGE (Burghes *et al* 1982a, Dunn and Burghes 1983a) and our system is shown diagrammatically in figure 4.13 except that the original casting tower has been replaced by one of the type described by Jones *et al* (1980). The Ultrograd scans a template of any desired gradient shape and controls a three-way valve to form the gradient. However, this electro-mechanical device is relatively expensive, will mix up to only three solutions and produces gradients as a function of time and not of volume. Microcomputer-controlled systems should be less expensive and give increased versatility (Dunn and Burghes 1983a). Recently, Altland has described a gradient mixer based on the computer-controlled cooperation of a series of stepmotor-driven burettes which can be used to form gradient polyacrylamide gels (Altland and Altland 1984a). The apparatus is shown in figure 4.14 and a sample computer output which documents the gradient which has been cast is shown in figure 4.15. Unfortunately, the stepmotor-driven burettes used in this system are rather expensive, but a gradient maker based on

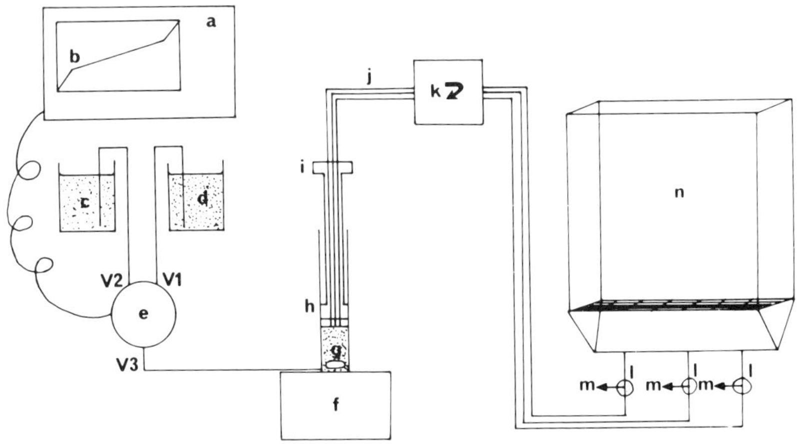

Figure 4.13 Apparatus for casting gradient polyacrylamide gels. (*a*) an LKB Ultrograd (*b*), fitted with the appropriate gradient template controls a three-way valve (*e*) for the dense (*c*) and light (*d*) acrylamide solutions. The solution in the mixing chamber (*h*) with its plunger (*i*) is mixed by a magnetic flea (*g*) and stirrer (*f*). Fluid flow is controlled by a peristaltic pump (*k*). All lines (*j*) are of narrow-bore tubing. The perspex casting tower (*n*) is fitted with three inlets controlled by three-way taps (*l*). Taken with permission from Burghes *et al* (1982a).

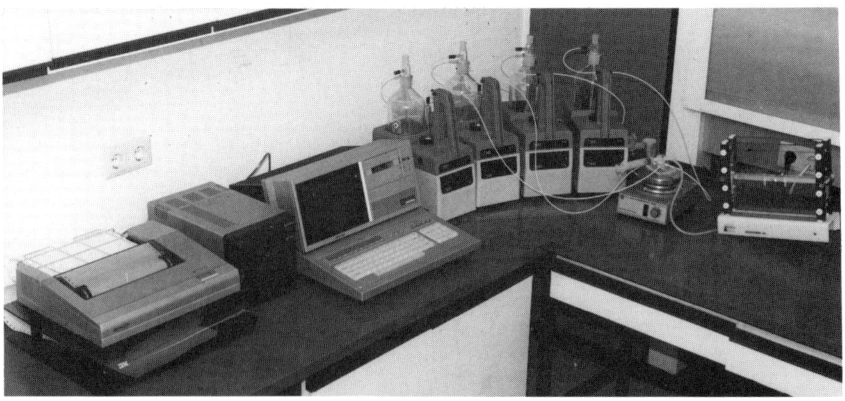

Figure 4.14 Computer-controlled stepmotor-driven burette system for mixing polyacrylamide gel gradients. From left to right: Computer, floppy disk station and matrix printer; 4 burettes A–Df; mixing chamber on magnetic stirrer; polymerisation stand with two separation cells to be filled from the bottom. Taken with permission from Altland and Altland (1984a).

Altland's design is commercially available from Desaga and it is likely more such devices will become available in the future. Such devices are very versatile and when also used to construct immobilised pH gradients (see §4.7.4) should significantly improve the resolution and reproducibility of 2D PAGE.

Gradient Control

Program: Polyacrylamide Gel Gradient

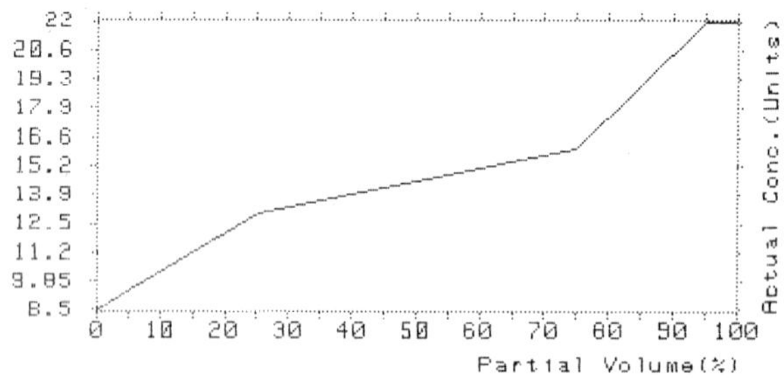

Data:

TV= 60 mL; BV= 10 mL; MV= .2 mL; TV/BV= 6 ;
LC= 8.5 Units; HC= 22 Units; Speed: 60 %
H1= 5 %; L1= 5 %; H2= 3 %; L2= 8 %;

Endpoints of lines

EP No.	VP(%)	C(U)	EP No.	VP(%)	C(U)	EP No.	VP(%)	C(U)	EP No.	VP(%)	C(U)
1	25	13.00	2	75	16.00	3	95	22.00	4	100	22.00

(EP=End Poi.; VP(%)=Part.Vol.; C(U)=Conc.in Units; TV=Tot.Vol.; BV=Bur.Vol.; MV=Mixing Vol.; LC=Low Conc.; HC=High Conc.; L1,H1,L2,H2 = Proportions of catalyst solutions 1 & 2 admixed to solutions of low and high conc. of gradient constituent

Figure 4.15 Final document of polyacrylamide gel gradient produced by the system shown in figure 4.14 including all data except the recipes for preparing the stock solutions for the reproduction of the gradient in any laboratory. Taken with permission from Altland and Altland (1984a).

The more sophisticated techniques of gradient formation are easily adaptable to the production of large batches of gradient gels which increase intergel reproducibility, at least within a given batch of gels (Anderson and Anderson 1978b, 1979, Burghes *et al* 1982a, Dunn and Burghes 1983a). These procedures can be used in conjunction with apparatus designed for the simultaneous IEF of multiple samples (Anderson *et al* 1978a; Burghes *et al* 1982a) and electrophoresis of batches of SDS gels (Anderson *et al*

1978b, 1979, Jones *et al* 1980). However, it is important to realise that these techniques do not fully overcome problems associated with batch to batch variability and with the exchange of comparable data between laboratories.

A problem which arises with multiple gel techniques is the handling of large numbers of SDS gels. This problem is alleviated if the gels are cast onto solid supports. It is possible to bind polyacrylamide gels to glass using silanes (Radola 1980), but it is more convenient for 2D PAGE to use specifically treated plastic backings (Burghes *et al* 1982a,b, Radola 1980). Suitable supports are commercially available from FMC Marine Colloids (Gel Bond PAG) and Serva (Gel-Fix). It should be noted that the vinyl bonds on the surface of these plastic sheets are sensitive to heat and light so that they should be stored under the appropriate conditions. Gels bound to silanised glass are stable in 20% TCA, whereas we found that gels became detached from Gel Bond PAG when overnight fixation was attempted (Dunn and Burghes 1983b). Thus, fixation of gels bound to Gel Bond PAG in strong TCA solutions should be carried out for as short a time as practicable. Gels bound to supports must be dried by special procedures if curling and cracking of the gels is to be avoided (Burghes *et al* 1983b, Dunn and Burghes 1983a). Recently a new material Acryl Aide (FMC Marine Colloids), an olefinic agarose derivative, has been developed as an alternative to small organic crosslinkers such as Bis. Polyacrylamide gels crosslinked with Acryl Aide and bound to Gel Bond PAG supports have been shown to be very resilient and can be easily dried, without vacuum, in an oven at 60°C (Nochumson and Gibson 1983).

4.9.2 Stacking

Most investigators have followed the original recommendation of O'Farrell (1975) and have used the discontinuous SDS gel system of Laemmli (1970). This system is based on the discontinuous (multiphasic) buffer systems of Ornstein (1964) and Davis (1964), but with the addition of SDS. A large-pore stacking gel is usually used to concentrate the proteins from the first-dimension IEF gel into narrow zones prior to their separation on a small-pore resolving gel. However, Garrels (1979) found that it was not essential to use a stacking gel as the protein bands within the first dimension gel are already concentrated and the non-restrictive IEF gel will itself act to concentrate the proteins.

It should be noted that the nature of stacking is somewhat modified in the presence of SDS. Since SDS-coated proteins have a constant charge to mass ratio (that is all proteins and their subunits have a uniform charge), they will migrate with the same mobility and thus will automatically stack. Moreover, as SDS–protein complexes are not titratable between pH 7 and pH 10, mobility is not variable within this range. It is, therefore, not necessary to have a discontinuity in pH and unstacking of the proteins will occur by the change in gel concentration (Wyckoff *et al* 1977). Another point to be considered is that SDS will act as a separate ion boundary to the Cl^- ions (Booth 1977). Allen *et al* (1984) have suggested that it is unnecessary to add SDS to the stacking or separating gel which would have the advantage of ensuring that conditions of polymerisation elaborated for the non-SDS case remain applicable to SDS-PAGE. However, it is likely that the more insoluble proteins often present within complex mixtures analysed by 2D PAGE require the continual presence of SDS within their environment to maintain them in a soluble state.

Dunn and Burghes (1983a) have suggested that urea may act to increase the viscosity of an IEF gel, thereby making it equivalent to a more restrictive gel and increasing the difficulty of protein elution from the first-dimension gel. Adaptation of the discontinuous buffer system to obtain a step voltage gradient (Booth 1977) might be advantageous in increasing the speed and efficiency of elution from first-dimension gels. The buffer concentration used for electrophoresis should be high enough to prevent the high levels of carrier ampholytes and proteins affecting pH. In addition, it should be remembered that any non-ionic detergent present will form mixed micelles with SDS, effectively reducing its mobility. All of these features may change stacking limits as compared with those established for standard SDS-PAGE.

4.9.3 Transfer between Dimensions

After equilibration, if this step is employed, the first-dimension IEF gel must be applied to the second-dimension SDS gel. It is essential to ensure good contact of the two gels over their entire length if good separations are to be obtained. Garrels (1979) reported that rod IEF gels could be applied directly to the surface of SDS slab gels without any agent to cement them in place and that during electrophoresis the IEF gels became compressed against and bonded to the surface of the SDS gels. Most investigators using rod IEF gels have found it necessary to fix the first-dimension gel in place with appropriately buffered agarose as originally recommended by O'Farrell (1975). However, caution should be exercised here as Spragg and his co-workers have found that agarose preparations can be contaminated with amino-acid containing impurities which can result in artefacts when sensitive detection methods such as silver staining are used (Amess *et al* 1985). As an alternative, rapidly polymerising acrylamide can be used to secure the IEF gel on top of the SDS slab (Klose and Feller 1981, Poehling and Nuhoff 1980). However, this method can result in loss of protein from the 2D map because of the heat denaturation during the rapid polymerisation and the direct polymerisation of the protein with acrylamide (Poehling and Neuhoff 1980). In contrast, agarose was found not to result in protein retention within the IEF gel as shown by staining with Coomassie blue

(Poehling and Neuhoff 1980). It should be noted that a considerable amount of protein is often found at the top of the second-dimension gel, particularly if sensitive detection methods such as autoradiography or silver staining are used. It is not known whether this is a very high molecular weight protein or insoluble, aggregated material (Dunn and Burghes 1983a). Using rod IEF gels considerable care must be exercised to minimise stretching of the gels during application to the second dimension as this can result in skewed bands. In addition, the resulting distortions of the 2D maps complicate the process of pattern matching between gels.

The problem of transfer between dimensions is somewhat different when the first-dimension IEF gel has been run as a horizontal slab using a flat-bed apparatus, as in the methods of Burghes *et al* (1982a) and Görg *et al* (1980, 1981). Here the samples are in the form of thin strips of gel bound to flexible plastic supports. As the gels are firmly adhered to a solid support they are relatively easy to handle and are not susceptible to stretching, but care must be exercised to ensure that the gels do not peel off the backing. When the second dimension is run in a vertical SDS gel slab (Burghes *et al* 1982a), we recommend that the top of the gel cassette is filled with equilibration buffer and the IEF strip is slid into place, avoiding entrapment of air bubbles, followed by sealing in place with agarose. When the second dimension is a horizontal SDS gel slab, the sample strip is placed, gel layer facing down, into a trench preformed in the SDS gel (Görg *et al* 1980, 1981).

4.9.4 Gel Size

In the original method of O'Farrell (1975), 13 cm long IEF gels were used in combination with 16.4 cm long SDS gels. In subsequent developments and modifications of this technology, few significant alterations have been made to the overall gel area available for separation. The resolution capacity of IEF gels depends on a number of factors including conductivity, gel length and voltage gradient (Allen *et al* 1984). The resolution of IEF gels of a given length can be improved by increasing the voltage gradient (Allen 1980, Läas and Olson 1981). However, this increase in resolution can only be gained at the expense of a considerable increase in Joule heating and, therefore, high field strengths can only be used if adequate cooling is available. This can be achieved using a flat-bed apparatus with an efficient cooling system, such as a Peltier cooling device, in combination with ultrathin-layer gels. Using 125 μm thick gels, voltage gradients of up to $700\ V\,cm^{-1}$ have been successfully used for IEF gels containing 4% ampholytes and cast on plastic supports (Allen and Arnaud 1983, Allen *et al* 1984). However, this approach is not possible using rod IEF gels, where field strengths are limited to 40 to $50\ V\,cm^{-1}$.

An increase in resolution in the second dimension of 2D PAGE can be achieved by an increase in the length of the SDS gel, in the same way that DNA sequencing gels can be improved (Ansorge and Garoff 1981). This has been demonstrated by Anderson and Peterson (1981) who used 28 cm gradient SDS gels to resolve complex mixtures into more than 150 proteins. Nickodem *et al* (1981) took advantage of this approach in a 2D PAGE system using rod IEF gels of normal size in the first dimension in combination with 29 cm long second-dimension SDS gels.

Recently, Young and his coworkers have advocated the use of large 2D gels, which they term 'giant' gels (Voris and Young 1980). Gel size in both dimensions was increased 2.5-fold over that used in the original O'Farrell (1975) system. Thus 32 cm IEF gels were used in combination with 40.64 cm long SDS gels, giving a six-fold increase in the effective gel area available for separation. The method together with details of the construction of the special apparatus required to run these large gels has recently been fully documented (Young *et al* 1983).

It is claimed that a substantial increase in resolution can be obtained with this system as measured by the number of spots detected (Voris and Young 1980). Using large gels, 5000 proteins have been detected in rat thymus cells (Young 1984) and as many as 10 000 species have been claimed to be found in human Hep cells (Klose and Zeindl 1984). Spots on the large gels are more discrete and rounded and many of the major spots seen on standard 2D gels, especially streaked regions, are resolved into multiple spots. This increase in resolution can be attributed to two factors. First, the large gels have a significantly higher sample capacity which improves the detectability of components present in low concentrations. Indeed, using thicker gels up to a hundred-fold increase in sample loading capacity can be achieved (Young *et al* 1983). This can also be achieved for radioactive samples by labelling the proteins to a higher specific activity. Secondly, and perhaps more importantly, the effects of autoradiographic spreading are minimised by the large gel area. This is of particular importance for higher-energy isotopes such as ^{32}P and ^{125}I, where autoradiographic spreading often seriously degrades resolution and the method is therefore useful for the study of phosphoproteins and iodinated proteins (Young 1984).

In contrast, other investigators have recommended the use of microscale systems for 2D PAGE (Poehling and Neuhoff 1980, Rüchel 1977). This approach is based on the fact that good resolution can be achieved in small, capillary IEF gels by the use of relatively high field strengths. These are used in combination with small SDS slab gels in the second dimension. The system described by Rüchel (1977) uses gels which are less than the size of a postage stamp. The 2D gels used by Neuhoff and his co-workers are $3 \times 3.5\ cm^2$ and they have documented the preparation and electrophoresis of this microgel system using both gradient and non-gradient techniques (Poehling and Neuhoff 1980). Microscale procedures have the advantages of reduced run times, shortened staining and destaining times if thin gels are used, and decreased cost due to the small consumption of reagents. However, these advantages are gained at the possible expense of resolution as the distances between spots on the final 2D map become reduced.

The optimum system would thus appear to be a combination of long, thin IEF gels run at a high field strength in combination with large, gradient SDS gels in the second dimension.

4.9.5 Molecular Weight Standards

When 2D gels containing a multiplicity of protein spots are to be compared it is essential to have a series of molecular weight standards as reference markers in the second, SDS dimension. These standards can be purified from mixtures of pure proteins of known molecular weight, which are electrophoresed along one edge of the SDS gels. A variety of such mixtures, both radioactive and non-radioactive, is commercially available. In theory, a series of crosslinked polymers of one protein of known molecular weight could be used (Inouye 1971). However, it is possible that these polymers would exhibit anomalous behaviour in SDS gels as they might not be exactly comparable to linear chains of the same molecular weight. Giometti *et al* (1980) have recommended the use of rat whole heart homogenate which is added to the agarose overlay used to secure the first-dimension IEF gels in place. These standards give a series of 80 bands, in the range 300 to 10 000 molecular weight, which stretch all the way across the gel. This homogenate, when stored in the presence of SDS and 0.7% agarose at $-20^{\circ}C$ or $-70^{\circ}C$, produces a stable protein pattern for over 12 months (Giometti *et al* 1980). The molecular weight markers discussed above are not suitable for use as pI markers in the first dimension. Thus, a separate set of markers must be used in each dimension. As an approach to this problem, Kurian *et al* (1981) have recommended the use of bacteriophage T4 coat proteins as a complete set of pI and molecular weights standards for 2D PAGE. Recently, Lubit (1984) has used fluorescent, dansylated proteins as internal standards for pI and molecular weight in 2D PAGE. Lastly, it should be noted that the presence of non-ionic detergent interferes with the use of Bromophenol blue tracking dye as a marker of the gel front for molecular weight estimations (Ogita and Markert 1979).

4.10 Visualisation, Detection and Analysis

After the 2D PAGE run is complete, gels are usually fixed prior to detection of the separated proteins. The best general fixative is trichloroacetic acid (TCA) which is most efficient at the 20% (w/v) level. Some investigators use sulphosalicylic acid (10–20%), mixtures of TCA and sulphosalicylic acid (10% of each) or acid alcohol solutions, but these procedures are generally less efficient than TCA alone (Frey and Radola 1982) and can result in significant loss of protein.

After fixation, the polypeptide spots can be visualised by protein staining using either a general purpose protein stain such as Coomassie brilliant blue R-250 or one of the more sensitive techniques such as silver staining. Alternatively, radio-labelled proteins can be detected by a suitable autoradiographic or fluorographic procedure. We do not propose to discuss these detection procedures here as we have recently reviewed them in some detail (Dunn and Burghes 1983b) and sensitive detection methods are fully discussed by Merril *et al* in chapter 7.

Once the polypeptide spots in 2D maps of complex protein samples have been visualised, one is left with the task of extracting qualitative and quantitative data from these complicated patterns. Simple visual inspection can at best provide only limited information about specific polypeptides. The large amount of data necessitates the use of sophisticated automated computer analysis systems (Dunn and Burghes 1983b). This topic will not be dealt with here as the computer analysis of 2D gels is discussed in detail by Spragg *et al* in chapter 8.

4.11 Heterogeneity and Artefacts

Various forms of post-translational modification activity can result in alteration of proteins so that microheterogeneity of these proteins is observed by 2D PAGE. Such modifications can result in charge alterations which will transform a single spot on a 2D map into a row or 'train' of spots (O'Farrell, 1975). Such changes can occur as a result of reactions such as phosphorylation, deamidation, acetylation and the addition of sialic acid residues. The situation is more complicated for glycoproteins as they can show anomalous behaviour in SDS gels. This can be partially overcome using borate buffers for the SDS dimension as borate ions will bind to the cis-hydroxyl group of the sugar and compensate for mobility changes (Poduslo 1981).

Similar charge and molecular weight changes can also be induced in samples as a consequence of various treatments, leading to artefactual heterogeneity of these proteins by 2D PAGE. However, these effects can usually be minimised provided reasonable care is exercised in handling the samples. Deamidation of asparagine and glutamine residues and oxidation of cysteine residues can readily occur, especially using procedures such as lyophilization (O'Farrell 1975). It should also be noted that many post-synthetic radio-labelling procedures, such as iodination and methylation, can introduce charge heterogeneity (Dunn and Burghes 1983b). The use of protease inhibitors can also result in protein modification (O'Farrell and O'Farrell 1977). Protein–protein and protein–ampholyte interactions can persist during IEF leading to artefactual spots in 2D PAGE, but such interactions should be minimised in the presence of both 8 M urea and detergent (Dunn and Burghes 1983a).

4.12 Estimates of Resolution

High-resolution 2D PAGE is generally used for the separation and analysis of complex protein mixtures. For example, data based on polysomal mRNA–DNA hybridisation experiments indicate that the informational RNA complexity in polysomes from a wide variety of mammalian cell types and tissues could encode between 5000 and 20 000 mRNA species. Although it is theoretically possible for several proteins to occupy the same position on a 2D gel, it has been found, in practice, that when two proteins are in similar locations they displace each other and form separate spots (McConkey 1979). If the analysis of such complex samples is to be attempted by 2D PAGE, it is essential to ascertain how many of these proteins can be resolved by the 2D PAGE procedure being used.

In his original 2D PAGE paper, O'Farrell (1975) attempted to assess the resolution capacity of his system as applied to *E. coli* proteins. Using one-dimensional IEF and SDS-PAGE about 70 and 100 bands respectively could be separated. Thus, a combination of the two procedures might be expected to be capable of resolving up to 7000 components. However, due to lateral diffusion which occurred during equilibration, transfer between dimensions and SDS-PAGE, spots were larger than expected on the final 2D maps so that the estimate of maximum resolution was reduced to 500 spots (O'Farrell 1975).

A major factor which can act to reduce resolution in 2D PAGE is the generally large range of relative abundancies of individual components within complex protein mixtures. Major proteins may account for as much as 3 to 5% of the total protein while less abundant species may account for 0.0001% or less. Bravo (1984) has calculated that if minor proteins are detected using fluorography, the major components will be overexposed by a factor of 10^4 to 10^5 times, producing a spot with an area at least 15 to 20 times larger than its minimal size. Such spreading, which also occurs using auto-radiographic and staining techniques, inevitably results in a considerable reduction in the resolving power of 2D PAGE. This problem can be partially overcome using pairs of weakly and strongly stained gels or short and long auto-radiographic exposures, which must then be concatenated, usually using a computer system, to obtain a total 2D map of the sample under investigation. Other factors, such as streaking of high molecular weight proteins and insolubility of components in the first dimension, also act to decrease resolution.

When these features are taken into account it is probably not surprising that using conventional 2D PAGE a maximum of about 1000 spots is normally observed on maps of total mammalian cell proteins. This finding has led some workers to suggest that there may be not more than 2000 physiologically significant primary gene products in a typical mammalian cell (Duncan and McConkey 1982). However, recent work based on the use of large or 'giant' 2D gels (Voris and Young 1980, Young *et al* 1983) has shown that as many as 5000 to 10 000 proteins can be resolved (Klose and Zeindl 1984, Taylor *et al* 1983, Young 1984), a figure now in the range predicted by hybridisation data.

Most estimates of the resolution capacity of 2D gels have been based on the rather subjective assessment of the number of spots obtained from complex protein mixtures. Recently, Taylor *et al* (1983) have devised an objective mathematical test to measure the theoretical resolution capacity of 2D gels. The test is essentially a ratio of areas and measures how many spots of a certain size can be packed into the pattern and still be resolved. Using this procedure, Taylor *et al* (1983) have calculated theoretical resolution capacities ranging from 15 000 spots for 18 cm × 18 cm gels to 50 000 spots for large 20 cm × 25.5 cm gels, again demonstrating the improved resolution capacity of large 2D gels. However, it is impossible for most, if not all, samples to achieve this level of resolution. The theoretical estimate assumes that spots are distributed uniformly across the final 2D pattern, but this ideal is not exhibited by most complex protein mixtures using the standard 2D PAGE procedure. This situation can be improved, however, by the use of gradient engineering in both dimensions so that a more uniform distribution of spots is obtained. The test also assumes that the protein spots are of a standard, 'typical' size. The large range of abundancies of components in most complex protein samples complicates this assumption, so that in practice a substantial variation in spot size can be observed. This is particularly significant for those proteins present in very high abundance as these spots can be artefactually enlarged by 'spreading' at the detection step (i.e. staining or auto-radiography). Taking these factors into account, Taylor *et al* (1983) have estimated that the number of components which can be resolved in practice will be reduced to 2000 and 6000 proteins for small and large format gels respectively. Thus, these estimates of resolution capacity of 2D PAGE based on mathematical models (Taylor *et al* 1983) are in concordance with those derived subjectively (Young 1984).

4.13 Applications

The unique ability of 2D PAGE procedures to separate simultaneously the components of complex protein mixtures has resulted in the application of this technique to many areas of biological, biochemical and biomedical research. The diversity of these applications make it impossible to review sensibly the several thousand publications in which 2D PAGE has been used. We have previously discussed some of the areas of application of 2D PAGE technology (Dunn and Burghes 1983b) and to get a flavour of the diversity of the uses of 2D PAGE the reader is referred to the recent books edited by Celis and Bravo (1984) and Neuhoff (1984) and to the proceedings of two scientific meetings devoted to 2D PAGE (*Clinical Chemistry* (1982) **28**, no 4, *Clinical Chemistry* (1984) **30**, no 12).

4.14 2D-PAGE and the Study of Human Genetic Disorders

From the foregoing discussion it is clear that there have been many recent advances in 2D PAGE technology including computer-controlled casting of gradients, binding of gels to plastic supports in both dimensions, immobilised pH gradients using the Immobiline system and large format 'giant' gels. All of these advances should make high-resolution 2D PAGE a more reliable and reproducible procedure. However, a question that still remains to be conclusively answered is whether 2D PAGE has the capacity to resolve all the proteins in a particular cell type and can changes be detected in protein maps due to the primary defects associated with human genetic diseases.

As we have discussed in §4.12 data based on physical mRNA-DNA hybridisation experiments suggest that any particular cell type can encode between 5000 and 20 000 mRNA species, with perhaps 30 000 to 50 000 genes in the total human genome. Although recent reports using large format gels indicate that the number of proteins resolved by 2D PAGE can be increased (Klose and Zeindl 1984, Taylor *et al* 1983, Young 1984), it still seems likely that a significant number of gene products can remain undetected by 2D PAGE.

What approaches can be made to overcome this problem? One approach may be to analyse the protein products of the *in vitro* translation of mRNA. As post-translational modification of proteins does not occur in such systems, problems associated with multiple spots on 2D maps due to protein microheterogeneity are minimised. In addition, any artefactual changes associated with differential metabolism of amino acid analogues should not occur. Perhaps more importantly this approach provides an opportunity to deplete samples of abundant mRNA species. In most cell types there is a large proportion of the total mRNA that contains very few mRNA species. This abundant component typically consists of fewer than 100 different mRNAs present in more than 1000 or even 10 000 copies per cell. It often corresponds to a major part of the mass, approaching 50% of the total mRNA. Upon *in vitro* translation the abundant mRNA will result in a few polypeptide spots present at high abundance in 2D maps, which makes resolution and detection of the less abundant species difficult. These less abundant proteins are specified by the other 50% of the total mRNA, which consists of up to 10 000 sequences, each represented by only a small number (perhaps < 10) of copies in the total mRNA. This problem can be approached by Cot fractionation of a DNA preparation, so that a short hybridisation time would be used to obtain the rapidly reassociating fraction representing high copy DNA. This fraction can be isolated by hydroxyapatite column chromatography and the isolated, high-abundance DNA sequences hybridised with the mRNA preparation of interest to deplete it of the abundant mRNA species. Even if such a DNA preparation were added directly to the *in vitro* translation system it could inhibit the translation of the corresponding abundant mRNA species in a manner analogous to hybrid arrest translation.

Another advantage of using mRNA preparations as the starting material for protein analysis by 2D PAGE is that the results can relate directly to cDNA experiments which are of increasing importance in the study of human genetic disorders. The Immobiline system for IEF using immobilised pH gradients is particularly suitable for 2D PAGE analysis of *in vitro* translation products, as this method can tolerate larger protein loads than traditional IEF techniques. It should also be noted that 2D PAGE should be an excellent procedure for characterising the protein products derived from DNA sequences cloned into expression vectors.

The approach discussed above should overcome one of the major problems associated with resolution of complex protein mixtures, namely the spreading of the highly abundant proteins at the stage of spot detection resulting in interference in the detection of the less abundant species. Alternative procedures that should be considered here, particularly with regard to samples such as serum and other body fluids, are immunodeletion (Dermer and Edwards 1983, Pearson and Anderson 1980) and pseudoligand affinity chromatography (Allen *et al* 1984, Arnaud *et al* 1979, Tracy *et al* 1982a).

A potential problem in the use of 2D PAGE for the study of genetic diseases is that the technique is insensitive to the detection of point mutations, for example, if an amino acid substitution occurs which results in no overall change in charge (i.e. pI). In addition, a further complication must be considered as it is now thought likely that many genetic disorders which have previously been assumed to be the result of a single alteration in a gene, are more likely to result from a variety of changes within a locus that produce clinically similar phenotypes. An example of this is the Lesch–Nyhan syndrome where five variants of hypoxanthine-guanine phosphoribosyl transferase have been detected and four of these determined (Wilson *et al* 1983). Thus the experimental design for the 2D PAGE analysis of such disorders is of prime importance. One approach to this problem (Dunn *et al* 1984) is to construct a so-called 'super-normal' sample from a mixture of normal samples all radio-labelled in the same way (e.g. with ^{35}S-methionine). Each normal sample, labelled with the other amino acid analogue (e.g. ^{75}Se-selenomethionine), is compared to the 'super-normal' sample by dual label 2D PAGE to check for any dilution effects of any particular spot. The 'super-normal' sample is then used for dual-label 2D PAGE comparison with each individual sample from diseased individuals. In this way it should be possible to detect either a single, specific change associated with the genetic disorder, or if the disease is polymorphic it would be expected to recognise a set of changes in a specific area of the gel with a number of individuals showing each change.

We are using another approach (Dunn *et al* 1984) in our studies of X-linked Duchenne muscular dystrophy (DMD) in which there are women who have balanced X:autosome translocations and who manifest clinical symptoms of DMD (Jacobs *et al* 1981). The breakpoint on the X-chromosome in these translocations is found in X_p21, although the precise location of the break is variable. In these individuals it is very unlikely that expression of the disease is the result of a point mutation. Rather, major disruption of the gene would be expected resulting in the absence of a gene product or in the synthesis of a very abnormal protein. This situation should be amenable to analysis by 2D PAGE and by comparing protein samples from these individuals with a 'super-normal' sample as discussed earlier.

Thus with the application of the technological advances discussed in this chapter to make 2D PAGE an easier and more reliable technique with increased resolution capacity combined with cDNA techniques may make it possible to analyse by 2D PAGE proteins of all abundance classes from any cell type. The use of sophisticated, automated computer systems for analysis of the resultant 2D maps should make high-resolution 2D page a very powerful technique for the investigation of a wide variety of biological, biochemical and biomedical problems.

Acknowledgments

Financial support from the Muscular Dystrophy Group of Great Britain, the Medical Research Council and Action Research is acknowledged. We would like to thank Dr Klaus Altland for his assistance and advice on the use of the Immobiline system. We are grateful to Mrs C Trand for skilfully typing the manuscript and to Mrs L White for her photographic expertise.

References

Allen R C 1980 *Electrophoresis* **1** 32–7
Allen R C and Arnaud P 1983 *Electrophoresis* **4** 205–11

Allen R C, Saravis C A and Maurer H R 1984 *Gel Electrophoresis and Isoelectric Focusing of Proteins* (Berlin: de Gruyter)
Altland K and Altland A 1984a *Electrophoresis* **5** 143–7
—— 1984b *Clin. Chem.* **30** 2098–103
Altland K, Banzhoff A, Hackler R and Rossmann U 1984 *Electrophoresis* **5** 379–81
Altland K and Rossmann U 1985 *Electrophoresis* **6** 314–25
Ames G F L and Nikaido K 1976 *Biochemistry* **15** 616–23
—— 1981 *Europ. J. Biochem.* **115** 525–31
Amess R, Fox J E and Spragg P S 1985 *Electrophoresis* **6** 186–8
An der Laan B and Chrambach A 1980 *Electrophoresis* **1** 23–27
Anderson D W and Peterson C 1981 *Electrophoresis '81* ed. R C Allen and P Arnaud (Berlin: de Gruyter) pp 41–8
Anderson N G and Anderson N L 1979 *Behring Inst. Mittl.* **63** 169–210
Anderson N G, Anderson N L, Tollaksen S L, Hahn H, Giere F and Edwards J 1979 *Anal. Biochem.* **95** 48–61
Anderson N L and Anderson N G 1978a *Anal. Biochem.* **85** 331–340
—— 1978b *Anal. Biochem.* **85** 341–354
Anderson N L and Hickman B J 1979 *Anal. Biochem.* **93** 312–20
Ansorge W and Garoff H 1981 *Electrophoresis '81* ed. R C Allen and P Arnaud (Berlin: de Gruyter) pp 635–46
Arnaud P, Galbraith R M, Chapuis-Cellier C, Galbraith G M P and Fundenberg H H 1979 *Protides of the Biological Fluids* **26** 649–53
Barret T and Gould H J 1973 *Biochim. Biophys. Acta* **294** 165–170
Basha S M M 1979 *Plant Physiol.* **63** 301–6
Bhakdi S, Knufermann H and Wallach D F H 1974 *Biochim. Biophys. Acta* **345** 448–57
—— 1975a *Progress in Isoelectric Focusing and Isotachophoresis* ed. P G Righetti (Amsterdam: North Holland) pp 281–91
—— 1975b *Biochim. Biophys. Acta* **394** 550–557
Bhatnagar Y N and Bellve A R 1978 *Analyt. Biochem.* **86** 754–60
Bianchi Bosisio A, Loehenstein C, Snyder R S and Righetti P G 1980. *J. Chromatogr.* **189** 317–30
Bonner W M, West M H P and Stedman J D 1980 *Eur. J. Biochem.* **109** 17–23
Booth A G 1977 *Biochem. J.* **163** 165–8
Booz M L and Travis R L 1981 *Phytochemistry* **20** 1773–1779
Bravo R, Bellatin J and Celis J 1981 *Cell Biol. Intl Rep.* **5** 93–6
Bravo R and Celis J E 1982 *Clin. Chem.* **28** 766–81
Bravo R 1984 *Two-dimensional Gel Electrophoresis of Proteins* ed. J E Celis and R Bravo (New York: Academic) pp 3–36
Breithaupt T B, Nystrom I E, Hodges D H and Babitch J 1978 *Anal. Biochem.* **84** 579–82
Bull H B, Breese K, Ferguson G L and Sevenson C A 1964 *Arch. Biochem. Biophys.* **104** 297–304
Burdett P, Lizana J, Eneroth P and Bremne K 1981 *Electrophoresis '81* ed. R C Allen and P Arnaud (Berlin: de Gruyter) pp 329–41
Burghes A H M, Dunn M J and Dubowitz V 1982a *Electrophoresis* **3** 354–63
Burghes A H M, Dunn M J, Statham H E and Dubowitz V 1981 *Electrophoresis '81* ed. R C Allen and P Arnaud (Berlin: de Gruyter) pp 295–308

—— 1982b *Electrophoresis* **3** 185–196
—— 1982c *Electrophoresis* **3** 177–85
Burghes A H M, Dunn M J, Witkowski J A and Dubowitz V 1983 *Electrophoresis '82* ed. D Stathakos (Berlin: de Gruyter) pp 371–80
Burghes A H M, Patel K and Dunn M J 1985 *Electrophoresis* **6** 453–61
Celis J E and Bravo R (ed.) 1984 *Two-dimensional Gel Electrophoresis of Proteins* (New York: Academic)
Charlionet R, Martin J-P, Sesboue, R, Madec P J and Lefebvre F 1979 *J. Chromatogr.* **176** 89–101
Chidakel B-E, Nguyen N Y and Chrambach A 1977 *Anal. Biochem.* **77** 216–25
Chignell D A and Wingfield P T 1974 *Fed. Proc.* **33** 1283
Creighton I E 1979 *J. Molec. Biol.* **129** 235–264
Cuono C B and Chapo G A 1982 *Electrophoresis* **3** 65–75
Cuono C B, Chapo G A, Chrambach A and Hjelmeland L M 1983 *Electrophoresis* **4** 404–7
Dale G and Latner A L 1969 *Clin. Chim. Acta* **24** 61–8
Davie J R 1982 *Analyt. Biochem.* **120** 276–81
Davie J R and Candido E P M 1978 *Proc. Natl. Acad. Sci. USA* **75** 3574–7
Davis B J 1964 *Ann. NY Acad. Sci.* **121** 404–36
Delincee H and Radola B J 1978 *Anal. Biochem.* **90** 603–23
Dermer G B and Edwards J J 1983 *Electrophoresis* **4** 212–8
Duncan R and Hershey J W B 1984 *Analyt. Biochem.* **138** 144–55
Duncan R and McConkey E H 1982 *Clin. Chem.* **28** 749–55
Dunn M J and Burghes A H M 1983a *Electrophoresis* **4** 97–116
—— 1983b *Electrophoresis* **4** 173–89
—— 1983c (unpublished results) Jerry Lewis Muscle Research Centre
Dunn M J, Burghes A H M and Patel K 1985 *Protides Biol. Fluids* **33** 479–82
Dunn M J, Burghes A H M, Patel K, Witkowski J A and Dubowitz V 1984 *Electrophoresis '84* ed. V Neuhoff (Weinheim: Verlag-Chemie) pp 281–4
Dunn M J, Burghes A H M, Witkowski J A and Dubowitz V 1985 *Protides Biol. Fluids* **32** 973–6
Edwards J J, Anderson N G, Tollaksen S L, van Eschenbach A C and Guevara J 1982a *Clin. Chem.* **28** 160–3
Edwards J J, Tollaksen S L and Anderson N G 1982b *Clin. Chem.* **28** 941–8
Emes A V, Latner A L and Martin J A 1975 *Clin. Chim. Acta* **64** 69–78
Fernandes P B, Nardi R V and Franklin S G 1978 *Analyt. Biochem.* **91** 101–14
Ferreira A and Eichinger D 1981 *J. Immunol. Meths.* **43** 291–9
Finlayson G R and Chrambach A 1971 *Anal. Biochem.* **40** 292–311
Franklin S G and Zweidler A 1977 *Nature* **266** 273–5
Frey M Atta M B and Radola B J 1984 *Electrophoresis '84* ed. V Neuhoff (Weinheim: Verlag-Chemie) pp 122–5
Frey M D and Radola B J 1982 *Electrophoresis* **3** 27–32
Gahne B and Juneja R K 1984 *Electrophoresis '84* ed. V Neuhoff (Weinheim: Verlag-Chemie) pp 285–8
Garrels J I 1979 *J. Biol. Chem.* **254** 7961–77
—— 1983 *Methods Enzymol.* **100** 411–23
Gelfi C and Righetti P G 1981 *Electrophoresis* **2** 213–219
—— 1984 *Electrophoresis* **5** 257–62

Gelsema W J, de Ligny C L and van der Veen N G 1978 *J. Chromatogr.* **151** 161–74
—— 1979 *J. Chromatogr.* **171** 171–81
Gianazza E, Artoni G and Righetti P G 1983a *Electrophoresis* **4** 321–6
Gianazza E, Astrua-Testori S, Giacon P and Righetti P G 1985b *Electrophoresis* **6** 332–9
Gianazza E, Celentano F, Dossi G, Bjellqvist B and Righetti P G 1984c *Electrophoresis* **5** 88–97
Gianazza E, Dossi G, Celentano F and Righetti P G 1983b *J. Biochem. Biophys. Methods* **8** 109–133
Gianazza E, Frigerio A, Tagliabue A and Righetti P G 1984a *Electrophoresis* **5** 209–16
Gianazza E, Giacon P, Sahlin B and Righetti P G 1985 *Electrophoresis* **6** 53–6
Gianazza E and Righetti P G 1984b *Electrophoresis '84* ed. V Neuhoff (Weinheim: Verlag-Chemie) pp 87–90
Gianazza E and Righetti P G 1980 *J. Chromatogr.* **193** 1–8
Giometti C S and Anderson N G 1981 *Clin. Chem.* **27** 1918–21
Giometti C S, Anderson N G, Tollaksen S L, Edwards J J and Anderson N L 1980 *Anal. Biochem.* **102** 47–58
Giometti C J, Danon M J, Anderson N G 1983 *Neurology* **33** 1152–6
Goldsmith M R, Rattner E C, Macy M, Koehler D, Balikov S R and Bock S C 1979 *Anal. Biochem.* **99** 33–40
Gonenne A and Ernst R 1978 *Anal. Biochem.* **87** 28–38
Goodwin G H, Wright C A and Johns E W 1981 *Nucleic Acids Res.* **9** 2761–5
Görg A, Postel W, Weser J and Westermeier R 1984 *Electrophoresis '83* ed. H Hirai (Berlin: de Gruyter) pp 525–32
Görg A, Postel W and Westermeier R 1981 *Electrophoresis '81* ed. R C Allen and P Arnaud (Berlin: de Gruyter) pp 259–270
Görg A, Postel W, Westermeier R, Gianazza E and Righetti P G 1980 *J. Biochem. Biophys. Methods* **3** 273–84
Groppi V E and Browning E T 1980 *Molec. Pharmacol.* **18** 427–37
Hari V 1981 *Anal. Biochem.* **113** 332–5
Hickman B J, Anderson N L, Willard K E and Anderson N G 1980 *Electrophoresis '79* ed. B J Radola (Berlin: de Gruyter) pp 341–60
Hirabayashi T 1981 *Anal. Biochem.* **117** 443–51
Hjelmeland L M, Nebert D W and Chrambach A 1978 *Electrophoresis '78* ed. N Catsimpoolas (Amsterdam: Elsevier) pp 29–56
—— 1979 *Anal. Biochem.* **95** 201–8
Hoffman N L and Dowben R M 1978 *Anal. Biochem.* **89** 540–9
Horst M N, Mahaboob S, Basha M, Baumbach G A, Mansfield E H and Roberts R M 1980 *Anal. Biochem.* **102** 399–408
Howard G A and Traut R R 1973 *FEBS-Letters* **29** 177–80
Iborra G and Buhler J M 1976 *Anal. Biochem.* **74** 503–511
Imada M, Hsieh P and Sueoka N 1978 *Biochim. Biophys. Acta* **507** 459–69
Imada M and Sueoka N 1980 *Biochim. Biophys. Acta* **625** 179–92
Inouye M 1971 *J. Biol. Chem.* **246** 4834–8
Jackle H 1979 *Anal. Biochem.* **98** 81–4
Jacobs P A, Hunt P A, Mayer P I T and Bart R D 1981 *Am. J. Hum. Genet.* **33** 513–18

Jones M I, Massingham W E and Spragg S P 1980 *Anal. Biochem.* **106** 446–9
Jonssen M 1980 *Electrophoresis* **1** 141–9
Kaderbhai M A and Freedman R B 1980 *Biochim. Biophys. Acta* **601** 11–21
Kadofuku T, Sato T, Manabe T and Okuyama T 1983 *Electrophoresis* **4** 427–31
Kaltschmidt E and Wittmann H G 1970a *Anal. Biochem.* **36** 401–12
—— 1970b *Proc. Natl. Acad. Sci. USA* **67** 1276–1282
Klose J 1975a *Humangenetik* **26** 231–43
—— 1975b *New Approaches to the Evaluation of Abnormal Embryonic Development* ed. D Neubert and H J Merkes (Stuttgart: Thieme) pp 375–87
Klose J and Feller M 1981 *Electrophoresis* **2** 12–24
Klose J and Zeindl E 1984 *Clin. Chem.* **30** 2014–20
Knopf U L, Sommer A, Kenny J and Traut R R 1975 *Molec. Biol. Rap.* **2** 35–40
Kuhn O and Wilt F H 1980 *Anal. Biochem.* **105** 274–80
Kurian P, Gersten D M, Suhocki P V and Ledley G 1981 *Electrophoresis* **2** 184–7
Läas T and Olsson I 1981 *Electrophoresis '81* ed. R C Allen and P Arnaud (Berlin: de Gruyter) pp 191–203
Laemmli U K 1970 *Nature* **227** 680–5
Latner A L and Emes A V 1975 *Progress in Isoelectric Focusing and Isotachophoresis* ed. P G Righetti (Amsterdam: North Holland) pp 223–33
Lubit B W 1984 *Electrophoresis* **5** 358–61
McConkey E M 1979 *Anal. Biochem.* **964** 39–44
MacGillivray A J and Rickwood D 1974 *Eur. J. Biochem.* **41** 181–90
Macko V and Stegemann H 1969 *Hoppe-Syler Z. Physiol. Chem.* **350** 917–19
Madjar J-J, Arpin M, Buisson M and Reboud J-P 1979a *Molec. Gen. Genet.* **171** 121–134
Madjar J-J, Michel S, Cozzone A J and Reboud J-P 1979b *Analyt. Biochem.* **92** 174–82
Manabe T, Hayama E and Okuyama T 1982a *Clin. Chem.* **28** 824–7
Manabe T, Kojima K, Jitzukawa S, Hoshino T and Okuyama T 1982b *Clin. Chem.* **28** 819–823
Manabe T and Okuyama T 1981 *J. Chromatogr.* **225** 65–71
Manabe T, Tachi K, Kojima K and Okuyama T 1979 *J. Biochem.* **85** 649–59
Manabe T, Takahashi Y and Okuyama T 1983a *Electrophoresis* **4** 359–62
Manabe T, Takahashi Y, Okuyama T, Maeda Y Y and Chihara G 1983b *Electrophoresis* **4** 242–6
Margolis J and Kenrick K G 1967 *Nature* **214** 1334–6
—— 1969 *Nature* **221** 1056–7
Marshall T 1984 *Electrophoresis* **5** 245–50
Marshall T, Williams K M and Vesterberg O 1985 *Electrophoresis* **6** 47–52
Martini O H W and Gould H J 1971 *J. Molec. Biol.* **62** 403–5
Means G E and Feeney R E 1968 *Biochemistry* **7** 2192–2201
Merril C R, Goldman D, Sidman S A and Ebert M H 1981 *Science* **211** 1437–8
Mets L J and Bogorad L 1974 *Anal. Biochem.* **57** 200–10
Neuhoff V (ed.) *Electrophoresis '84* (Weinheim: Verlag-Chemie)
Nguyen N Y and Chrambach A 1976 *Anal. Biochem.* **74** 145–53
—— 1977 *Anal. Biochem.* **82** 226–35
—— 1980 *Electrophoresis* **1** 14–22
Nickodem V M, Trus B L and Rau J E 1981 *Proc. Natl. Acad. Sci. USA* **78** 4411–15

Nochumson S and Gibson S G 1983 *Electrophoresis '82* ed. D Stathakos (Berlin: de Gruyter) pp 177–82
O'Farrell P H 1975 *J. Biol. Chem.* **250** 4007–21
O'Farrell P H and O'Farrell P Z 1977 *Meths. Cell Biol.* **16** 407–20
O'Farrell P Z, Goodman M M and O'Farrell P H 1977 *Cell* **12** 1133–42
Ogita Z-I and Markert C L 1979 *Anal. Biochem.* **99** 233–241
Olsson I and Läas T 1981 *J. Chromatogr.* **215** 273–378
Ornstein L 1964 *Ann. NY Acad. Sci.* **121** 321–49
Orrick L R, Olson M O J and Busch H 1973 *Proc. Natl. Acad. Sci. USA* **70** 1316–20
Panyim S and Chalkley R 1969 *Arch. Biochem. Biophys.* **130** 337–46
Pearson T and Anderson N L 1980 *Anal. Biochem.* **101** 377
Perdew G H, Schaup H W and Selivonchick D P 1983 *Anal. Biochem.* **135** 453–5
Peters K E and Comings D E 1980 *J. Cell Biol.* **86** 135–55
Peterson J L and McConkey E M 1976 *J. Biol. Chem.* **251** 548–54
Piperno G, Huang B, and Luck D J L 1977 *Proc. Natl. Acad. Sci. USA* **74** 1600–4
Poduslo J F 1981 *Anal. Biochem.* **114** 131–9
Poehling H M and Neuhoff V 1980 *Electrophoresis* **1** 90–101
Radola B J 1980 *Electrophoresis* **1** 43–56
Ramasamy R, Spragg S P, Jones M I and Amess R 1985 *Electrophoresis* **6**, 43–6
Rangel-Aldao R, Kupiec J W and Rosen O M 1979 *J. Biol. Chem.* **254** 2499–508
Raymond S and Nakamichi M 1964 *Anal. Biochem.* **7** 225–32
Raymond S and Weintraub L 1959 *Science* **130** 711
Righetti P G 1979 *J. Chromatogr.* **173** 1–5
—— 1983 *Isoelectric Focusing: Theory, Methodology and Application* (Amsterdam: Elsevier)
Righetti P G and Drysdale J W 1976 *Isoelectric Focusing* (Amsterdam: North Holland)
Righetti P G, Gianazza E and Bjellqvist V 1983 *J. Biochem. Biophys. Meths.* **8** 89–108
Righetti P G and Macelloni G J 1982 *J. Biochem. Biophys. Meths.* **5** 1–15
Rosen A, Ek K and Aman P 1979 *J. Immunol. Med.* **28** 1–11
Rosenblum B B, Hanash S M, Yew N and Neel J V 1982 *Clin. Chem.* **28** 925–31
Rosenmann E, Kreis C, Thomson R G, Dobbs M, Hamerton J L and Wrogemann K 1982 *Nature* **298** 563–5
Rothe G M and Purkhanbaba H 1982 *Electrophoresis* **3** 33–42
Rubin R W and Milikowski C 1978 *Biochim. Biophys. Acta* **509** 100–10
Rubin R W and Penny N S 1983 *Anal. Biochem.* **131** 520–4
Rüchel R 1977 *J. Chromatogr.* **132** 451–68
Sanders M M, Groppi V E and Browning E T 1980 *Anal. Biochem.* **163** 157–65
Saravis C A and Zamcheck N 1979 *J. Immunol. Meths* **29** 91–6
Scheele G A 1975 *J. Biol. Chem.* **250** 5375–85
Serwer P and Hayes S J 1982 *Electrophoresis* **3** 80–5
Shapiro A L, Vinuela E and Maizel J V 1967 *Biochem. Biophys. Res. Commun.* **28** 815–20
Shackelford D A, Mann D L, Van Rood J J, Ferrara G B and Strominger J L 1981 *Proc. Natl. Acad. Sci. USA* **78** 4566–70
Shackelford D A and Strominger J L 1980 *J. Exp. Med.* **151** 144–65

Siemankowski R F, Giambalvo A and Dreizen P 1978 *Physiol. Chem. Phys.* **10** 415–434
Sinclair J and Rickwood D 1981 *Gel Electrophoresis of Proteins* ed. B D Hames and D Rickwood (Oxford: IRL Press) pp 189–218
Singer B S, Morrissett H and Gold L 1978 *Anal. Biochem.* **85** 224–9
Slater C G 1965 *Fed. Proc.* **24** 225
Smithies O and Poulik M D 1956 *Nature* **177** 1033
Stegemann H 1970 *Angew. Chem.* **82** 640–1
Steinberg R A, O'Farrell P H, Friedrich U and Coffino P 1977 *Cell* **10** 381–91
Suria D and Liew C C 1974 *Can. J. Biochem.* **52** 1143–53
Svensson H 1961 *Acta Chem. Scand.* **15** 325–41
—— 1962 *Acta Chem. Scand.* **16** 456–66
—— 1967 *Protides Biol. Fluids* **15** 515–22
Taylor J, Anderson N L and Anderson N G 1983 *Electrophoresis* **4** 338–46
Thompson B J, Burghes A H M, Dunn M J and Dubowitz V 1981 *Electrophoresis* **2** 251–8
Thompson B J, Dunn M J, Burghes A H M and Dubowitz V 1982 *Electrophoresis* **3** 307–314
Tollaksen S L, Edwards J J and Andersen N G 1981 *Electrophoresis* **2** 155–60
Tracy R P, Currie R M, Kyle R A and Young D S 1982a *Clin. Chem.* **28** 900–7
Tracy R P, Currie R M and Young D S 1982b *Clin. Chem.* **28** 908–14
Tuszyncki G P, Buck C A and Warren L 1979 *Anal. Biochem.* **35** 224–229
Ui N 1973 *Ann. NY Acad. Sci* **209** 198–209
Vanderhoff J W, Micale F J and Krumrine P H 1977 *Sep. Pur. Methods* **6** 61–87
Vesterberg O and Hansen L 1978 *Biochim. Biophys. Acta* **534** 369–73
Vlasuk G P and Walz F G 1980 *Anal. Biochem.* **105** 112–20
Voris B P and Young D A 1980 *Anal. Biochem.* **104** 478–84
Weber K and Osborn M 1969 *J. Biol. Chem.* **244** 4406–12
Westermeier R, Postel W and Görg A 1981 *Electrophoresis '81* ed. R C Allen and P Arnaud (Berlin: de Gruyter) pp 281–7
Wilson J M, Young A B and Kelley W N 1983 *New Engl. J. Med.* **309** 900–10
Wyckoff M, Rodbard D and Chrambach A 1977 *Anal. Biochem.* **78** 459–82
Young D A 1984 *Clin. Chem.* **30** 2104–8
Young D A, Voris B P, Maytin E V and Colbert T A 1983 *Methods Enzymol.* **91** Part 1, 190–214
Zweidler A 1978 *Methods Cell Biol.* **17** 223–65

Chapter 5

Immunoelectrophoretic Methods

by
P M H Heegaard and T C Bøg-Hansen

5.1 Introduction

The techniques of separating antigens and identifying them specifically by their reaction with a suitable antibody have been given the common name immunoelectrophoresis. Under this name hides a number of very different methods, each method being furthermore equiped with an almost unlimited number of variants. To contribute to the description of this vast field of techniques one must therefore, for the sake of clarity, confine oneself to a coherent group of techniques within the heading immunoelectrophoresis. We have thus chosen to deal only with techniques in which the reaction of antigen with antibody takes place under electrophoresis (of either antigen or antibody) and in which this reaction leads to the formation of a precipitating complex of antigen and antibody. This is apart from the introductory paragraph of historical background where some related techniques are mentioned. By this delineation of the subject we hope to be able to describe adequately some of the potentials of the method and to stress the applicability of the basic principles.

5.2 Historical Background

The immunoelectrophoretic method introduced in 1953 (Grabar and Williams 1953), referred to as classical immunoelectrophoresis, was a great improvement in resolving protein mixtures. It combined for the first time a zone electrophoretic separation (based mainly on the free electrophoretic mobilities of the antigens) with antigenic detection (detection by diffusion of the separated antigens into antibodies, themselves diffusing from a reservoir in the gel). As always, the combination of two parameters greatly improved the resolving power of the method. By this method the presence in serum of a population of hitherto unrecognised alpha-1- and beta-1-proteins was shown, hiding each other in the conventional zone electrophoresis. The original method is still often recommended (Arquembourg 1975, Clausen 1981, Feinstein 1976, Kohn 1983). Later, the method was miniaturised, cutting down the amount of expensive antibody required (Scheidegger 1955).

Another early method for immunological characterisation of antigens by their precipitation with antibodies in gels is the double immunodiffusion method of Ouchterlony (1948), only suitable for pure antigens or demanding the use of monospecific antibodies and based solely on diffusion. Another approach was that of Ressler (1960a), who electrophoresed antigens into antibody-containing agar gels and achieved some precipitation of the sample components. He later combined a separating step (starch gel electrophoresis, Smithies 1955) with an immunological reaction by forcing the separated antigens into an antibody-containing agar gel by electrophoresis (Ressler 1960b). His results were not promising, but the approach was a logical forerunner of the important techniques developed later by Laurell.

Laurell used the same combination of electrophoresis and immunoprecipitation as in Ressler's second technique (Ressler 1960b) but optimised the amounts of antigen and antibody and performed the experiments in agarose (Laurell 1965). His results were promising and even indicated the method's potential for being quantitative. He stressed the basic principle of electrophoresis being superior to diffusion for the immunoprecipitation step by the former limiting the time needed for reaction, thereby increasing the resolving power by maintaining the initial resolution obtained by the first separating electrophoresis. He used goat antibodies against human serum proteins and noted that rabbit antibodies might also be used but horse antibodies could not. He termed the technique antigen–antibody crossed electrophoresis. Clarke and Freeman (1967) further refined this method into a truly quantitative one, the surface area of the precipitates being proportional to the amount of antigen in the precipitate and inversely proportional to the amount of antibody of this specificity per cm^2 of gel (this last figure is often somewhat erroneously called the antibody concentration). Today the method is most often referred to as crossed immunoelectrophoresis or two-dimensional immunoelectrophoresis.

Laurell also developed one of the other main methods in immunoelectrophoresis, termed rocket immunoelectrophoresis or electroimmunoassay (Laurell 1966, 1967). Again, it was a refinement of the principle first introduced by Ressler (1960a), the protein in question being electrophoresed directly into the antibody-containing gel causing the formation of rocket-shaped precipitates. By measuring the heights of these rockets it is possible to precisely and sensitively quantify the protein, even when it is present in a complex mixture, by the use of a mono-specific antibody. Laurell approximated the heights of the precipitates to the antigen–antibody ratio (being in reality proportional to the surface area of the precipitate, just as in crossed immunoelectrophoresis) and obtained reproducible results. Thus, this method was a supplement to the quantitative technique of Mancini *et al* (1964), which depended upon diffusion of the protein into an antibody-containing gel, demanding much more time and yielding less consistent results than rocket immunoelectrophoresis.

For quantification of antigens, another approach is to mix the antigen(s) in question into a gel and then electrophorese the material from this gel into another gel containing the antibody (Krøll 1968, 1983b). The precipitates thus formed have the shape of lines, with a distance from the origin of the antibody-containing gel which is proportional to the amount of antigen mixed into the first gel. By this method it is easier to measure more antigens at any one time.

Another principle of quantitative immunoelectrophoresis is counter current immunoelectrophoresis (crossover electrophoresis, immunoelectrosmophoresis) (Lang and Haan 1957, Wadström 1983). This technique demands opposite mobilities of antigen and antibody, which normally means that it takes place in gels of a high electroendo-osmosis (see §5.4.3). The antigen and the antibody are simply loaded into two wells in plain agarose and the current applied so that antigen and antibody will meet each other in between the wells and give rise to precipitation there. The position of this precipitate (distance from wells) is indicative of the amount of antigen in relation to the amount of antibody. The technique is a supplement to other nonqualitative techniques like ELISA (Enzyme Linked ImmunoSorbent Assay) and RIA (RadioImmunoAssay) for measuring antigen amounts. It demands a monospecific antibody of high titre.

A very elegant and straightforward variant of rocket immunoelectrophoresis is fused rocket immunoelectrophoresis devised by Svendsen and Rose (1970) primarily to give relative quantitative information on specific proteins during a fractionation procedure. In this method polyspecific antibodies are used so that antigenic heterogeneity of any one fraction can be analysed in one step. This is very convenient when one is concerned with the purification of, for example, a specific protein from a complex mixture (see §5.6.7).

The methods described above remain the basis of all the later uses of immunoelectrophoresis as defined in the introduction. Later improvements have been concerned with increasing the resolving power and sensitivity, and with characterisation procedures during the electrophoretic procedure (see §5.4.5)

5.3 Theory

Under the conditions of free electrophoretic flow in nonsieving anticonvection media, the only parameter determining the mobility of a certain molecule is the number of ionisable groups and the pK values of these groups, both to be compared with the size of the whole molecule: in short, the charge density at the pH given. This is also the condition for electrophoresis of antigens in agar or agarose but here two additional factors are of importance, namely diffusion and electroendo-osmosis (EEO). Diffusion will always tend to distribute the molecule homogeneously in a gel and is roughly inversely proportional to the size of the molecule.

EEO is a phenomenon particular to matrices carrying fixed charges. In the case of agar and certain types of agarose these charges are contributed by pyruvate and sulphate groups, bound to the sugar backbone (immobilised negative charges, Araki 1956, see §5.4.3). This forces the cations and the neutral constituents of the buffer to move towards the cathode. This is of great importance for the resolving power of immunoelectrophoresis. By purifying agar to different degrees of anionic substitution, it is possible to make agarose of controlled EEO (see below and table 5.1). IgG molecules from most species which have been electrophoresed at weakly alkaline pH in gels of this type are supplied with a net mobility of zero or slightly cathodal by the EEO-force, while in this system, for example most other serum proteins from man, will have anodal mobility (without EEO a bigger number of the IgG molecules would be moving anodally). Thus, it is ensured that all antigens meet almost every antibody, having the pronounced electrophoretic heterogeneity of IgG in mind (e.g. Grubb 1972)

When an antigen meets its antibody counterpart, a complex is formed. This complex will be soluble in the gel matrix and able to move corresponding to its net charge, but with lower mobility than the free antigen. With the addition of more antigen molecules bigger soluble complexes will be formed and eventually a nonsoluble, immobile precipitate will be deposited in the gel. This happens in comparison with test-tube precipitation (Heidelberger experiment) in the zone of antigen excess (Laurell and McKay 1981). Before this does happen some precipitation will have occurred but this precipitate is not stationary in the electrophoresis medium (see figure 5.2), and once formed, the final precipitate is both stationary in the gel, even after prolonged electrophoresis, and not dissolved again by addition of more antigen (if e.g. rabbit antibodies are employed). This is in contrast to titration of antibody with antigen in a test-tube which shows no precipitate formation in excess of antigen. The ratio of antigen to antibody critical for precipitation varies with the antigen–antibody system employed, depending on the antigen size, the number of antigenic determinants carried and the degree of charge difference between the two components. The size of the precipitate is determined by the antigen/antibody mass ratio while the total amounts determine the density of precipitates. It has been hypothesised that agarose gel structure is important in precipitate stabilisation in gels but the phenomenon is not yet fully explained (Cann 1975, Laurell and McKay 1981). In cellulose acetate (see below), it has been claimed that precipitation occurs in the zone of excess antibody (Chang and Inglis 1975).

Suggestive of the precipitates' structure are the cases of nonspecific entrapment of other molecules in the precipitate lattice often observed (see below, enzymes, lectins), the non-interaction of different precipitates composed of immunologically nonrelated antigens (see §5.6.3 antigen identification and figure 5.6) plus also the special cases of specific adsorption of proteins on the free combining sites of molecules immobilised in the precipitate (see amplifying reactions and figure 5.8) and the formation of 'false spurs' in line immunoelectrophoretic methods (see below). This, apart from the purely quantitative aspect of the size of the precipitate, gives qualitative information on the nature of the complex, i.e. which protein is actually bound. This contributes a lot to the resolution and the facility of interpretation of the method.

5.4 General Conditions and Guidelines

5.4.1 Antibodies

As mentioned briefly a number of animals are suitable for immunisation with for example human antigens and subsequently using their antibodies in immunoelectrophoretic systems. The species most often used now are rabbit, sheep, goat and swine, and mouse for monoclonal antibodies. The use of monoclonal antibodies in these techniques is not yet very widespread, but might in the future become so, remembering that this special kind of antibody does not precipitate normal antigens in immunelectrophoretic systems, thereby not excluding their use as a first dimensional ligand in crossed immunoelectrophoresis or a local absorption ligand in line immunoelectrophoresis (see below and figure 5.5). Multiple epitope replicas on the antigen are demanded for precipitation by monoclonal antibodies. The exact scheme of immunising a rabbit for the production of polyclonal (monospecific or polyspecific) antisera can be found in Harboe and Ingild (1973, 1983), the rabbit being, in general, the chosen animal. The animal should be chosen judging from its response to antigens from the species in question (also with regard to any interspecies cross reactivity, see below), its ease of bleeding, the cost of keeping the animal, the productive lifetime of the animal, the quality of its antibodies (titre and precipitating ability), the usefulness of the IgGs in the immunoelectrophoretic system employed and the ease of purifying the antibody from serum.

Essential features of antibodies will be discussed below and these comprise specificity, purity and electrophoretic behaviour as well as titre, affinity, avidity and precipitating ability, the last four characteristics being very much interdependent. Specificity is a function of the heterogeneity of the immunogen solution injected, but the process of purifying an antibody after immunisation by either liquid or solid phase absorption, with specific proteins removing specific antibody specificities, is well known and of great value. To test the monospecificity of a given antibody one must always use the crude starting material as the antigen. If absorbing, it is worthwhile to remember that excess antigen in the antibody solution might give rise to a number of artefacts (described below, and see for example figure 5.6) in the immunoelectrophoretic picture. A special feature is the cross reactivity of an antibody with antigens from more than one species. As a very general rule, crossreacting antibodies are most easily obtained by immunising an animal which is evolutionarily far from the animal delivering the immunogen and from the other animal with the antigen one wishes to compare it with (e.g. Crumpton 1975, Jensenius *et al* 1981).

Complete purity of an antibody preparation is, in the case of rabbit antibodies, not necessary in order to obtain good results in immunoelectrophoresis. The main problem using the crude antiserum is some background staining, due to a certain amount of nonreacting immunoglobulins (most other possible nonreacting proteins are washed out of the plate during electrophoresis by their anodic mobility). In the case of unfractionated rabbit serum, however, the lipoprotein content especially contributes to the background staining (Harboe and Ingild 1983). Also, in certain systems the lipoproteins may give rise to artefacts (Carey 1975). To ensure lower background, and thereby higher sensitivity, antibodies could be purified from crude rabbit serum by the method of Harboe and Ingild (1973) which employs ammonium sulphate precipitation, ion exchange and dialysis, resulting in an IgG fraction with 95% recovery (Blirup-Jensen 1982). As this will also contain 'nonreactive' molecules ('nonsense' antibodies) and non-inhibited proteases (plasmin (Bjerrum and Bøg Hansen 1976)) further purification may be desirable. It can be performed by pseudospecific affinity chromatography (protein A) (Hjelm *et al* 1972) yielding completely pure IgG, or by specific affinity chromatography using the antigen in question immobilised for solid phase absorption (immunosorption) (Rouslahti 1976). This last method ideally will only yield IgG reacting with the antigen in question. This approach is also worthwhile when considering if a poly- or oligospecific antibody is to be rendered monospecific (solid phase absorption, mentioned above); drawbacks are the chance of losing high affinity antibodies and the risk of denaturing the antibodies recovered by the drastic methods demanded for their elution from the antigen column. These problems can be circumvented by using this method only to remove contaminating activities.

Also the behaviour of a certain species' IgG under electrophoretical immunoprecipitation is important when choosing an animal and treating the antiserum. As mentioned horse IgG does not always seem well suited for electrophoresis, and the precipitate may dissolve in antigen and antibody excess (Week 1973, Williams 1971). Still, some results have been obtained using antibodies from this species (Krøll 1969a). Avian antibodies demand as a general rule high concentrations of salt to be precipitated (Williams 1971).

For conventional use, antibodies from rabbit, goat, sheep and swine are most often chosen. The four parameters used to describe the precipitation behaviour of an antibody in an electrophoretic system are affinity, avidity, titre and precipitating ability.

The affinity of antibodies is very variable, is normally found within the wide range of dissociation constants from 10^4 to 10^{10} M (Kabat 1980), and describes the reaction of one IgG molecule with one epitope, which is (ideally) the case for immune complexes in antigen excess.

Avidity is a somewhat confusing term often operationally defined and differing from system to system, but this parameter basically describes the functional affinity of an antibody preparation, i.e. the total affinity constant for the whole antibody molecule interacting with the antigen (e.g. Roitt 1980). The time needed for this complex formation becomes important (dependent on the activation energy of all the individual transition states). Antibodies of low avidity can therefore be used in slow systems, as e.g. Mancini quantitation, better than in systems demanding quicker reactions as e.g. rocket immunoelectrophoresis, where high avidity antibodies are demanded.

Both avidity and affinity are important for the titre of the antibody, which is defined as the amount of antigen being precipitated by one volume of the antibody solution. Here, additional factors are important, such as the amount of nonprecipitating species in the solution. Again, the measurement of titres gives very different results from method to method (Harboe and Ingild 1973), and can only be regarded as a relative guideline for the amount of antibody to use. Generally the lowest possible amount of antibody should be used to decrease background staining and enhance sensitivity. As a general rule, injections with a low amount of antigen give rise to a low titre antibody of high avidity (Danielsen 1983).

Precipitating ability (by some, set equal to avidity (Harboe and Ingild 1973)) is a term totally dependent upon the system used for precipitation and this feature (as well as the other three features) can vary enormously from one individual animal to another (e.g. Ganrot 1972), hence one justification of the old rule of always immunising more than one animal when producing antibodies. In addition different antigens may, generally speaking, vary enormously with respect to their ease of precipitation. Differences between precipitate morphology from antibody to antibody are also observed, but with the same antibody species this morphology characteristic will mostly depend more on the nature of the antigen being precipitated. In general, precipitation in gels can be enhanced 'artificially' by the addition of polymers to the gel (PEG 6000 (polyethyleneglycol) mol wt 6000), dextrans, see below) (Ceska 1968, Hellsing 1973, Kostner and Holasek 1972, Ramlau and Bjerrum 1977) or by modifying the antigens, by e.g. desialylation (Brogren and Bøg-Hansen 1975b).

Antibodies can be stored for years with relatively small loss of activity (2% a year) at 4°C in 0.1 M NaCl and 15 mM azide, while freezing is not recommended (Harboe and Ingild 1973). Solid phase affinity purified antibodies are claimed to be less stable than antibodies in a crude IgG-fraction (Laurell and McKay 1981).

5.4.2 Antigens

Almost any substance foreign to the animal will be immunogenic, i.e. able to induce the production in the animal of specific antibodies against the injected substance. In general, proteins should be bigger than 5000 daltons (D) and polysaccharides more than 70000 D, but this is a rule with many exceptions (Crumpton 1975). Moreover, low molecular weight antigens (haptens, hormones) can be rendered immunogenic by coupling to carrier proteins. Poor immunogens can be rendered good either by changing to another species, modifying the antigen chemically or optimising the composition of the mixture injected (by adjuvants). There is a correlation between 'foreigness' and crossreactivity of antibody produced but it is not linearly connected with differences in the amino acid sequence of the antigen (Atazzi 1975). A small, yet important point is that antigens are still often immunogenic when complexed with IgG from the same animal, making possible the method of cutting out agarose-enclosed precipitates directly from the gel and immunising with the whole precipitate to obtain specific antibodies against the antigen in the complex (Vestergaard 1975). The use of complexed antigens is even claimed to enhance the response (Crowle *et al* 1972). Generally, proteins should be between 3×10^4 D and 2×10^6 D and have two or more antigenic sites per molecule. Also they should be kept in their native state and solubilised during electrophoresis. The exclusion limit of the gel (see below) as well as the pH and the ion content in the system must be considered too in the set-up of the immunoelectrophoresis. As pH is often not a variable due to the demand for immobility of the IgGs employed, an antigen not fit for this pH must be altered (ionic detergents, chemical modification). Alternatively, a gel with another EEO must be chosen or the antibodies might be modified (see below).

5.4.3 Gels

Gels used for immunoelectrophoresis are now normally agarose type gels (see table 5.1) although some still recommend the use of agar and cellulose

Table 5.1 Types of agarose employed for immunoelectrophoretic techniques

Type				*Electro endo-osmosis*	*Gelling temperature (°C)*	*Strength 1.5% gel ($g\,cm^{-2}$)*	*Max. % SO^4*
Litex	*Pharmacia*	*Biorad*	*LKB*				
HSA	A	Low	M	−0.13	42	1200	0.45
HSB				−0.07	42	2400	0.30
HSC	C	Zero	L	−0.02	42	1000	0.30
HSLF			Z	0	41–3	700	0.25
LSL				−0.13	36	2400	0.09
LSM				−0.17	36	2000	0.35
LSA	B	High	H	−0.25	36	1900	0.60
LSHE				−0.35	36	1600	0.25
		Low gelling		−0.10	37	500	0.15
			Isogel	0	41–3	1600	0.10
IBF							
A 37				−0.17	36–8	800	0.30
A 37 HAA				−0.19			
A 37 NA				−0.17	36–8	1000	0.30
Marine Colloids							
Seakem	*Sigma*						
LE	I			−0.1–0.15	36	1000	0.35
ME	II			−0.16–0.19	36	1000	0.35
HE	III			−0.23–0.26	36	650	0.35
HEEO	IV			−0.30(min.)	36	650	0.35
HGT	V			−0.10(max.)	42	800	0.30
HGT(P)	VI			−0.10(max.)	42	1000	0.15
Isogel	VII			0	—	700	0.10
Seaprep	IX			−0.05(max.)	8–17	75(2%)	0.10
Nusieve	VII			−0.15(max.)	35	150	0.15

acetate (Jurd 1981, Kohn 1976). Agarose is a naturally occurring, linear polysaccharide with beta-D-galactopyranose-1-3-(3,6-anhydro)-alpha-L-galactopyranose-1-4 as the repeating unit ('agarobiose') (Araki 1956) substituted methyl groups, sulphate and pyruvateketal, the latter two causing the electroendoosmosis (EEO) (see §5.3). Upon heating and subsequent cooling of this polysaccharide in aqueous solutions, it gellifies as bundles of left-handed double helices with big pores in between resulting in a uniform and transparent gel (Arnott *et al* 1974) with an exclusion limit of 20×10^6 (Pluzek 1981) (see table 5.1), freely penetrable even to antigen–

antibody complexes. The agarose is normally used in a concentration of 0.7–1.2% in the electrophoresis buffer. The EEO is an important parameter and is easily measured by comparing the migration of dextran (uncharged) with that of albumin at standard conditions (pH 8.6, in 0.75 M Veronal, 1% agarose, 6 V cm^{-1} for 90 minutes (Wieme 1965)), the sign for cathodal migration of dextran being negative. By taking off sulphate and pyruvate from the agarose in a controlled manner one can produce agaroses of well defined EEO (Lönnderdal and Låås 1976). For most immunoelectrophoretic techniques, agarose of medium EEO (around −0.13) is used. High EEO agarose is needed for counter current immunoelectrophoresis (see §5.2), and zero EEO agarose for procedures involving isoelectric focusing (see below, 'Hybrid Techniques'). Generally, the sulphate content should be kept below 0.2% (see table 5.1), and in some cases even lower (see below). Moreover, agarose is non-toxic, has a good mechanical strength, is very easy to handle, can be produced with controlled gelling temperatures (see table 5.1) and has a great positive influence on the formation of stationary antigen–antibody precipitates with the advantageous features described above (Laurell and McKay 1981). The correlation between EEO and pH (for applications at unusual pH values) has been investigated for different agarose gel types (Pluzek and Bjerrum 1982). For special purposes, other gel types may be used, although (see below), except for the second-dimensional immunological characterisation, agar or agarose is almost always used. Agarose interacts only in special cases with antigens. Examples of this are notably galactose specific lectins and enzymes (Kint and Leroy 1985, Tomita *et al* 1971) and certain heparin binding proteins from human serum (Complement factor C1q, antithrombin III) and other sources (lactoferrin, pancreatic elastase) by adsorption of the basic proteins to the immobilised anionic groups (Johansson 1972, Laurell and McKay 1981). In these cases low sulphate agarose or cellulose acetate is recommended for immunoelectrophoresis. Gels are poured on glass plates or polyester (GelBond, Renn and Evans 1976) at a thickness of 1–2 mm, the solvent being the electrophoresis buffer. Low percentage of agarose is employed when background staining is a problem, higher percentages can be employed if large volumes of antiserum are used. A detailed procedure is found in Axelsen *et al* (1973, 1983a):

(i) All types of detergents can be added to gels for solubilisation or for investigations of amphiphilicity (Bhakdi and Bjerrum 1983, Bjerrum 1983c); (ii) ligands, lectins (Bøg-Hansen *et al* 1975, Bøg-Hansen and Brogren 1975), immobilised lectins (Bøg-Hansen 1973); and (iii) polymers can be added to enhance the precipitation in the gel as mentioned above.

The morphology of the gel and the mild conditions of the whole system make staining of precipitates very easy and fast and makes it possible to use histochemical methods to stain precipitated enzymes which still retain their biologiocal activity since the antibody binding does not itself destroy this activity (Arnon 1975). Enzymes can be found precipitated directly, coprecipitated or entrapped in other precipitates (Augustinsson 1968, Brogren and Bøg-Hansen 1975a). The use of cellulose acetate as a medium for electrophoresis is well established (Kohn 1976) and is also described for all common immunoelectrophoretic variants, as for example classical immunoelectrophoresis *ad modum* (Grabar and Williams 1953, Kohn 1976), rocket immunoelectrophoresis (Schanbacher and Smith 1974), crossed immunoelectrophoresis (with agarose as the second-dimension gel) (Kint and Leroy 1985), counter current immunoelectrophoresis and immunofixation (Chang and Inglis 1975) as well as the more elaborate techniques like counterflow immunoisotachophoresis of Abelev and Karamova (1984). Precipitation in cellulose acetate is claimed to take place in the zone of excess antibody (Chang and Inglis 1974) in contrast to the situation in agarose (Laurell and McKay 1981). Cellulose acetate as an electrophoresis medium when compared with agarose has the advantages of—

(i) being totally chemically inert showing no nonspecific interactions,
(ii) making analysis of agarose-interacting proteins possible (galactose-specific lectins (Kint and Leroy 1985), lactoferrin (Schanbacher and Smith 1974)),
(iii) being easy to cool very efficiently,
(iv) large samples and short electrophoresis times are possible (Kroll 1969a).

The disadvantages include:

(i) contact between several cellulose acetate sheets is more difficult to obtain,
(ii) consistent patterns are also not obtained easily,
(iii) the sensitivity of precipitation is not as great as in agarose.

The use of cellulose acetate will not be dealt with in detail in this review.

5.4.4 Electrophoretic Conditions

Generally, the conditions of electrophoresis should aim at keeping the mobility of the antibody IgGs as near to zero as possible, at supplying antigens with a different charge to the antibodies and at speeding up the whole procedure. The latter involves considering the size and the number of antigenic determinants of the antigens analysed (smaller antigens are 'faster' than bigger ones, the bigger the charge difference and the higher the number of antigenic sites the faster the immunoprecipitation upon electrophoresis). Consistent results should of course be aimed at too. To obtain this, low ionic strength buffers are used, making possible the application of high voltages and the creation of distinct precipitates (ionic strengths between 0.01 and 0.03 are recommended for most systems (Weeke 1973)) without creating excessive heat. The ionic strengths used cannot be too low as this will cause loss of buffering capacity (which is of course essential in order to obtain constant results) or demand large volumes of (circulating) buffer. Moreover, the ionic strength should be high enough to keep all antigen and antibody molecules in solution. In ordinary medium-EEO agarose (EEO around −0.13, see table 5.1) employing rabbit antibodies, a weakly alkaline pH will give an average migration of zero for antibody molecules (Grubb 1972). Buffers commonly employed for work at standard conditions are listed in table 5.2. When other pH values are desired (e.g. in the study of complex formations between antigens at different pHs (Pluzek 1982)) or when a better resolution of the pattern is required (Smyth *et al* 1976) or when an antigen is alkali labile and in order to reduce unwanted enzymatic activity (Verbruggen 1975a, Verbruggen and Baeck 1976) or to make immobile proteins mobile (Ramlau and Bjerrum 1977), the change to agarose of another EEO is one way to obtain nonmoving antibodies (Pluzek 1982). Another method is to modify the antibodies by, for example, carbamylation (Bjerrum *et al* 1973), where antibodies are

Table 5.2 Commonly employed buffers for immunoelectrophoresis at standard conditions

Laurell 1965	*Weeke 1973*	*Svendsen 1973*
Veronal, Na-Veronal, 0.075 M Ca-lactate, 0.001 M pH: 8.6, *I*: 0.05	Veronal, 0.0043 M Na-Veronal, 0.0179 M pH: 8.6, *I*: 0.02 Spec. conductivity (5 times concentrated stock solution): 7 mS (23°C)	Veronal, 0.0054 M Na-Veronal, 0.0282 M Glycine, 0.1300 M Tris, 0.1000 M pH: 8.8, *I*: 0.08
Bjerrum et al 1974	*Monthony et al 1978*	*Brogren and Bøg-Hansen 1975a, b, Svendsen 1979, Svendsen et al 1983*
Glycine, 0.100 M Tris, 0.038 M pH: 8.7	Tris, 0.0800 M Tricine, 0.0235 M Ca-lactate, 0.00034 M pH: 8.6, *I*: 0.02 Spec. Conductivity: 1.109 mS (20°C)	Veronal, 0.0243 M Tris, 0.0732 M Ca-lactate, 0.00034 M pH: 8.6, *I*: 0.02 Spec. Conductivity: 1.247 mS (20°C) (Monthony 1978)

I = ionic strength
Veronal = diemal = Barbitone = barbital = 5,5 diethylbarbituric acid (pKa = 7.4)

reacted with isocyanate converting the amino-groups of the antibodies to carbamylamino-groups, lowering the pI of IgG to a desired pH, e.g. pH 5.0. This procedure does not alter the titre of antibodies very much (Bjerrum *et al* 1974). Another way of modifying antibodies is to react them with carbodiimide followed by nucleophile attack (elimination of COOH groups (Grubb 1974)) raising the mean pI to 10.3. These modifications might be necessary when antigen and antibody have similar mobilities at the pH employed. As rule of thumb, pIs of antibodies and antigens respectively should be more than two pH units apart (Laurell and McKay 1981).

Addition of calcium ions to the buffer may give sharper precipitates (Laurell 1972b). It is frequently necessary to lower the speed of the electrophoresis when a precipitating step is performed since antibodies of low avidity may not be able to precipitate quickly enough. In the case of rocket immunoelectrophoresis, procedures have been worked out with antibodies of high avidity so as to speed up the procedure by running the electrophoresis at a very high voltage (Glad 1983), obtaining electrophoresis times comparable with cellulose acetate procedures (2–3 h). It is important to control the thickness of the agarose gels employed, as thin gels will cause a bigger voltage drop than thicker gels at the same current, giving rise in some systems (rocket immunoelectrophoresis, see §5.5.1) to faults in the relative quantitation of the same antigen from plate to plate. For the same reason the homogeneity of the field across the plate should also be ensured. The temperature of the electrophoresis should be kept constant but not too low compared with room temperature, as this will often cause undesired condensation of water onto the plates.

5.4.5 Visualisation, Sensitivities and Limits

Visualisation and precipitation are two important parameters in the description of the resolving power and the sensitivity of an immunoelectrophoretic system, and they both have their limits in terms of the smallest total amount of antigen needed.

Because of the advantageous features of agarose gels described above (big pores, squeezability) and immunoprecipitates therein (immobility, accessibility of precipitated antigens), general staining of antigens is very easy and fast and intensive washing/pressing procedures can be performed to rinse the gel of unprecipitated antigens and soluble immune complexes (Weeke 1973) without damaging the precipitates. If staining is not wanted (e.g. when the aim is to cut out precipitates), precipitates can often be detected by direct visual inspection of the wet gel by light or dark incident field illumination (Krøll 1981). For staining of proteins, virtually any protein stain may be used. The aim is normally to obtain the highest sensitivity possible, so Coomassie brilliant blue R 250 (Kenacid blue) in ethanol/water has long been the stain of choice being three times more sensitive than amidoblack 10B (Wieme 1959, Johansson 1972) (see table 5.3). Also other dyes are in use (Ponceau S, nigrosin and Coomassie PAGE (Kohn 1983)), but the Coomassie brilliant blue R 250 is the stain which is used by far the most. Prerequisites for sensitive visualisation are the use of the lowest possible amount of antigen and antibody (observing of course the optimal ratio between the two), extensive washings, thin gels and high avidity antibodies. Recently the new method of silver staining for general staining with increased sensitivity towards proteins has been applied to proteins precipitated in agarose gels. It is claimed to be up to eight times as sensitive as the usual Coomassie stain (Porro 1982) (see table 5.3). This way of staining has also found use in other electrophoretic systems in agarose, e.g.

Table 5.3 Sensitivity in immunoelectrophoresis. Limit for the amounts of antigen

rockets = rocket immunoelectrophoresis; crossed = crossed immunoelectrophoresis; 'anti-alpha-2 macroglobulin-anti-do' = a system employing alpha-2-macroglobulin as antigen and monospecific anti-alpha-2-macroglobulin as antibody. When nothing else is indicated, the agarose is of medium EEO and non-sieving. References: (1) Stevens *et al* 1981, (2) Svendsen 1979, (3) Johansson 1972, (4) Porro 1982, (5) Broe and Ingild 1983, (6) Kostner and Holasek 1972, (7) Grieninger and Granick 1975, (8) Darcy 1972, (9) Saravis and Bonacker 1970, (10) Nørgaard Pedersen *et al* 1983, (11) Kindmark and Thorell 1972, (12) Axelsen 1983, (13) Heegaard *et al* 1984a (15) Brogren 1975a, (16) Laurell 1972, (17) Ramlau and Bjerrum 1977, (18) Prause *et al* 1978, (19) Bjerrum *et al* 1982, (20) Laurell and McKay 1981, (21) Krøll 1981, (22) Andersen *et al* 1982.

Limit for precipitation

Unmodified basic systems

System: 1% agarose: 0.3 ng (2)
1% agarose: 0.2 ng (20)

Modified systems

System: 1% agarose + 2% (w/v) PEG 6000, rockets,
alphafetoprotein-anti-do: 0.2 ng (5)
crossed: 0.2–1.0 ng (5)
1% agarose + 5% (w/v) PEG 6000, rockets,
alphafetoprotein-anti-do: 0.3 ng (10)
alpha-2 macroglobulin-anti-do: 1.0 ng (17) (12)
ceruloplasmin-anti-do: 0.6 ng (17)
alpha-1 antichymotrypsin-anti alfa-do: 0.4 ng (17)

1% agarose + 4% Dextran T 70, crossed,
human serum-anti-do: 5 times increase in sensitivity compared to normal procedure (6)

addition of tannic acid, sensitivity as with dextran (7)

Table 5.3 continued

Limit for visualisation

Dark field illumination:
1% agarose, line immunoelectrophoresis: 50 ng cm^{-1} (21)

General stains

System: 1% agarose, rockets and crossed,
amidoblack 10 B: 30 ng (3)
Coomassie brilliant blue R 250: 5–10 ng (2)
10 ng (20)
10–20 ng (3)

1% agarose, rockets, Coomassie brilliant blue R 250
bovine albumin-anti-do: 6.5 ng (4)
pneumococcal capsular polysaccharide 6A-anti-do: 0.4 ng (4)
mannose (inositol-P)2-ceramid-anti-do: 10–20 ng (19)

1% agarose + 5% PEG 6000, rockets, Coomassie brilliant blue R 250
alpha-2 macroglobulin-anti-do: 4 ng (17)
alpha-1-antichymotrypsin-anti-do: 8 ng (17)

1% agarose (85% HSA + 15% LSR (Litex)), rockets in capillaries, Coomassie brilliant blue R 250,
alpha-2 macroglobulin-anti-do: 0.1–1.0 ng (18)

1% agarose, rockets, silver stain,
bovine albumin-anti-do: 0.8 hg (4)
pneumococcal capsular polysaccharide 6A-anti-do: 0.1 ng (4)

Direct staining for enzymes

System: 1% agarose, crossed, staining for esterases by fast red +
alfa- and beta-naphthylacetate in acetone,
cholinesterase-anti-do: 0.10–0.15 ng (15)

Amplified visualisations

System: agarose rockets amidoblack
carcinoembryonic antigen-anti-do (goat) + anti-Goat-IgG-overlay: 10 ng (8)

Cellulose acetate, rockets, Ponceau S, Nigrosin,
Hepatitis associated antigen-anti-do (human) +
guinea pig complement + anti guinea pig complement antibody (rabbits): increased sensitivity (9)

Enzyme
1% agarose + 2% PEG 6000, rockets and crossed,
Horse radish peroxidase labelled second antibody (swine)-overlay,
alphafetoprotein-anti-do, anti-anti-do-HRP,

Table 5.3 continued

3-amino-9-ethyl-carbazol: 0.2 ng (precipitation limit, see above) (5)
By Autoradiography 0.8% agarose, rockets, overlay with ^{125}I-labelled primary antibody after precipitation with cold antibody CRP-anti-do + 125I-anti-do: 0.1 ng (11)
1% agarose + 5% PEG 6000, crossed and rockets, ^{125}I-labelled antigen in intermediate gel + cold antigen coprecipitates, alphafetoprotein + ^{125}I-alphafetoprotein-anti-do: 0.3 ng (10) (precipitation limit)
1% agarose + 5% PEG 6000, ^{125}I-labelled + cold primary antibody in gel, crossed and rockets, ferritin-anti-do + ^{125}I-anti-do: Precipitation limit reached (antigen amount not known) (22)
1% agarose + 5% PEG 6000, ^{125}I-labelled second antibody (swine) second electrophoresis at pH 5, rockets, alpha-2 macroglobulin-anti-do + ^{125}I-anti-anti-do-elph.: 0.5 ng (12)
1% agarose + 5% PEG 6000, ^{124}I-labelled second antibody (swine) second electrophoresis at pH 5, rockets, alpha-2 macroglobulin-anti-do + ^{125}I-anti-anti-do-elph.: 1.0 ng (17) ceruloplamin-anti-do, ^{125}I-anti-anti-do-elph.: 0.6 ng (17) alpha-1-antichymotrypsin-anti-do + ^{125}I-anti-anti-do-elph.: 0.4 ng (17)
0.75% agarose + 1% PEG 6000, overlay with ^{125}I-labelled protein A, rockets, *Salmonella minnesota* core glycolipid antigen-anti-do: 0.125 ng (1)
By Fluorography: 1% agarose + 5% PEG 6000, ^{14}C-labelled second antibody (swine)- overlay, sodium salicylate enhancement, rockets, alpha-2-macroglobulin-anti-do + ^{14}C-anti-anti-do-overlay: 0.05 ng (13); glycophorin-anti-do + ^{14}C-anti-anti-do-overlay: 0.10 ng (13)

isoelectric focusing (Vesterberg and Gramstrup-Christensen 1984) and zone electrophoresis (Willoughby and Lambert 1983) giving typically 10–30 fold increases in sensitivity compared with conventional Coomassie brilliant blue R 250 staining.

To increase specifically the sensitivity of visualisation one can take advantage of special characteristics of the antigen in question—such as enzymatic activity (Brogren and Bøg-Hansen 1975a), binding of monoclonals (Bjerrum 1984, see figure 5.5), binding of IgE by allergens (Løwenstein 1978), figure 5.8)—and often in this way dramatically lower the visualisation limit (up to 50 times, see e.g. Bøg-Hansen 1983a). Specific visualisation in immunoelectrophoresis can be performed for carbohydrates on glycoproteins and glycolipids, checking that the stain does not interact with the agarose itself or the antibody molecules as these are also glycoproteins containing 3–12% sugar (Wasserman and Carra 1977). Suitable stains that have been used are: Schiff-reagent (Crowle 1961), lectins (Bøg-Hansen 1973, 1983, Bøg Hansen *et al* 1973, 1983), lipids in lipoproteins (Sudan black B, oil red, Uriel 1964), and glycolipids (Sudan black plus Triton X 100, Bjerrum *et al* 1982, Bjerrum and Gerlach 1982), enzymes stainable after precipitation (esterases (Augustinsson 1968, Uriel 1964, Brogren and Bøg Hansen 1975a), ceruloplasmin (oxidase activity (Uriel 1964)), xanthine oxidase (Nielsen and Bjerrum 1977), ATPase, 5′-nucleotidase (Gomori 1952), haemoglobin (peroxidaseactivity) (Uriel 1964)). This is only a short list, and alternatives exist to label a certain compound before precipitation and visualisation (see below). Another approach is to amplify the precipitates in the sense of increasing the 'signal-to-noise' ratio in the system. This has been done to a large extent by using radioactive or enzymatic markers and/or secondary antibodies (or protein A (Stevens *et al* 1981)) applied postelectrophoretically, directly or by coprecipitation with 'cold' antibodies (Kinmark and Thorell 1972, Nørgaard-Pedersen and Gaede 1975, Ramlau and Bjerrum 1977, Andersen *et al* 1982, see table 5.3). Labelling can also be performed with other affinity systems as e.g. the biotin–avidin system (biotinylated second antibody and HRP-avidin (Lihme and Heegaard 1984)).

Using these methods, the precipitation limit of the system, in terms of the total amount of antigen, might be reached and the threshold for detection becomes more a question of this than the sensitivity of the visualisation method. Nothing general can be said about this limit, as it depends heavily on the features of the antibody employed (Ramlau and Bjerrum 1977), the antigen–antibody system investigated and the type of immunoelectrophoresis and the presence of precipitation-enhancing factors (PEG, dextrans). In these systems, PEG 6000 in concentrations of 2–5 w/v% is often employed, and the limit for precipitation in terms of the antigen amount is typically around 200 pg (Laurell and McKay 1981). Another way is to miniaturise the system as has been done with capillary rocket immunoelectrophoresis (Prause *et al* 1978) or to increase the amount of sample applied (Krøll 1976).

5.5 Contemporary Methods

As a manual for the contemporary use of immunoelectrophoretic methods Axelsen (1983a) should be consulted concerning the general procedures.

Other basic reviews are Axelsen (1975), Axelsen and Bock (1972), Axelsen *et al* (1973b) and Laurell (1972a). The main types of these techniques in use today will be considered briefly here, along with their limitations and potentials.

5.5.1 Semiquantitative Methods

When an immunoelectrophoretic method is termed quantitative it is always meant to be relative to a known standard. That is, the system must always be calibrated with dilutions of a known sample of the antigen in question. This limitation taken into account, the applicability of immunoelectrophoresis is, as will be shown, almost unlimited. We will describe here in more detail the following basic immunoelectrophoretic methods:

Rocket immunoelectrophoresis (Laurell 1966)
Crossed immunoelectrophoresis (Laurell 1965, Clarke and Freeman 1967)
Tandem crossed immunoelectrophoresis (Krøll 1968)
Intermediate gel technique (Svendsen and Axelsen 1972)
Fused rocket immunoelectrophoresis (Svendsen and Rose 1970)

Rocket Immunoelectrophoresis, Electroimmunoassay

A most useful alternative to the single radial immunodiffusion (Mancini 1947) and nephelometry (see Ingild 1983), the original procedure of Laurell (Laurell 1966), is used almost without modification today (Laurell 1972b, 1981, Axelsen and Bock 1983) (see introduction for procedure and figure 5.1). Only if the antigen in question is of a low molecular weight (below 30 D) or exhibits extensive electrophoretic heterogeneity might single radial immunodiffusion be considered. Nephelometry is inferior both to rocket immunoelectrophoresis and single radial immunodiffusion in specificity (Ingild 1983). After electrophoresis (overnight at a low voltage or 2–3 h at a higher voltage) and staining, the heights of standards are measured as a first approximation to their areas and correlated to the known content of antigen. Measures must be taken to ensure that the precipitation has been completed (see, e.g. figure 5.2 and Axelsen *et al* 1983). However, this is normally always the case after overnight electrophoresis. The content in the unknown samples of the antigen in question can then easily be calculated by measuring the rocket heights. The advantage is that the ease with which the rocket heights can be measured makes it also much quicker than tedious calculations of precipitates' surface areas (see below). A series of standard dilutions of the antigen must be included on all plates as the calibration curve can change from plate to plate. The curve is seldom linear; the inhomogeneous distribution of antibodies developing under electrophoresis due to EEO is one factor ruling out linearity. It has been suggested that heights squared might give more linear

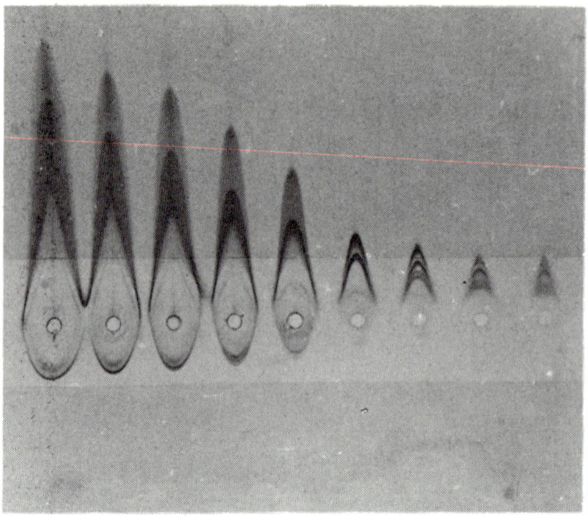

a

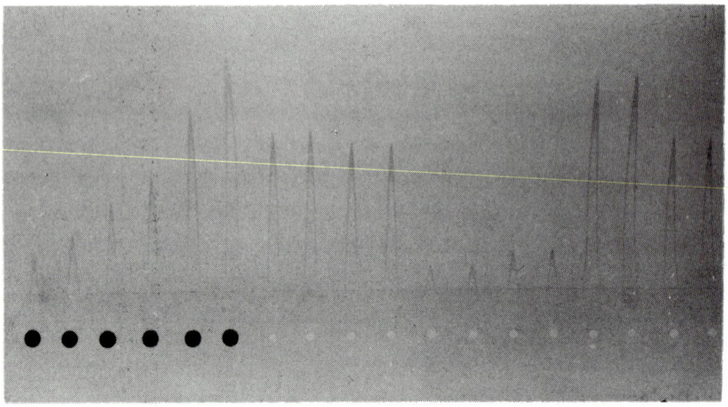

b

Figure 5.1 Rocket immunoelectrophoresis. Rocket immunoelectrophoresis of mouse serum with (*a*) antimouse IgG from rabbits and (*b*) anti-mouse alpha-2 macroglobulin from rabbits as antibodies.

(*a*) A serial dilution of mouse serum pool is applied in 5 μl in the empty middle gel. Upper gel, 3.0 μl anti IgG per square centimetre gel; lower gel, 1.5 μl anti IgG cm^{-2} gel. Anode at top. Electrophoresis was performed overnight at 2V cm^{-1} using Tris/Veronal ionic strength 0.02 and pH 8.6, anode at top. Agarose was 1 (w/v)% Litex type HSA in the Tris/Veronal buffer. Staining by conventional Coomassie brilliant blue procedure. The rockets formed are indicative of a heterogeneous charge protein migrating quite slowly with pronounced side to side diffusion and both anodically and cathodically. Several distinct subclasses are visible.

(*b*) Serial dilution of mouse serum pool applied in 5 μl: 0.01, 0.02, 0.04, 0.05, 0.1 and 0.5 μl undiluted (marked by dots). The rest of the plate contains individual mouse sera (0.05 μl applied in 5 μl) in double applications. The application gel is empty (no antibodies) and the upper gel contains 16.7 μl anti-alpha-2 macroglobulin cm^{-2} gel. Anode at top. To the gel is added 2 (w/v)% PEG 6000, conditions apart from this are as in (*a*). Sharp, well-defined rockets are formed indicative of a homogeneous high molecular weight quickly migrating protein showing practically no diffusion during the period of electrophoresis. Antibodies were produced at the Protein Laboratory.

standard curves in a plot against antigen amounts (Mueller and Potter 1979). It is important to ensure exact antigenic identity of sample components and standards as well as the buffer composition of the two should be equal. Moreover, to retain the basic correlation between height and antigen amount, all samples should be applied in the same volume, the same size of application well, and the diffusion of samples in the time between application and start of electrophoresis must be avoided, e.g. by applying the samples while the current is on. Also, the field strength from one end of the plate to the other should be invariable (same thickness of gel, exactly fitting wicks). Determinations are most accurately done with rockets between 20 and 50 mm high. Observing these rules the reproducibility can be as good as 99% (Axelsen and Bock 1983a), with typical errors in the 2–5% range (Laurell and McKay 1981). Two principal patterns obtained by ordinary rocket immunoelectrophoresis are shown in figure 5.1. Samples applied in an agarose gel without any antibody included increase the sensitivity of the rocket imunoelectrophoresis, a phenomenon so far unexplained (Broe and Ingild 1983). All amplification procedures can be used with rocket immunoelectrophoresis too, and detergents can be included for amphiphilic antigens (Bjerrum 1983c), and the principle can be extended to other affinity systems, as for example (denatured) glycoproteins into lectins (Bøg-Hansen *et al* 1977), or lectins into glycoproteins (Bøg-Hansen and Nord 1974).

The line procedure (Krøll 1983b) can be used with polyspecific anti-sera and for preparative purposes (Danielsen 1983) (see below). The reversed system (Axelsen and Svendsen 1973) where antibodies are forced into antigen-containing gel at a low pH can be used to screen different antibody preparations. Rocket line immunoelectrophoresis (Krøll 1983c) combines the simultaneous quantitation of more antigens at one time with characterisation of their mutual identity or non-identity as revealed by a common antibody or other binding activities. 'Local absorption' under the electrophoresis is a very elegant method for quantitation and characterisation (of interaction with antibodies, lectins and other ligands) in one step (Krøll 1983b) (see figure 5.6).

Crossed Immunoelectrophoresis, Two-dimensional Immunoelectrophoresis

Crossed immunoelectrophoresis for quantitative purposes (Clarke and Freeman 1967, Høiby and Axelsen 1983) is a very versatile technique and a lot of variants, stressing either the qualitative or the quantitative aspects, have been described (see below). It is a two step technique which consists of, in the standard version, a zone electrophoretic first step ('first dimension') in an agarose gel (at a relatively high voltage) followed by another electrophoresis, perpendicular to the first, forcing the separated antigens out of the first dimension gel and into an antibody-containing, 'second dimension' gel (at a relatively low voltage) in principle like Laurell's 'crossed electrophoresis' (Laurell 1965, see figure 5.2). The higher the voltage for the first dimension separation the sharper the zones obtained (Bjerrum 1983c). In the standard set-up, the technique uses the same buffer, the same type and concentration of agarose and the same electrophoresis apparatus for the first and the second dimensions (Høiby and Axelsen 1983). Thus, the technique is very easy to practice and demands only a limited amount of work (maximum 15 min per plate (personal experience)). The crucial step in the practical performance is the transfer of gel from the first dimension plate to the second dimension plate; this transfer should be completely quantitative, performed within reasonable time (half an hour) and contamination must be avoided. The second dimension electrophoresis should take place immediately after transfer to avoid diffusion of bands. When it is ensured that the mobility of the antibodies used is zero, both cathodally and anodally moving antigens can be detected and quantitated. A peculiarity quite often seen is that an antigen moves anodally in the first dimension but forms a cathodal precipitate in the second dimension, indicating that the complex of IgG and the antigen moves differently from the antigen itself.

The procedure for quantitation in crossed immunoelectrophoresis is obviously not as simple as that for rocket immunoelectrophoresis as here there is no correlation found between heights and antigen amounts, one has to measure the areas of the precipitates. This discipline has given rise to a number of manual or semi-automatic, ingeneous methods (counting squares, planimetry, weighing paper) and the most refined ones can give standard deviations as small as 1% (Høiby and Axelsen 1983). It must be remembered that the system—as the rocket immunoelectrophoretic system—should be calibrated with a standard antigen. An important point in obtaining reproducible measurements is to have a well defined 'baseline' from where to measure areas. A way of obtaining this has been described employing monospecific antibodies in the second dimension gel (Hansen *et al* 1984).

The sensitivity of crossed immunoelectrophoresis is comparable to or even better than that of rocket immunoelectrophoresis (Høiby and Axelsen 1983), see table 5.3. When a monospecific antibody is available, however, the rocket immunoelectrophoresis is quicker and easier to perform and measure, especially if a large number of samples are to be compared.

The big advantage of crossed immunoelectrophoresis is its potential for molecular characterisation, identification and resolution of complex mixtures, and the technique has accordingly been applied to a number of very different problems (see below). In the technique of crossed line immunoelectrophoresis (Krøll 1969, 1983d) the identification by the crossed procedure is combined with the quantitation, titre-determination or local absorption performed in the line part. Precipitates in the gel can be

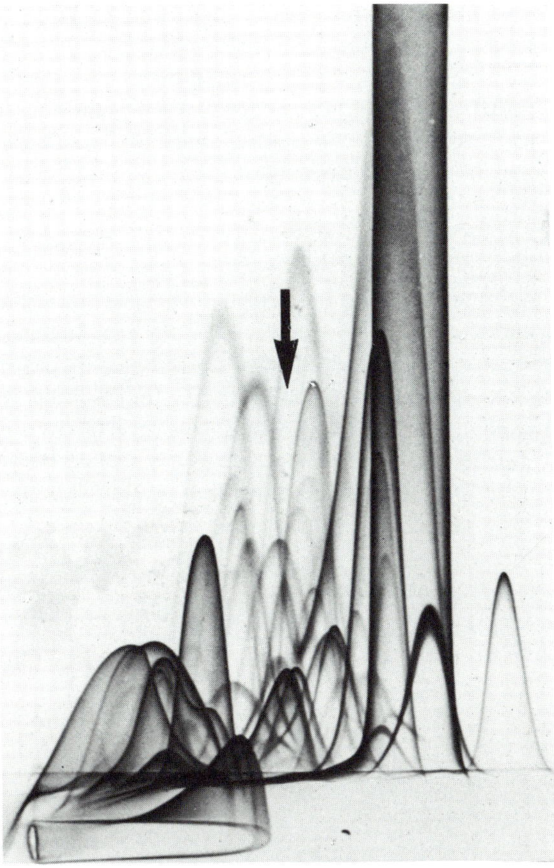

Figure 5.2 Crossed immunoelectrophoresis. Crossed immunoelectrophoresis oı 1 μl human serum pool. First dimension at 10 V cm^{-1} for 1 h, anode at right. Second dimension was run overnight at 2 V cm^{-1} anode at top. The second dimension gel contained 16.7 μl anti total human serum protein antibody from rabbits cm^{-2} gel. The gel was 1 (w/v)% agarose type HSA (Litex) in 0.02 ionic strength 8.6 pH Tris Veronal buffer. In this case a specific antibody is included in the first dimension gel. This is anti human haptoglobin from rabbits at a concentration of 25 μl per square centimetre gel. This causes haptoglobin to precipitate in the first dimension and disappear from its standard position (arrow) in the second dimension gel. The plate furthermore shows that the haptoglobin rocket is not 'finished' within the 1 h and thus not 'stationary', the rocket point being turned at right angles upon turning the electric field 90° from the original direction (see also Axelsen *et al* 1983). Visualisation by Coomassie brilliant blue R 250. Antibodies were produced at the Protein Laboratory.

characterised, either directly in the plate (enzymes, secondary antibodies, see above) or after the precipitate has been cut out (by SDS-PAGE (Norrild 1983), two-dimensional gel electrophoresis (Pødenphant *et al* 1983)), the latter technique demanding radioactive labelling of antigens (small amounts). An indispensable requirement for the successful application of crossed immunoelectrophoresis is a good antibody as the whole method relies entirely on what the antibody used 'sees'.

Tandem Crossed Immunoelectrophoresis

A number of variants from the basic principles concern the simultaneous quantification and the establishment of an antigenic relationship between two or more antigens and the identification of a specific protein in the mixture. Tandem crossed immunoelectrophoresis (Krøll 1968, Krøll 1983a) is designed to explore the antigenic relationship (identity, non-identity or partial identity) between two antigens towards one reference antibody (the same information obtained by the old double diffusion technique

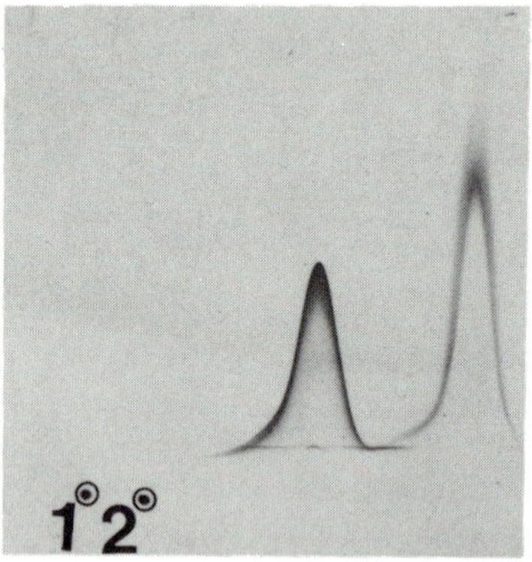

Figure 5.3 Tandem crossed immunoelectrophoresis. Tandem crossed immunoelectrophoresis of (first well) 4 μl unmodified human serum albumin (from Sigma) and (second well) 4 μl 4-iso-cyano-toluene-coupled human serum albumin. First dimension anode was at right and electrophoresis was performed for 20 min at $10 \mathrm{V\,cm^{-1}}$ followed by the second dimension electrophoresis (anode at top) at $2 \mathrm{V\,cm^{-1}}$ overnight. Antibody was $1.25\ \mu\mathrm{l\,cm^{-2}}$ gel anti-human serum albumin (DAKO, type 001) and buffer had ionic strength 0.02 pH 8.6 Tris-Veronal. Agarose was Litex type HSA at 1 (w/v)% in the electrophoresis buffer. Staining by Coomassie brilliant blue R 250 (standard procedure).

A distinct change in precipitate morphology and first dimension mobility is observed upon modifying the albumin, and a reaction of partial identity between this and the unmodified albumin can be seen indicating the destruction of antigenic determinants by the modification. Figure by courtesy of Dr K S Danielsen (The Protein Laboratory).

(Ouchterlony 1948)), i.e. to determine the distribution of antigenic determinants between the two antigens. The technique is identical to the basic crossed immunoelectrophoresis, but two application wells are used one after the other in the first dimension gel to contain the two antigens being compared. The distance between these wells should allow the precipitates eventually formed to touch each other, see figure 5.3. The degree of overlapping determines the morphology of the final combined precipitate. Three principal patterns can arise indicative of three distinctly different relationships between the two antigens towards the reference antibody

(i) Precipitates 'cross over' without interaction like two different precipitates: this indicates that no antigenic determinants are shared by the two antigens
(ii) Precipitates fuse to form a multipeak precipitate: this indicates that the two antigens are completely alike to the antibody i.e. the antibody sees the same determinants on both antigens.
(iii) Precipitates fuse, but exhibit also 'spurs' inside the precipitate; indicates that the antibody sees some common antigenic determinants in addition to some determinants specific for each antigen.

The second situation can occur, e.g. between native proteins and their precursors if the antibody employed is raised against the native species (which is often smaller than the precursor). Situations (i), (ii) and (iii) are called reactions of complete, no and partial identity, respectively (Bock and Axelsen 1972).

Intermediate Gel Technique

This technique (Svendsen and Axelsen 1972, Axelsen 1983b) is suited both for identifying specific antigens and for establishing cross reactivities of several antibodies towards a single protein. The modification in this technique takes place in the second dimension of a crossed immunoelectrophoresis: part of the gel is replaced by a gel containing another antibody. Antigens will then be precipitated in this intermediate gel according to the activity of the antibody (see figure 5.4). If for example a monospecific antibody is placed in an intermediate gel, only one precipitate will be seen here, while any activity toward this special antigen in the rest of the second dimension gel will not cause further precipitation as the antigen is already precipitated. If the titre of the intermediate gel antibody is not sufficiently high, different degrees of second dimension retentions of the corresponding precipitate are seen, from the above mentioned complete precipitation in the intermediate gel to the reaction of 'inward feet' (Axelsen 1983b). This last indication of activity in an antibody preparation towards an antigen is very sensitive and can detect cross reactivities not detectable in any other non-amplified systems (Heegaard *et al* 1984b). The method of intermediate gel immunoelectrophoresis can be extended to other immunoelectrophoretic techniques and is often successfully applied to for example fused rocket immunoelectrophoresis (see figure 5.7). The degree of retention is always dependent upon the relation between the 'concentration' of the second dimension antibody and intermediate gel antibody against the 'common' antigens, see figure 5.4.

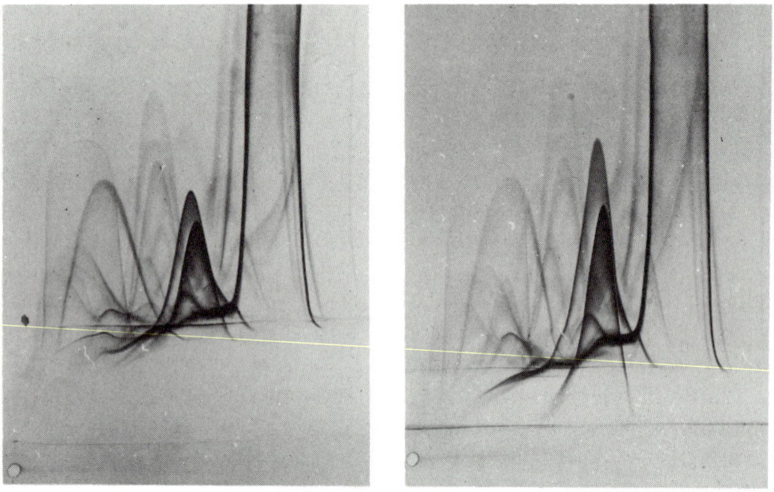

a *b*

Figure 5.4 Crossed immunoelectrophoresis with intermediate gel. (*a*) Crossed immunoelectrophoresis of 1 μl mouse serum (Balb/c, male) employing a 2 cm wide intermediate empty gel in the second dimension. First dimension electrophoresis was performed for 1 h at 10 V cm^{-1}, anode at the right. In the upper section of the second dimension gel, rabbit antibody against whole mouse serum was incorporated in a concentration of 2.4 μl cm^{-2}. Second dimension electrophoresis was at 2V cm^{-1} overnight, anode at top. Electrophoresis buffer was Tris-Veronal, 0.02 ionic strength pH 8.6. Gel was agarose (type HSA, Litex) at a concentration of 1 (w/v)%. (*b*) As (*a*), but having incorporated in the intermediate gel 28.6 μl cm^{-2} specific rabbit antimouse alpha-2 macro- globulin antiserum. A selective precipitation of this protein as compared to (*a*) is observed in the intermediate gel accompanied by a decrease of the precipitate height in the upper second dimension gel. Antibodies were prepared at the Protein Laboratory.

Any content of antigen in an intermediate gel will cause line precipitates to develop in the second dimension gel by their reaction with the corresponding antibody. This phenomenon is very often observed with certain liquid phase absorbed commercial monospecific antibodies. It can also be used voluntarily to show the fusing/non-fusing (identity/non-identity) of the line from an antigen-containing intermediate gel with the second dimension precipitate (Krøll 1969, Krøll 1983d) in the technique of crossed line immunoelectrophoresis.

Ligand Interactions

Ligand interactions may be studied with the ligand in the first dimension of a crossed immunoelectrophoresis, in the intermediate gel of a crossed immunoelectrophoresis (see figure 5.5), by incubating the whole plate (any immunoelectrophoretic method) post-electrophoretically with the ligand, by co-electrophoresing the whole complex at a suitable pH or performing local absorption in a line procedure (see figure 5.6). Generally, when studying ligand interactions, more information about the antigen is obtained when the interaction is allowed to take place during the first dimensional separation as here partly binding subfractions of the antigen for example will be clearly separated. The prerequisite for this is that the mobility of the complex is different from the mobility of the free antigen. Because of the antigenic identity between the different subfractions, they will all be clearly precipitated by the antibody in the second dimension at their different positions. This separation into microheterogeneous populations does not, of course, take place after the second dimension precipitation of the whole population. When choosing between these methods and the methods of allowing the interaction between ligand and antigen to take place in an intermediate gel, that is during the second dimension electrophoresis of a crossed immunoelectrophoresis or as a local absorption in a line procedure (figure 5.6), it should be kept in mind that the nature of the ligand and the receptor–ligand interaction might prohibit the use of one or more of these alternatives. Also the reaction might not be possible during electrophoresis or with the precipitate-bound antigen. In general, such problems can be overcome and it is often worthwhile making this effort with regard to the gain in information obtained by immunoelectrophoretic systems, using only a very small amount of antigen (examples will be given in §5.6.4).

Fused Rocket Immunoelectrophoresis

The fused rocket procedure (Harboe and Svendsen 1983, Svendsen and Rose 1970) could also be called quantitative, but it is normally used only to characterise qualitatively the content of fractions from a fractionation procedure for the antigens of interest. Aliquots of fractions are applied close together in an antibody-free gel and allowed to diffuse into the gel for some time (typically 1 h at room temperature) before they are electrophoresed into an antibody-containing gel. It is thus a one-dimensional technique, the electrophoresis taking place overnight at a relatively low voltage. In this way all fractions containing an antigen are delineated by a precipitate showing the presence of this antigen in the different fractions, and an elution profile is obtained for each antigen, see figure 5.7. This variant of the rocket immunoelectrophoresis is especially well suited to specifically follow the fate of a multitude of antigens during a fractionation procedure (figure 5.7). As mentioned above, an intermediate gel can be

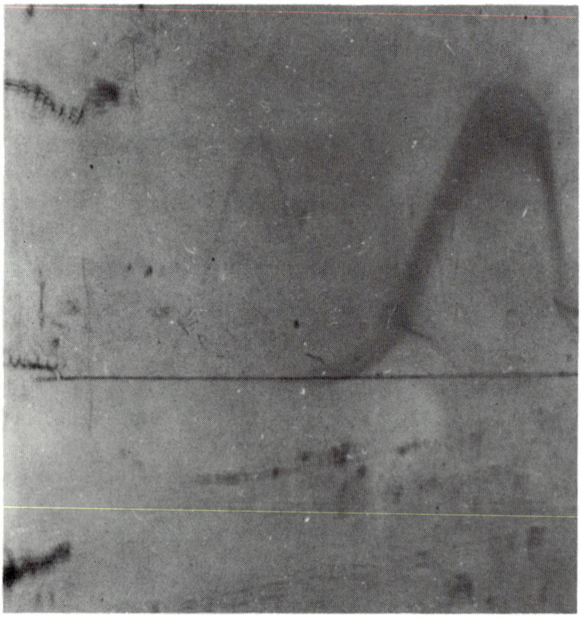

a

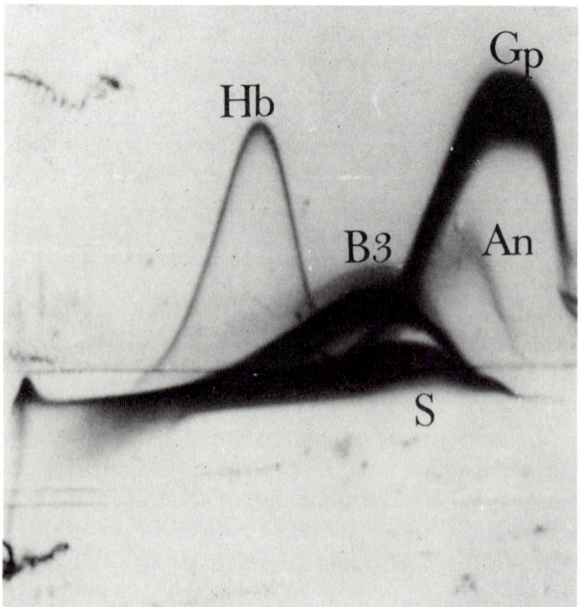

b

employed to specifically follow the fate of one antigen in relation to the whole mixture. Moreover it can be used in the presence of ampholytes, detergents and nucleotides (Bjerrum and Bøg-Hansen 1976). So it is in every respect superior to the standard methods used to follow the fractionation of for example proteins, i.e. E280 measurements, activity assays etc.

5.6 Applicability of Immunoelectrophoretic Methods

The applicability of immunoelectrophoretic methods has proven to be enormous when considering the widespread use of both basic and modified systems and these two in combination with various other methods for both analytical and preparative purposes. In most cases antigens can be analysed without modifications of the basic systems. In short, the following demands have to be met before methods can be used: (i) protein should be soluble (or rendered soluble); (ii) antibody employed should be precipitating; (iii) amount of protein should be big enough to ensure visualisation (with or without amplification of sensitivity); (iv) protein should have the right

---

Figure 5.5 Immunoelectrophoresis with ligand interaction. Detection of binding specificity of monoclonal antibodies by crossed immunoelectrophoresis with intermediate gel.
(*a*) Crossed immunoelectrophoresis of 20 μg Triton X-100 solubilised human erythrocyte membrane proteins applied in 10 μl with an intermediate gel containing 150 μl of the supernatant of a mouse hybridoma cell culture and rabbit anti-human erythrocyte membrane protein antibody (DAKO A 104), 6 $\mu l\,cm^{-2}$ of gel as second dimension antibody. The immunoelectrophoresis was performed in the presence of Triton X-100 (0.1% (v/v)) (Bjerrum and Bøg-Hansen 1976) and after electrophoresis, the plate was carefully pressed and washed in 0.1 M NaCl + 0.1% (v/v) Triton -100 (procedure performed two times) followed by overnight incubation in PBS containing 0.1% (w/v) bovine serum albumin, 0.1% (v/v) Triton X-100 and 15 μM Na_3N plus horse radish conjugated rabbit anti-mouse-IgG-antibody (DAKO P 260) (1 : 2000 (v/v)). After pressing and washing as above and drying at room temperature the plate was stained with 0.04% (w/v) 3-amino-9-ethyl-carbazole and 0.012% (w/v) hydrogen peroxide.
(*b*) As (*a*) except that the plate after electrophoresis was subjected to the normal Coomassie brillant blue R 250 staining procedure. Both plates have 1st dimension anode at the right and second dimension anode at the top.
In (*a*) bound mouse-IgG is seen specifically in two human erythrocyte membrane protein precipitates, namely glycophorin (gp) and haemoglobin (hb), revealed by the incubation with the HRP conjugated second antibody and staining for enzymes. This binding was confirmed by the corresponding ELISA experiments on purified antigens. (*b*) shows the identity of proteins precipitated: Glycophorin (Gp), ankyrin (An), Band 3 protein (B3), spectrin (S) and haemoglobin (Hb). Figure by courtesy of Dr O J Bjerrum, The Protein Laboratory.

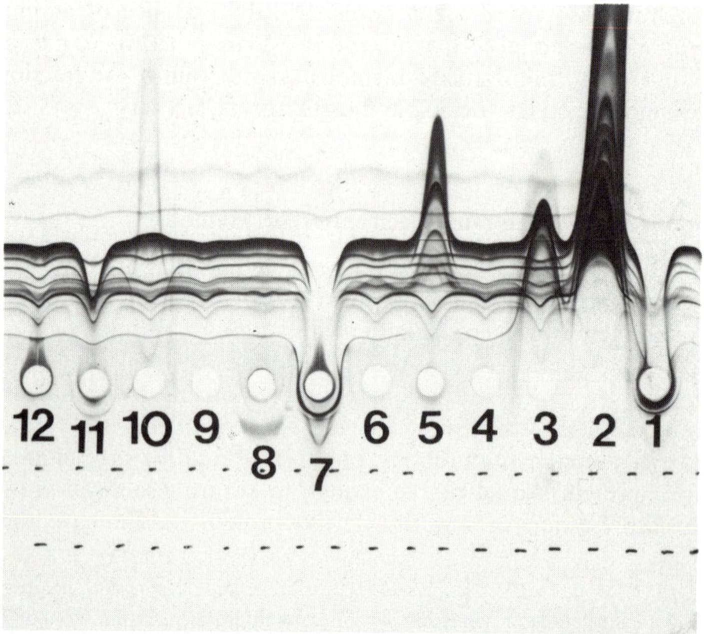

Figure 5.6 Immunoelectrophoresis with ligand interaction. Line immunoelectrophoresis with local absorption. Line immunoelectrophoresis with a line of 100 μl rat cerebrospinal fluid in the part of the gel delineated by dotted lines. The following antibodies are placed in the wells: 1, anti-rat-alpha-1 lipoprotein; 2, anti-rat-alpha-2 acute phase protein; 3, anti-rat-cerebrum; 4, anti-guinea pig immunoglobulins; 5, anti-mouse-IgA; 6, anti-mouse-IgM; 7, anti-mouse-alpha-1-lipoprotein; 8, anti-mouse-alpha-1-serum proteins; 9, anti-human-IgG; 10, anti-human-alpha-1-antitrypsin; 11, anti-human-alpha-1-lipoprotein; 12, anti-human-orosomucoid (all antibodies produced in rabbits at the Protein Laboratory). Gel was 1% (w/v) HSA (Litex) in standard Tris/veronal buffer: pH, 8.6; *I*, 0.02; all antibodies were applied in 10 μl and the second dimension antibody was 8.75 $\mu l\ cm^{-2}$ gel anti-rat serum-proteins (produced in rabbits in the laboratory). Electrophoresis was performed at 2V cm^{-1} overnight (anode at top) and followed by staining with conventional Coomassie brilliant blue R 250 staining.

This plate allows for detection of crossreactions between several antigens and one antibody, between several antibodies and one antigen, the type of crossreaction, specificity and relative titre of antibodies. Also content of absorption antigens in the antisera is detected (as 'normal' rockets).

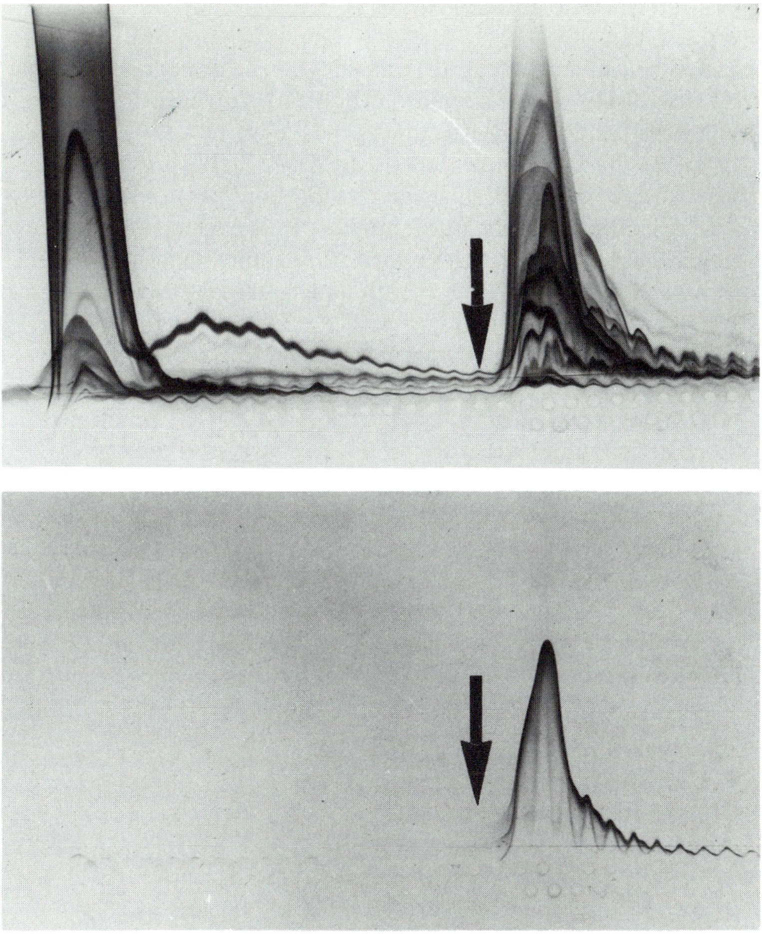

a

b

Figure 5.7 Fused rocket immunoelectrophoresis. Fused rocket immunoelectrophoresis of fractions from an affinity chromatographic separation of 0.5 ml human serum pool on 2.5 ml con A coupled to Sepharose 4B. 5 μl of each fraction was applied in the lower gel (empty) and allowed to diffuse for 1 h at room temperature before electrophoresis was performed at 2V cm^{-1} overnight (anode at top). The upper gel contained anti-total human serum protein antibody in (*a*) and anti-human haptoglobin in (*b*). The plates were stained after electrophoresis by the conventional Coomassie brilliant blue R 250 procedure.

The arrow indicates the start of column elution with the specific displacer alpha-methyl-D-glucopyranoside (0.5 M), and as can be inferred from (*b*), haptoglobin is only found in these fractions, indicative of its strong and homogeneous binding to the con A. By inspection of (*a*), it seems that several serum proteins are heterogeneous with respect to binding to con A, that is they contain both binding and non-binding molecules, as their precipitate is found in both the fractions containing non-bound material and in those containing bound proteins. Certain proteins even show intermediate binding characteristics. (Modified from Nilsson and Bøg-Hansen 1980.)

physical/chemical properties (pI outside the neutral area, size between 3×10^4 and 2×10^6 (Laurell and McKay 1981). A list of examples of these uncomplicated cases is given below.

Moreover, immunoelectrophoretic methods have proven to be most versatile with respect to further characterisation of antigens. This can be achieved either by combining the immunochemical recognition with another chromatographic or electrophoretic method other than the normal non-sieving size/charge dependent separation in agarose (before, simultaneously or after precipitation), the so called 'hybrid methods', or by a physical/chemical or biological activity-dependent characterisation of the antigens, as these antigen characteristics are most often unaffected by the procedures employed. The latter approach can also be performed before, during or after immunoelectrophoresis. In short, in these cases, immunoelectrophoresis adds a specific recognition of antigens to the characterisation performed. A few examples of this too will be given below.

As general reviews for the methodological approach to immunoelectrophoresis Axelsen 1975, 1983, Axelsen and Bock 1972, Axelsen *et al* 1973, Laurell 1972a and Verbruggen 1975b should be consulted.

Systems can be arranged according to the degree of modification. It should be remembered that the methods can be combined.

Basic Systems

Rocket immunoelectrophoresis
Crossed immunoelectrophoresis
Fused rocket immunoelectrophoresis

Elaborated Systems

Tandem crossed immunoelectrophoresis
Line immunoelectrophoresis
Crossed line, rocket line (local absorption) immunoelectrophoresis
Intermediate gel crossed immunoelectrophoresis

Modified Systems

Component modified
Antigen: carbamylation, aggregation, iodination, desialylation, etc
Antibody: carbamylation, protein A binding, second antibody binding, iodination, etc
Gel modified: see table 5.1, EEO change, percent employed, mix with other gel types
Added Factors: PEG 6000/dextrans, antigen-interacting substances (ligands, free or immobilised, affinity gels etc.
Buffer (see table 5.2)
Visualisation (see table 5.3)

Combination
- Hybrids
- Agarose
- Other gels
- Agarose and other matrices
- Two step procedures

In the following overview the types mentioned above will be dealt with but with reference to the research problems encountered.

5.6.1 Types of Antigens Applied—Without Modifications

Originally designed for the investigation of serum proteins in man (Clarke and Freeman 1967, Ganrot 1972, 1974, Laurell 1966), crossed immunoelectrophoresis has been used without modifications to analyse serum proteins from a number of other species; e.g. mice (Crowle and Miller 1981, Heegaard *et al* 1984b), rats (Add-el-Fattah *et al* 1981), nude mice (Brogren *et al* 1977), as well as other body fluids (cerebrospinal fluid from guinea pig (Boiesen *et al* 1980), human urine and human cerebrospinal fluid (Krøll 1976, Laurell 1972, Lenger *et al* 1984), to mention a few. Other major applications concern allergens (Løwenstein 1978) (see §5.6.4 and figure 5.8), microbial antigens (Axelsen 1973, 1976), often establishing cross-reactivities by intermediate gel techniques: *Neisseria gonorrhoea* (Smyth 1976), *Mycobacteria* (Roberts *et al* 1972), *Pseudomonas aeruginosa* (Høiby 1975), *Neisseria meningitidis* (Hoff and Høiby 1978a), virus proteins (polio virus capsid proteins (van Wezel *et al* 1984)) and plant proteins (beta-amylases from barley grain (Hejgaard 1976), peanut proteins (Neucere and Cherry 1975), glutamate dehydrogenases from alfalfa pollen (De Vienne *et al* 1984)).

5.6.2 Types of Antigens Applied—Involving Modifications

Membrane antigens can be analysed by immunoelectrophoresis in general only if special solubilisation procedures are employed (Bjerrum 1977, 1983a, Bjerrum *et al* 1981, Bjerrum and Bøg-Hansen 1976). Moreover, during electrophoresis, these antigens must be kept in solution and this is most often achieved by adding detergents to the system. This method, termed ‘detergent immunoelectrophoresis’, causes certain technical problems and is thoroughly dealt with by Bjerrum (1983c). This principle is of use in all basic methods like crossed and rocket immunoelectrophoresis and the following have all been analysed: e.g. lipopolysaccharides (Back *et al* 1976), glycolipids (mannose-(inositol-P)2-ceramide) (Bjerrum and Gerlach 1982), lipid haptens (galactocerebroside, cardiolipin) (Niedick 1978), maltose/glucoamylase from intestinal microvilli of pig (Sørensen *et al* 1982), human kidney brush border membrane proteins (Scherberich *et al* 1974), peptidases from pig intestinal brush border (Noren *et al* 1977), membrane proteins from the human erythrocyte (Bhakdi *et al* 1976b, Bjerrum and Bøg-Hansen 1976), bovine fat globule membrane proteins (Nielsen and Bjerrum 1977) and algal proteins (Plumley and Schmidt 1983). Some more special cases using detergent immunoelectrophoresis are described below.

Other proteins not directly applicable have been subjected to immunoelectrophoretic analysis. These include human serum IgM which was quantitated in agarose of EEO = 0 at pH 10.3 using modified antibodies of mean pI 10.3 (Grubb 1974) by rocket immunoelectrophoresis. The isoenzymes of subtilopeptidase A from *Subtilisin carlsberg* were analysed at pH 6.5 by crossed immunoelectrophoresis (Verbruggen 1975a, Verbruggen and Baeck 1976). This pH also reduced autoproteolysis. Succinylated lipomannan from *Micrococcus lysodeikticus* was analysed in the presence of Triton X-100 (Owen and Salton 1975) and lactoperoxidase from leucocyte lysosomes was analysed in agarose of low EEO (cationic protein, Johansson and Hjérten 1974). Antibodies which had developed in patients' sera against *Neisseria meningitidis* could be detected simply by using the sera as antiserum in crossed immunoelectrophoresis and comparing with a reference antiserum produced in rabbits (Hoff and Høiby 1978a, b, c). Proteins present only in trace amounts can be analysed in the basic systems when sensitive visualisation is employed (see table 5.3).

5.6.3 Characterisation of Antigens

5.6.3.1 Quantitation

As mentioned before, quantitation can be performed by rocket immunoelectrophoresis as well as by crossed immunoelectrophoresis and variants thereof. Non-monospecific antibodies can be used to quantitate an antigen in a mixture by rocket immunoelectrophoresis if the antigen in question can be identified and recognised or selectively stained: arylesterase (Brogren and Bøg-Hansen 1975a); rat liver malic enzyme (Frenkel 1974); identification by crossed line immunoelectrophoresis and sudan black staining (Dang 1984); precipitation by lectins instead of antibodies (Bøg-Hansen *et al* 1977, Owen *et al* 1977, Owen and Salton 1976). Denaturation of antigens cannot be consistently followed by rocket immunoelectrophoresis, as it may give rise to opposite results, that is, either heightening of the rockets (by destroying the determinants) or lowering them (total denaturation of a certain population of antigen molecules, Bjerrum 1983b). By using a monospecific antibody there are seldom any problems in quantitating antigens and a few examples that can be mentioned are: lipoprotein A constituents by rocket immunoelectrophoresis (Curry *et al* 1976), human IgG and IgM at pH 5 (Bjerrum *et al* 1973), human IgM at pH 10.3 (Grubb 1974) and human IgGs at pH 8.6 (carbamylation of antigens, Weeke 1968). Weeke suggested the use of carbamylated transferrin as a reference antigen when using crossed immunoelectrophoresis for quantitation (Weeke 1970). Crossed immunoelectrophoresis and planimetry (Høiby and Axelsen 1983) were used to measure arylesterase in mouse serum (Bøg-Hansen *et al* 1978c). Subfractions of human orosomucoid, each having different degrees of reactivity with concanavalin A, were also quantitated by planimetry using an incorporated baseline to ensure more accurate measurements (Hansen *et al* 1984) in crossed affino-immunoelectrophoresis (Bøg-Hansen *et al* 1975), see below.

5.6.3.2 Chemical–Physical Characterisation

A number of the so called hybrid techniques have been of use in the physical–chemical characterisation of various antigens. Typical examples of techniques combining the ordinary second dimension precipitation of a crossed immunoelectrophoresis with types of first dimension separations other than just zone-electrophoresis in agarose include

(i) In agarose gels: high % agarose, sieving of big molecules (Crowle and Miller 1981); SDS in 2% agarose (Siefring, 1978), zero EEO agarose, isoelectric focusing (Erwal 1976).
(ii) In other gels: polyacrylamide: isoelectric focusing (Skude and Jeppson 1972; Smyth 1983), isotachophoresis (Brogren 1977), SDS-PAGE (directly or indirectly Kessler 1981, Owen 1976). Sephadex: gel filtration (Nielsen 1975), phenylsepharose, hydrophobic interaction chromatography (Bjerrum 1978, 1983b), AlOH chromatography (Weeke *et al* 1975), ion exchange chromatography (Løwenstein *et al* 1976).
(iii) In agarose equipped with certain ligands, free or immobilised (see §5.5.1 and below).

These 'hybrid' methods constitute powerful alternatives to the basic procedures and are often of value in designing corresponding chromatographic experiments, as well as often replacing a specific characterisation of precipitates after immunoelectrophoresis.

pIs of antigens can be established in different ways using the immunoelectrophoretical approach, most simply by inclusion of an ion exchange gel in the first dimension procedure of a crossed immunoelectrophoresis. Jirka and Blanický (1973) analysed purified albumin and Zn-alpha-2-glycoprotein from human serum by combining isoelectric focusing in polyacrylamide rods with subsequent electrophoresis into antibody-containing agarose gels, the rod being simply incorporated into the agarose after the isoelectric focusing. The areas of precipitates which developed could be measured by planimetry and related to the amount of antigen in the band. Terminal complement complex proteins from target sheep erythrocyte membranes were subjected to isoelectric focusing in polyacrylamide rods and precipitated by electrophoresis into Triton X-100 and antibody-containing agarose (Bhakdi *et al* 1976a). Another technique established the pattern for plasma membrane proteins from lymphoid cells transformed by *Simian* virus 40 by isoelectric focusing in polyacrylamide slabs containing urea and Triton X-100 followed by washings (removing urea and ampholytes), whereafter the slab was placed on top of the antibody-containing agarose gel and electrophoresis was performed (Schmidt-Ullrich *et al* 1975). General procedures for combining isoelectric focusing with immunoprecipitation ('crossed immunoelectrofocusing') have been explained by Smyth (1983) and Söderholm *et al* (1975) using 'laying on' techniques, focusing in thin layer polyacrylamide gels and investigating diphtheria toxins, rat liver microsomal proteins, human serum proteins, and *Candida albicans* cytoplasmic antigens. The method is claimed to be particularly useful for the investigation of genetic variants and charge microheterogeneity of single proteins.

Molecular weight determinations can also be carried out, a rough estimate being readily obtainable by the 'barrier electroimmunodiffusion' (Crowle *et al* 1974). Here crossed immunoelectrophoresis is performed, having first incorporated a 'barrier gel' consisting of 4% agarose, which acts as a sieving matrix in the first dimension. This 'barrier' therefore has some impact on the first dimension migration of big proteins like alpha-2-macroglobulin and alpha-1-lipoprotein of human serum. Thin layer gel filtration can also be combined with a second dimension immunoelectrophoretic precipitation as suggested by Nielsen (1975) in an investigation of human serum proteins, but the most accurate analyses are obtained using combinations of sodium dodecylsulphate polyacrylamide gel electrophoresis (SDS-PAGE) and immunoprecipitation (Converse and Papermaster 1975). In this technique, for the analysis of both soluble proteins and membrane proteins, an intermediate gel containing a nonionic detergent (in this case Lubrol PX) is essential, replacing the SDS (anionic detergent) from the polyacrylamide gel thus avoiding the SDS in the second dimension procedure. SDS, if present might cause serious disturbance of the patterns (see Bjerum 1983c). In this crossed system, peak heights were found to correlate with the amount of protein in the band. Kessler (1981) simply washed SDS out of the polyacrylamide slab before immunoelectrophoresis. Fixed and stained slabs were run in parallel to ensure localisation of the bands. Antigens were pheromone induced *Streptococcus faecalis* proteins extracted by Triton X-100. Quantitation was performed by cutting out squares from the stained band in the polyacrylamide slab and subjecting this to rocket immunoelectrophoresis by placing the piece of acrylamide in a well and covering it with agarose. Others (Bjerrum *et al* 1975) have used a preparative PAGE step in the investigation of human erythrocyte membrane proteins and certain human serum proteins, combined with a second dimension immunoprecipitation by cutting out bands and resolubilising these with Berol EMU 043 or Triton X-100 (nonionic detergents). Incorporation of one of these non-ionic detergents throughout the gel normalised the precipitation pattern which had been heavily influenced by the presence of SDS.

The reverse approach (Norrild 1983, Norrild *et al* 1977) involves cutting out precipitates from agarose gels after crossed immunoelectrophoresis and thereafter performing SDS-PAGE. Precipitates were radioactively labelled and the SDS-PAGE was followed by autoradiography. Human serum proteins, *Herpes simplex* virus type I proteins and human erythrocyte membrane proteins were analysed in this way. The SDS-PAGE part of this two-step technique could be performed even after Coomassie brillant blue R-250 staining of the immunoplate, using the dye at 0.01% (w/v) in water. The gel-enclosed precipitate was solubilised for SDS-PAGE by ultrasonic treatment in phosphate buffer containing mercaptoethanol and SDS.

Isotachophoresis in thin layer polyacrylamide gels was combined with a second dimension immunoprecipitation in agarose in a laying on technique (Brogren 1979). Human serum and bovine thymic extracts were analysed, mainly to predict the outcome of preparative isotachophoresis and to analyse the protein mixtures by isotachophoresis.

Hydrophobicity of proteins and glycolipids (from human erythrocyte membranes, thrombocytes, serum, bovine milk fat globules and yeast plasma membranes) was investigated in a crossed immunoelectrophoretic system with detergent (Triton X-100) and hydrophobic dye (Sudan black) included in the agarose gel (Bjerrum *et al* 1982). Amphiphilic proteins were hereby detected by their binding during electrophoresis of this dye. The dye had no effect on the solubility of the antigens. Another approach is to soak the immunoplate after electrophoresis with radiolabelled non-ionic detergent e.g. Triton X-100. In this case binding by hydrophobic parts of proteins precipitated in the gel is revealed by the corresponding staining on the gel after autoradiography (Bjerrum *et al* 1981). The method of charge shift immunoelectrophoresis (Bhakdi *et al* 1977), developed from a method employing the same principle for purified proteins upon electrophoresis in agarose (Helenius and Simons 1977), was employed in the investigation of the amphiphilicity of human serum aryl esterase (Bøg-Hansen *et al* 1978b) and in a study of human serum proteins and human erythrocyte membrane proteins (Bhakdi *et al* 1977). In this system three first dimension electrophoreses are run in parallel in the presence of non-ionic, non-ionic plus anionic and nonionic plus cationic detergents respectively. A shift in first dimension mobility from one first dimension to the other indicates binding of detergent and thus amphiphilicity of the protein. In this case serum arylic esterase was found to be amphiphilic and furthermore it was inhibited by the presence of detergents (Bøg-Hansen *et al* 1978b). Alpha-1-lipoprotein was thus established as the only amphiphilic serum protein in man (Bhakdi *et al* 1977). Jørgensen (1977) used this system to analyse synaptosomal membrane proteins. Hydrophobicity can also be analysed in another crossed immunoelectrophoretic system using agarose mixed with alkyl- or aryl-substituted Sepharose as a first dimension gel. Amphiphilic proteins will interact with this type of matrix and their first dimension mobility will change accordingly (Bjerrum 1978). This also suggested the inclusion of a hydrophobic matrix in an intermediate gel. This method, applied in the above study of human serum proteins, erythrocyte membrane proteins and human intestinal brush border aminopeptidase, was also used to predict the outcome of hydrophobic interaction chromatography of human serum proteins (Bjerrum *et al* 1980).

The proportionality of sialic acid substitution in human serum glycoproteins and the free electrophoretic mobility of these proteins was investigated by Bordas *et al* (1981) simply by comparing rocket immunoelectrophoretic quantitation with quantitation by single radial immunodiffusion of native and partially desialylated proteins, the latter method being independent of electrophoretic mobility, thus obtaining a relative expression of the contribution of bound salic acid to the mobility of the protein. Crossed immunoelectrophoresis of partially desialylated proteins can also be used to get an impression of sialic acid substitution. A big shift in the first dimension migration was seen for alpha-1-acid glycoprotein from mouse serum upon *Clostridium perfringens* desialylation in this system (Heegaard *et al* 1984b). Classical immunoelectrophoresis (Grabar and Williams 1953) was employed in a study of the degradation of orosomucoid from human serum by a number of degradative enzymes (Krotoski 1983). With this method non-participation of the sugar moieties of this heavily glycosylated protein in antigenicity was established as has been done by others before (e.g. Brogren and Bøg-Hansen 1975b). The two dimensional structure of human ceruloplasmin, as indicated by its antigenic structure followed by different fractionation procedures and crossed, crossed line and fused rocket immunoelectrophoresis, was investigated by Løwenstein and coworkers (1975). They were able to propose a model of this protein using only these immunoelectrophoretic methods in combination with automatic amino acid analysis, molecular weight determinations with Sephadex G 150, affinity chromatography with con A Sepharose and hydroxyapatite chromatography.

5.6.4 Detection of Biological Activity: Binding of Ligands

Biological activity expressed in the binding of certain macromolecular or smaller ligands has been utilised and investigated in a big number of modifications of basic immunoelectrophoretic systems. The following may be listed as examples:

(i) Lectins–glycoproteins (Bøg-Hansen 1983a, Bøg-Hansen *et al* 1983), monoclonal antibodies (Bjerrum 1984) (see figure 5.5), ions (Ramlau and

Bock 1979), substrate inhibitor (Bleyl and Peichl 1980, Hagen 1981a, b), virus–virus receptor protein (Milgrom *et al* 1967).
(ii) Ankyrin and Band 3 of human erythrocyte membranes (Pluzek and Bjerrum 1982), neurotransmitter–receptor protein and ^{125}I-neurotoxin (Mattson *et al* 1979), ions (Takesue 1975), alpha-1-lipoprotein and arylesterase (Brogren and Bøg-Hansen 1975a).
(iii) IgE–allergen (Løwenstein 1978), Triton X-100-hydrophobic proteins (Bjerrum *et al* 1981).

These few examples have been divided as follows:

(i) interaction taking place in the first dimension of a crossed immunoelectrophoresis, the ligand being included in the first dimension gel;
(ii) interaction taking place before and allowed to be maintained during electrophoresis at a suitable pH;
(iii) interaction taking place after the electrophoresis on precipitate-bound proteins has finished.

Generally speaking, one can allow interaction to take place before (ii), during (i) or after (iii) electrophoresis. During electrophoresis in two-step procedures, interaction can either take place in the first dimension or in the second dimension. A reaction taking place after electrophoresis demands a specific visualisation procedure. On the other hand, sensitivity can be increased this way. Thus, depending upon the aim of the analysis the binding ability of different antigens has been investigated in different modifications of the basic systems (membrane proteins, Bjerrum *et al* 1981; glycoproteins, Bøg-Hansen 1979, 1983a, Bøg-Hansen *et al* 1983c; allergens, Løwenstein 1978).

The binding of various glycoproteins to a large number of lectins has been thoroughly studied and used in immunoelectrophoretic systems (Bøg-Hansen 1979, 1983a, b Bøg-Hansen *et al* 1983). The lectin concanavalin A (con A), specific for certain oligosaccharides containing mannose and/or glucose (Baenziger and Fiete 1979, Goldstein *et al* 1965, Ogata *et al* 1975) was employed in the study of human serum proteins in a crossed immunoelectrophoretic system in an intermediate gel. Con A was utilised either as a free lectin or an immobilised one, bound to Sepharose (analog to the intermediate gel crossed immunoelectrophoresis, see §5.5.1; Bøg-Hansen 1973, Bøg-Hansen and Brogren 1975). This approach and the other method of placing the lectin in the first dimension of a crossed immunoelectrophoresis were used as a very powerful means of revealing serum glycoprotein sugar-based microheterogeneity (Bøg-Hansen *et al* 1975). This was because any binding of con A through the sugar part of the glycoprotein could be revealed as different degrees of first dimension migration retardation corresponding to—

(i) the total affinity of the molecule for the lectin and

(ii) the number of lectin molecules bound per molecule of protein.

These two characteristics are dependent on the number and the type of sugars attached to the amino acid backbone and variations in the binding of lectins within a population of seemingly homogeneous molecules therefore reflect differences in their glycosylation. The precipitates obtained are still a measure of the amount of protein present and these methods have later been used in a large number of clinical studies on cancer and inflammation, analysing especially the so-called acute phase reactants of human serum (Kushner 1982) and rat serum (Koj *et al* 1982). This was in order to refine the measurements and thus improve the diagnosis power (Cooper and Stone 1979, Hansen *et al* 1984, Smith *et al* 1979) of procedures used to follow the development of and status of inflammation (Garnot 1974) and diagnosis of cancer (Baskies 1980). The uses of these methods can be divided into three general areas

(i) determination of lectin–glycoprotein binding constants (Bøg-Hansen and Takeo 1980), which has also been done by a line immunoelectrophoretic procedure using local absorption (wheat germ agglutinin and glycophorin, Heegaard *et al* 1983);
(ii) prediction (with some inherent limitations, Nilsson and Bøg-Hansen 1980) of the outcome of affinity chromatographic experiments (Bayard and Kerckaert 1980, Bøg-Hansen 1978a);
(iii) quantitation of subpopulations for clinical purposes (e.g. Hansen *et al* 1984, Raynes 1982).

Monoclonal antibodies have also been used in immunoelectrophoretic techniques as an intermediate gel labelling technique (see figure 5.5) (Bjerrum *et al* 1984). As monoclonal antibodies are normally non-precipitating, they cannot be used directly, but the approach mentioned here might be of use in characterising the epitope heterogeneity of a population of molecules.

Allergens can be detected in the technique of crossed radioimmunoelectrophoresis (CRIE, see figure 5.8, Løwenstein 1978), where a mix of potential allergens (e.g. pollen, house dust mites etc) is subjected to crossed immunoelectrophoresis with antibodies against this mixture. After electrophoresis, plates are incubated with the patient's serum and thereafter with radioactively labelled rabbit anti-human IgE. Upon autoradiography any protein allergic to the patient will be revealed on the film through the binding of the patients's IgE to the antigens in the precipitate (Løwenstein and Weeke 1976).

The principles of detecting ligand-binding activity described above have been extended to several cases of biologically active proteins as, for example, membrane receptors (Bjerrum *et al* 1981) (demanding inclusion of detergent in the gels as described above). Also, in this case, purification schemes have been proposed on the basis of these experiments (Bjerrum and

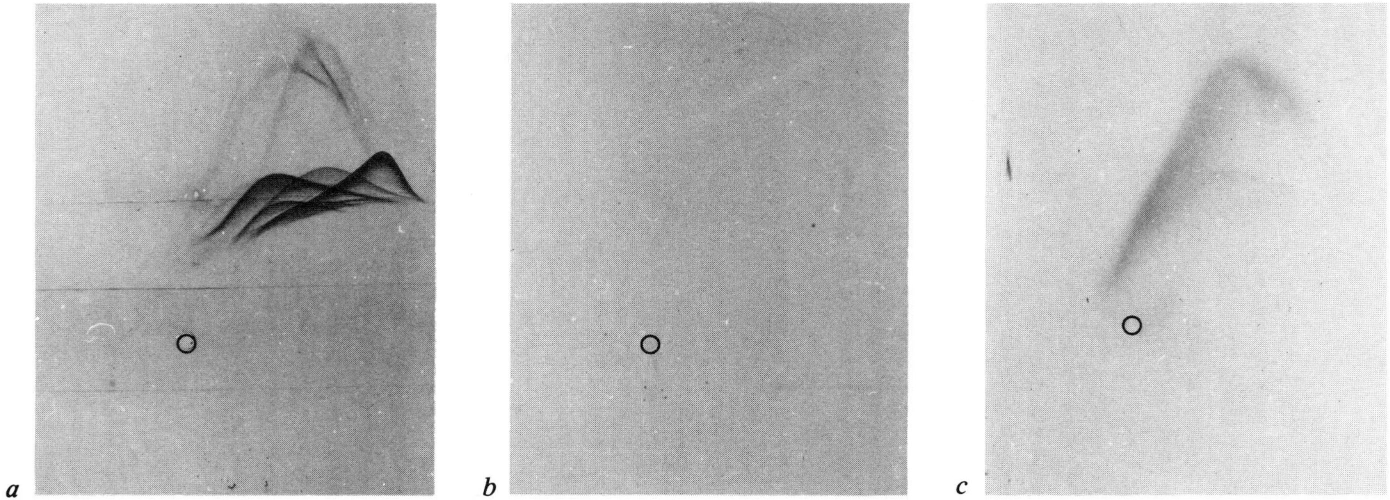

Figure 5.8 CRIE. Crossed radio immunoelectrophoresis of a 60 μg freeze dried extract of purified *Dermatophagoides pteromyssimus* (house dust mite). The first dimension was performed for 25 min at 10 V cm^{-1}, anode at right, and the second dimension at 2V cm^{-1} overnight, anode at top, both in agarose type HSA (Litex) at 1 (w/v)% in Tris-Veronal pH 8.6, ionic strength 0.02. The upper second dimension gel contained 5.7 $\mu l\, cm^{-2}$ and the lower second dimension gel 3.3 $\mu l\, cm^{-2}$ of rabbit antibodies against *Dermatophagoides pteromyssimus* culture extract. After electrophoresis plate (*b*) was incubated with normal human serum, plate (*c*) with an allergic patients serum and plate (*a*) was stained with Coomassie brilliant blue R 250. (*b*) and (*c*) were left soaking in the sera overnight in a phosphate buffer at pH 7 at room temperature, and then both were incubated with ^{125}I-labelled anti-human IgE produced in rabbits overnight in pH 7 phosphate buffer at room temperature.

After this, autoradiography at $-20°C$ revealed binding of the patient's IgE to precipitate-bound allergens (see (*c*)), while the control ((*b*), normal sera) showed no binding activity. This general procedure is found in Løwenstein 1978. (Figure: courtesy of Dr P Lind, The Protein Laboratory.)

Gianazza 1983). Another approach was used by Pluzek and Bjerrum (1982) where a pH dependent association in the sample (human erythrocyte membrane proteins) was shown to take place between ankyrin and band 3 protein, resulting in a soluble complex which was detected by the antibodies. By changing the pH, one precipitation 'arc' (the complex) could be split into two; the two proteins, now dissociated, were precipitating separately. This approach is only of use in cases where both components of the interacting system are recognised by the antibody employed and the complex is soluble. Moreover, the approach only yields qualitative information (complex or no complex observed).

5.6.5 Detection of Biological Activity: Enzymes

Biological activity is most directly exploited in the detection of specific enzymes by their enzymatic activity (Brogren and Bøg-Hansen 1975a, Owen and Smyth 1977). This is possible because most enzymes retain their activity even after precipitation by antibodies in agarose gels. As the amount of protein which is able to be detected is much lower with this method than by general protein stains (see table 5.3, Brogren and Bøg-Hansen 1975a) and specific visualisation is ensured, these methods are very useful.

Generally enzymes can either be precipitated as such by a corresponding antibody, coprecipitated via another protein to which it is associated and to which precipitating antibodies are present, or 'entrapped' non-specifically in a dense immunoprecipitate (Brogren and Bøg-Hansen 1975a). Histochemical methods for detection of enzymatic activity have often been directly applied to agarose-enclosed enzymes (Gomori 1952, Nielsen and Bjerrum 1977). In normal human serum a high density lipoprotein-associated aryl esterase has been detected by crossed immunoelectrophoresis and rocket immunoelectrophoresis. The serum was stained with 5-chloro-O-toluidine (fast red TR salt (Sigma)) and alpha- and beta-naphthyl acetate as co-substrates, giving a diazocompound as the product, resulting in a bordeaux colouration of the above enzyme and an orange colouration of another enzyme, cholinesterase. The latter enzyme was not stained by Coomassie dye (amount precipitated too low) (Brogren and Bøg-Hansen 1975a, Back *et al* 1976). Human serum lipoprotein-associated arylesterase was found (by charge shift crossed immunoelectrophoresis (see §5.5.1)) to be an amphiphilic protein, inhibited by detergents (Bøg-Hansen 1978b). Amylases from barley and from bacteria were studied by anti-alpha- and beta-barley-amylase antibodies from rabbits by *in situ* immunabsorption in the well of an agarose gel zone electrophoresis (Daussant and Carfantan 1975). Microbial glycosidases were investigated by Every and Ashworth (1974). Urinary carboxylic esterases were detected by alpha-naphthyl acetate and 5-chloro-O-toluidine (Bøg-Hansen *et al* 1973). In this study the presence of both antibodies and con A did not exclude the specific enzymatic staining in the crossed immunoelectrophoretic system used. In the serum of mice, lipopolysaccharide (LPS) was shown by planimetry to induce arylesterase production (thereby classifying arylesterase in this species as an acute phase reactant, Kushner 1982) (Bøg-Hansen *et al* 1977). In this study arylesterase was found, furthermore, to coprecipitate totally with high density lipoprotein in contrast to the pattern in human serum. Other examples include studies on rat liver microsomal enzyme polymorphy upon treatment with various detergents and proteases, giving rise to a number of apparently different antigens having retained or lost their enzymatic activity. Conclusions could then be drawn regarding the arrangement of native enzymes within the microsomal membranes studied (Raftell and Blomberg 1974). In an extension of this study, detergent-solubilised rat liver microsomal enzymes were analysed by electrofocusing and immunoprecipitation (by the method of Söderhom *et al* 1975) revealing pI heterogeneity of these enzymes (Berzins *et al* 1976).

5.6.6 Cross reactivity

Reviewed by Axelsen 1973, Axelsen and Bock 1973, 1983b and Bock and Axelsen 1983, crossed tandem immunoelectrophoresis, rocket line immunoelectrophoresis and intermediate gel crossed immunoelectrophoresis constitute very powerful means of detecting even the slightest cross reactivities of, e.g. two antigens, towards a common antiserum. The degree of cross reactivity in this case depends on the relation between the antigen producer and the antibody producer; this has been studied with mammals (Kolos and Kellen 1972) and avian species as the producers of antibodies against mammalian antigens (Jensenius *et al* 1981). As a rule the more distant the antibody producer is from the antigen suppliers, the fewer the determinants which are immunogenic and the larger the crossreactivity that is found.

Microbial antigens have been studied to a large extent, using rabbit antibodies. These have been found to be very useful (Axelsen 1973), giving well defined precipitates for a large number of microbial antigens. Membrane proteins from *Enterobacteriae* (eleven strains) were studied in crossed immunoelectrophoresis (Hofstra *et al* 1980) and from *Neisseria meningitidis* and 27 other species in crossed immunoelectrophoresis, tandem crossed immunoelectrophoresis and crossed line immunoelectrophoresis (Hoff and Høiby 1978c) in both cases using rabbit antisera. The double diffusion method of Ouchterlony (still in use for this purpose, e.g. Hofstra *et al* 1981) was compared to tandem, line and crossed line immunoelectrophoresis and rocket line immunoelectrophoresis with *in situ* (local) antibody absorption for baboon albumin, human serum albumin and rat albumin using rabbit antibodies (Axelsen and Bock 1973). *In situ* absorption in rocket line and crossed line immunoelectrophoresis were found to be especially well suited for the analysis of complex antigen mixtures (see figure 5.6).

5.6.7 Preparative Purposes

For preparative purposes, immunoelectrophoresis can also be a method of choice, when purity is more mandatory than quantity. This is the case for purifying antigens to produce antibodies, and for this an agarose-enclosed precipitate can be used directly as immunogen, choosing the same species for the antibody production as the one supplying the antibody for the immunoelectrophoresis. In this way the animal will only respond to the antigen in the complex (the amount of antigen obtained by this technique is usually in the order of μg). As pointed out by Crowle *et al* (1972), even if very small quantities of antigen are recovered (20–100 ng), the orientation of the antigen in the precipitate seems to enhance the immune response compared with that of the 'free' antigen (also observed by Krøll 1976). Also, subsequent 'boosting' of the animal after immunisation with a minute amount of precipitate-bound antigen with a mixture of proteins containing the antigen in question, can give rise to production of antibodies of the specificity desired. Precipitates can be cut out of the wet gel (illuminated by incident light) containing only the immunoprecipitated protein. In this way, one can use a polyspecific or crossreacting antibody, reactive against the antigen in question for the production of a monospecific antibody (Shivers and James 1967, Krøll and Anderson 1976, Harboe and Closs 1983). This of course demands knowledge of the identity of the precipitates on the plate. This method was also used by Bradwell *et al* (1976), who subjected only partly purified antigen to crossed immunoelectrophoresis against an oligospecific antibody. From the precipitates, pure thyroxine binding protein was recovered and used for the production of a monospecific antibody. The freshness and small quantity of antigen is stressed as being responsible for the high-titred, monospecific antibody production.

5.7 Advantages and Drawbacks

To summarise, the immunoelectrophoretic techniques described here have quite a few advantageous features. They can simultaneously characterise a multitude of antigens in a mixture with respect to:

electrophoretic mobility, microheterogeneity, isoelectric points, relative amounts, behaviour upon various fractionation procedures, molecular size, tissue specificity, subcellular localisation, affinities for ligands, enzymatic activity, hydrophobicity, biological activity, nature of antigen (lipoprotein, glycoprotein, etc.), homology with other antigens, species specificities and complex formations.

This can all be done at mild or controlled denaturing conditions, without purifying the antigen, with a big resolution, and done very rapidly for any antigen which is or can be rendered immunogenic. Drawbacks are few, but the prominent one which must be mentioned is that these methods demand a precipitating antibody—this is not always as simple as it sounds. Moreover, all of the methods presented can only tell what the antibody sees, which means that the interpretation of results might sometimes be very difficult and/or demand another (better) antibody. For separation, characterisation and specific detection with very high sensitivity, the only comparable method today seems to be the electroblotting procedure (see e.g. Bjerrum and Larsen 1983, Gershoni and Palade 1983).

References

Abd el F Hah M, Scherer R, Fonad F M and Ruhenstroth-Bauer G 1981 *Cancer Res.* **41** 2198–555

Abbelev G I and Karamova E R 1984 *Anal. Biochem.* **142** 437–44

Andersen M M, Lihme A O F and Bøg-Hansen T C 1982 *Lectins, Biology, Biochemistry and Clinical Biochemistry* vol. II ed. T C Bøg-Hansen pp 475–85 (Berlin: de Gruyter)

Atazzi M Z 1975 *Immunochem.* **12** 423–38

Araki C 1956 *Bull. Chem. Soc. Japan* **29** 543–4

Arnon R 1975 *in* Sela M vol. I 1973 87–159

Arnott S, Fulmer A, Scott W E, Dea E M C, Moorhouse R and Rees D A 1974 *J. Mol. Biol.* **90** 269–84

Arquembourg P C 1975 *Immunoelectrophoresis* (Basel: Karger)

Augustinsson K B 1968 *Biochim. Biophys. Acta* **159** 197–200

Axelsen N H 1973 *Infect. Immun.* **7(6)** 949–60

—— 1975 ed. *Quantitative Immunoelectrophoresis, New Developments and Applications* (*Scand. J. Immunol.* **4** suppl. 2) (Oslo: Universitetsforlaget)

—— 1976 *Scand. J. Immunol.* **5** 177–90

—— 1983a ed. *Handbook of Immunoprecipitation-in-gel Techniques* also *Scand. J. Immunol.* **17** suppl. 10 (Oxford: Blackwell)

—— 1983b, *in* Axelsen 1983a pp 141–9

Axelsen N H and Bock E 1972 *J. Immunol. Meth.* **1** 109–22

—— 1973 *J. Immunol. Meth.* **2** 393–401

—— 1983a *in* Axelsen 1983a pp 103–6

—— 1983b *in* Axelsen 1983a pp 177–82

Axelsen N H, Bock E, Larsen P, Blirup-Jensen S, Svendsen P J, Pluzek K, Bjerrum O J and Ramlau J 1983 *in* Axelsen 1983a pp 87–96

Axelsen N H, Krøll J and Weeke B 1973 ed. *A Manual of Quantitative Immunoelectrophoresis, Methods and Applications* also *Scand. J. Immunol.* **2** suppl. 1 (Oslo: Universitetsforlaget)

Axelsen N H and Svendsen P J 1973 *in* Axelsen 1973 155–7

Back U, Møller B B and Bøg-Hansen T C 1976 *The Lancet* July 24 188–91

Baenziger J U and Fiete D 1979 *J. Biol. Chem.* **254** (7) 2400–7

Baskies A M, Chretien P B, Weiss J F, Makuch R W , Beveridge R A, Catalona W J and Spiegel H E, 1980 *Cancer* **45** 3050–60

Bayard B and Kerckaert J P 1980 *Biochem. Biophys. Res. Comm.* **95** 777–794

Berzins K, Blomberg F, Kjellgren M, Smyth C and Wadström T 1976 *FEBS-Lett.* **61** (1) 77–80

Bhakdi S, Bhakdi-Lehnen B and Bjerrum O J 1977 *Biochim. Biophys. Acta* **470** 35–44

Bhakdi S and Bjerrum O J 1983 *in* Axelsen 1983a pp 313–23

Bhakdi S, Bjerrum O J and Knüffermann H 1976b *Biochim. Biophys. Acta* **446** 419–431

Bhakdi S, Ey P and Bhakdi-Lehnen B 1976a *Biochim. Biophys. Acta* **419** 445–7

Bjerrum O J 1977 *Biochim. Biophys. Acta* **472** 135–95

—— 1978 *Anal. Biochem.* **89** 331–48

—— 1983a *Electroimmunochemical Analysis of Membrane Proteins* (Amsterdam: Elsevier)

—— 1983b *in* Axelsen 1983a 303–12

—— 1983c *in* Bjerrum 1983a pp 4–44

Bjerrum O J, Bhakdi S, Bøg-Hansen T C, Knufermann H and Wallach D F H 1975 *Biochim. Biophys. Acta* **406** 489–504

Bjerrum O J, Blirup-Jensen S and Larsen P 1980 *Prot. Biol. Fluids* **27** 775–80

Bjerrum O J and Bøg-Hansen T C 1976 *Biochemical Analysis of Membranes* ed. A H Maddy (London: Chapman and Hall) pp 378–426

Bjerrum O J and Gerlach J H 1982 *Proc. Lipidforum* 7–15

Bjerrum O J, Gerlach J H, Bøg-Hansen T C and Hertz J B 1982 *Electrophoresis* **3** 89–98

Bjerrum O J, Ingild A and Løwenstein H 1974 *Immunochem.* **11** 797–802

Bjerrum O J and Gianazza E 1983 *in* Bjerrum 1983a pp 127–55

Bjerrum O J, Ingild A, Løwenstein H and Weeke B 1973 *Clin. Chim. Acta* **46** 337–43

Bjerrum O J and Larsen K P 1983 *in* Tschesche 1983 pp 79–124

Bjerrum O J, Ramlau J, Bock E and Bøg-Hansen T C, 1981 *Membrane Receptors* ed. S Jacobs and P Cuatrecasas (London: Chapman and Hall) pp 116–56

Bjerrum O J 1984 *unpublished* The Protein Laboratory, Copenhagen University

Bleyl H and Peichl H 1980 *in* Radola 1980 pp 733–42

Blirup-Jensen S 1982 personal communication, The Protein Laboratory, University of Copenhagen

Bock E and Axelsen N H 1972 *J. Immunol. Meth.* **2** 75–87

—— 1983 *in* Axelsen 1983 pp 183–7

Bøg-Hansen T C 1973 *Anal. Biochem.* **56** 480–8

—— 1979 *Les Colloques de l'INSERM* **86** 399–416

—— 1983a *in* Axelsen 1983a pp 243–53

—— 1983b *Solid Phase Biochemistry, Analytical and Synthetic Aspects* ed. W H Scouten (New York: Wiley) pp 223–51

Bøg-Hansen T C, Bhakdi S, Back U and Hanel K H 1978b *Prot. Biol. Fluids*, Proc. 25th Coll. ed. H Peeters pp 251–4

Bøg-Hansen T C, Bjerrum O J and Brogren C H 1977 *Anal. Biochem.* **81** 78–87

Bøg-Hansen T C, Bjerrum O J and Ramlau J 1975 *in* Axelsen 1975 pp 141–8

Bøg-Hansen T C and Brogren C H 1975 *in* Axelsen 1975 pp 135–9

Bøg-Hansen T C, Brogren C H and Rostgaard M 1973 *J. Int. Res. Comm.* **1** 9
Bøg-Hansen T C, Krogh H H and Back U, 1978c *FEBS-Lett.* **93** (1) 86–90
Bøg-Hansen T C and Nord M 1974 *J. Biol. Education* **8** (3) 167–73
Bøg-Hansen T C, Prahl P and Løwenstein H 1978a *J. Immunol. Meth.* **22** 293–307
Bøg-Hansen T C and Takeo K 1980 *Electrophoresis* **1** 67–70
Bøg-Hansen T C, Teisner B and Hau J 1983 *in* Tschesche 1983 pp 125–48
Boiesen M, Krüger C and Bøg-Hansen T C 1980 *Natural Toxins* ed. D Eaker and T Wadström (Oxford: Pergamon) pp 355–60
Bordas M C, Biou D R, Feger J M, Durand G, Joziasse D H and van den Eijnden D H 1981 *Clin. Chim. Acta* **116** 17–24
Bradwell A R, Burnett D, Ramsden D B, Burr W A, Prince H P and Hoffenberg R, 1976 *Clin. Chim. Acta* **71** 501–16
Broe M K and Ingild A 1983 *in* Axelsen 1983 pp 255–8
Brogren C H 1977, *Electrofocusing and Isotachophoresis* ed. B J Radola and D Graeslin (Berlin: de Gruyter) pp 549–58
Brogren C H 1979 *Electrophoresis* ed. B J Radola and D Graeslin (Berlin: de Gruyter) pp 549–58
Brogren CH and Bøg-Hansen T C 1975a *in* Axelsen 1975 37–51
—— 1975b *in* Axelsen 1975 pp 59–65
Brogren C H, Warren H S, Nielsen E and Rygaard J 1977, *Proc. 2nd. Int. Workshop on Nude Mice* (Stuttgart: Fischer) pp 157–65
Cann J R 1975 *Immunochemistry* **12** 473–6
Carey C, Wang C-S and Alaupovic P 1975 *Biochim. Biophys. Acta* **401** 6–14
Ceska M 1968 *Immunology* 15 837–43
Chang C H and Inglis N R 1975 *Clin. Chim. Acta.* **65** 91–7
Clarke H G M and Freeman T 1967 *Prot. Biol. Fluids* **14** 503–9
Clausen J 1981 *Immunochemical techniques for the identification and estimation of macromolecules* 2nd edn (Amsterdam: Elsevier)
Converse C A and Papermaster D S 1975 *Science* **189** 469–72
Cooper E H and Stone J 1979 *Adv. Cancer Res.* **30** 1–44
Crowle A J 1961 *Immunodiffusion* (New York: Academic)
Crowle A J, Atkins A A and Revis G J 1974 *J. Immunol. Meth.* **4** 173–88
Crowle A J and Miller J L 1981 *J. Immunol. Meth.* **43** 15–28
Crowle A J, Revis G J and Jarrett K 1972 *J. Immunol. Comm.* **1 (4)** 325–36
Crumpton M J 1975 *in* Sela 1975 vol. II pp 1–78
Curry M D, Alaupovic P and Suenram C A 1976 *Clin. Chem.* **22 (3)** 315–22
Danielsen E M 1983 *in* Bjerrum 1983a pp 157–69
Dang Q Q, Camare R and Douste-Blazy P 1984 *Electrophoresis 1984* ed. V Neuhoff (Weinheim: Verlag-Chemie) pp 289–90
Darcy D A 1972 *Clin. Chim. Acta* **38** 329–37
Daussant J and Carfantan N 1975 *J. Immunol. Meth.* **8** 373–82
De Vienne D, Duval M F, Savina F and Daussant J 1984 *C. R. Acad. Sci., Paris* **299** III (11) 475–80
Erwall K 1976 *J. Immunol. Meth.* **12** 103–15
Every D and Ashworth J M 1974 *Biochem. J.* **143** 785–7
Feinstein A 1976 *Chromatographic and Electrophoretic Techniques* vol II 4th edn, ed. I Smith (London: Heinemann)
Frenkel R 1974 *Arch. Biochem. Biophys.* **162** 392–400

Ganrot K 1974 *Scand. J. Clin. Lab. Invest.* **34** 75–81
Ganrot P O 1972 *in* Laurell 1972 pp 67–70
Gershoni J M and Palade G E 1983 *Anal. Biochem.* **131** 1–15
Glad C 1983 *in* Axelsen 1983a pp 205–8
Goldstein I J, Hollerman C E and Smith E E 1965 *Biochemistry* **4** 876–83
Gomori G 1952 *Microscopic Histochemistry, Principles and Practice* (Chicago Press)
Grabar P and Williams C A 1953 *Biochim. Biophys. Acta* **10** 193–4
Grieninger G and Granick S 1975 *Proc. Natl Acad. Sci. USA* **72** 5007–11
Grubb A O 1972 *Immunochemistry* **0** 635–45
—— 1974 *J. Immunol.* **112** (4) 1420–5
Hagen I, Brosstad F, Solumn N O and Korsmo R 1981a *J. Lab. Clin. Med.* **97** 213–20
Hagen I, Brosstad F, Gogstad G O, Solumn N O and Korsmo R 1981b *Thromb. Res.* **27** 77–82
Hansen J E S, Lihme A O F and Bøg-Hansen T C 1984 *Electrophoresis* **5** 196–201
Harboe M and Closs O 1983 *in* Axelsen 1983a pp 353–9
Harboe N M G and Ingild A 1973 *in* Axelsen *et al* 1973 pp 161–4
—— 1983 *in* Axelsen 1983a pp 345–51
Harboe N M G and Svendsen P J 1983 *in* Axelsen 1983 pp 107–12
Heegaard P M H, Christensen U and Bjerrum O J 1983 *Lectins, Biology, Biochemistry, Clinical Biochemistry* vol. III ed T C Bøg-Hansen and G Spengler (Berlin: de Gruyter) pp 387–96
Heegaard N H H, Hebsgaard K and Bjeerum O J 1984a *Electrophoresis* **5** 263–9
Heegaard P M H, Bienvenu J, Sabbagh I, Delvil A, Laurent P and Bøg-Hansen T C in 1984b *Marker Pɩoteins in Inflammation* vol II ed P Arnaud, J Bienvenu and P Laurent (Berlin: de Gruyter) pp 539–48
Hejgaard J 1976 *Physiol. Plant.* **38** 293–9
Helenius A and Simons K 1977 *Proc. Natl. Acad. Sci. USA* **74** 529–32
Hellsing A 1973 *Prot. Biol. Fluids* **21** 579–83
Hjelm H, Hjelm K and Sjöquist J, 1972 *FEBS-Lett.* **28** 73–6
Hoff G E and Høiby N 1978a *Acta Path. Microbiol. Scand.* §C **86** 1–9
—— 1978b *Acta Path. Microbiol. Scand.* §C **86** 123–129
—— 1978c *Acta Path. Microbiol. Scand.* §B **86** 87–92
Hofstra H and Dankert J 1981 *J. Gen. Microbiol.* **125** 285–92
Hofstra H van Tol M J D and Dankert J 1980 *J. Bacteriol.* 143(1) 328–37
Høiby N 1975 *Acta Path. Microbiol. Scand.* §B **83** 433–42
Høiby N and Axelsen N H 1983 *in* Axelsen 1983a pp 125–34
Ingild A 1983 *in* Axelsen 1983a pp 41–56
Jensenius J C, Andersen I, Hau J, Crone M and Koch C 1981 *J. Immunol. Meth.* **46** 63–8
Jirka M and Blanicky P 1973 *Biochim. Biophys. Acta* **295** 1–7
Johansson B G 1972 *in* Laurell 1972 pp 7–19
Johansson B G and Hjertén S 1974 *Anal. Biochem.* **59** 200–13
Jorgensen O S 1977 *FEBS-Lett.* **79** 42–44
Jurd R D 1981 *Gel Electrophoresis of Proteins: a Practical Approach* ed. B D Hames and D Rickwood (IRL press Ltd) pp 229–48
Kabat E A 1980 *Meth. Enzymol.* **70** 3–49

Kessler R E 1981 *Anal. Biochem.* **116** 129–32

Kindmark C O and Thorell J I 1972 *in* Laurell 1972 pp 49–53

Kint J A and Leroy J G 1985 *Lectins* vol. IV ed. J Breborowitz and T C Bøg-Hansen (Berlin: de Gruyter)

Kohn J 1976 *Chromatographic and Electrophoretic Techniques* vol. II 4th **edn** ed. I Smith (London: Heinemann) pp 90–137

Kohn J 1983 *Electrophoretic Techniques* ed. C F Simpson and M Whittaker (New York: Academic) pp 119–48

Koj A, Dubin A, Kasperczyk H, Bereta J and Gordon A H 1982 *Biochem. J.* **206** 545–53

Kolos O T and Kellen J A 1972 *Rev. d'Immunologie* **36** (3) 103–8

Kostner G, Holasek A 1972 *Anal. Biochem.* **46** 680–3

Krøll J, 1968 *Scand. J. Clin. Lab. Invest.* **22** 112–4

—— 1969 *Scand. J. Clin. Lab. Invest.* **23** 227–30

—— 1969a *Scand. J. Clin. Lab. Invest.* **23** 55–60

—— 1976 *J. Immunol. Meth.* **13** 333–9

—— 1981 *Meth. Enzymol.* **73b** 52–7

—— 1983a *in* Axelsen 1983a pp 135–9

—— 1983b *in* Axelsen 1983a pp 151–63

—— 1983c *in* Axelsen 1983a pp 165–9

—— 1983d *in* Axelsen 1983a pp 171–3

Krøll J and Anderson M M *J. Immunol. Meth.* **13** 125–30

Krotoski W A and Weimer H E 1983 *Int. J. Peptide Res.* **21** 155–62

Kushner I 1982 *CRP and the Plasma Protein response to Tissue Injury* (*Ann. NY Acad. Sci.* **389**) ed. I Kushner J E Volanakis and H Gewurz pp 39–48

Lang N and Haan J 1957 *Int. Arch. Allergy* **10** 305–16

Laurell C B 1965 *Anal. Biochem.* **10** 358–61

—— 1966 *Anal. Biochem.* **15** 45--52

—— 1967 *Prot. Biol. Fluids* **14** 499–502

—— 1972a ed. *Electrophoretical and Electro-Immunochemical Analysis of Proteins* (*Scand. J. Clin. Lab. Invest.* **29** suppl. 124) (Oslo: Universitetsforlaget)

—— 1972b *in* Laurell 1972a pp 21–37

Laurell C B and McKay E J 1981 *Meth. Enzymol.* **73B** 338–69

Laurell C B and Persson U 1973 *Biochim. Biophys. Acta* **310** 500–7

Lenger K, Blunk M and Warelka K *in* 1984 *Electrophoresis 1984* ed. V Neuhoff (Weinheim: Verlag-Chemie) pp 422–5

Lihme A O F and Heegaard P M H 1984 unpublished, The Protein Laboratory, University of Copenhagen

Lönnerdal B and Lààs T 1976 *Anal. Biochem.* **72** 527–32

Løwenstein H 1975 *Int. J. Peptide Res.* **7** 1–9

—— 1978 *Progress Allergy* **25** 1–62

Løwenstein H, Markussen B and Weeke B 1976 *Int. Archs. Allergy Appl. Immunol.* **51** 48–67

Løwenstein H and Weeke B 1976 *Scand. J. Immunol.* suppl **3** 59–65

Mancini G, Vaerman J P, Carbonara A O and Heremans J F 1964 *Prot. Biol. Fluids* **11** 370–3

Mattson E, Ramlau J and Bock E 1979 *J. Neurochem.* **32** 301–9

Milgrom F, Barron A L and Loz U 1967 *Virology* **33** 144–9

Monthony J F, Wallace E G and Allen D M 1978 *Clin. Chem.* **24** 1825–7

Mueller U W and Potter J M 1979 *Anal. Biochem.* **100** 51–3

Neucere N J and Cherry J P 1975 *Peanut Science* **2** 66–72

Niedick B 1978 *Immunochemistry* **15** 11–2

Nielsen C S 1975 *in* Axelsen 1975 pp 101–5

Nielsen C S and Bjerrum O J 1977 *Biochim. Biophys. Acta* **466** 496–509

Nilsson M and Bøg-Hansen T C 1980 *Prot. Biol. Fluids* **27** 599–602

Norén O, Sjöström H, Svensson B, Jeppesen L, Staun M and Josefsson L 1977 *Peptide Transport and Hydrolysis* Ciba Foundation symposium 50 pp 177–97

Nørgaard-Pedersen B and Gaede P 1975 *Scand. J. Immunol.* **4** suppl. 2 19–24

Nørgaard-Pedersen B, Toftager-Larsen K and Axelsen N H 1983 *in* Axelsen 1983a 259–63

Norrild B 1983 *in* Axelsen 1983a pp 279–82

Norrild B, Bjerrum O J and Vestergaard B F 1977 *Anal. Biochem.* **81** 432–41

Ogata S, Muramatsu T and Kobata A 1975 *J. Biochem. (Tokyo)* **78** 687–96

Ouchterlony Ö 1948 *Acta Path Microbiol. Scand.* **25** 186

Owen P, Oppenheim J D, Nachbar M S and Kessler R E 1977 *Anal. Biochem.* **80** 446–457

Owen P and Salton M R J 1975 *Proc. Natl. Acad. Sci. USA* **72** (9) 3711–5

—— 1976 *Anal. Biochem.* **73** 20–6

Owen P and Smyth C J 1977 *Immunochemistry of Enzymes and their Antibodies* ed. M R J Salton (New York: Wiley) pp 147–202

Plumley F G and Schmidt G W 1983 *Anal. Biochem.* **134** 86–95

Pluzek K 1981 *XIth Int. Seaweed Symp.* ed. T Levring (Berling: de Gruyter) pp 711–8

Pluzek K and Bjerrum O J 1982 *Biochem. Biophys. Meth.* **6** 261–5

Pødenphant J, Bock E, Thymann M and Gozes I 1983 *in* Axelsen 1983a pp 283–17

Porro M, Viti S, Antonio G and Saletti M 1982 *Anal. Biochem.* **127** 316–321

Prause J U, Sjöström H, Norén O and Josefsson L 1978 *Anal. Biochem.* **85** 564–71

Radola B J 1980 ed. *Electrophoresis '79* (Berlin: de Gruyter)

Raftell M and Blomberg F 1974 *Eur. J. Biochem.* **49** 31–9

Ramlau J and Bjerrum O J 1977 *Scand. J. Immunol.* **6** 867–71

Ramlau J and Bock E 1979 *Proc. 3rd Int. Symp. Affinity Chromatography and molecular interactions INSERM 86* ed. J M Egly pp 147–74

Raynes J 1982 *Biomedicine* **36** 77–86

Renn D W and Evans E 1976 *Anal. Biochem.* **71** 588–93

Ressler N 1960a *Clin. Chim. Acta* **5** 359–65

—— 1960b *Clin. Chim. Acta* **5** 795–800

Roberts D B, Wright G L, Affronti L F and Reich M 1972 *Infect. Immun.* **6** 564–73

Roitt I 1980 *Essential Immunology* 4th edn (Oxford: Blackwell)

Rouslahti E 1976 ed. *Immunoadsorbents in Protein Purification* (Oslo: Universitetsforlaget)

Saravis C A and Bonacker L 1970 *Nature* **228** 61–2

Schanbacher F L and Smith K L 1974 *Anal. Biochem.* **59** 235–47

Scheidegger J J 1955 *Int. Arch. Appl. Allergy Appl. Immunol.* **7** 103–10

Scherberich J E, Falkenberg F W, Mondorf A W, Müller H and Pfleiderer G 1974 *Clin. Chim. Acta* **55** 179–197

Schmidt-Ullrich R, Wallach D F H and Hendricks J 1975 *Biochim. Biophys. Acta* **382** 295–310

Sela M 1973 and 1975 *The Antigens* vol. I–II ed. (New York: Academic)
Siefring G E 1978 *Biochemistry* **17** 2558–604
Shivers C A and James J M 1967 *Immunology* **13** 547–54
Skude G and Jeppson J O 1972 *in* Laurell 1972a
Smith C J, Kelleher P C, Belanger C and Dallaire C 1979 *Brit. Med. J.* **1** 920–1
Smithies O 1955 *Biochem. J.* **61** 629–41
Smyth C J 1983 *in* Axelsen 1983a pp 233–41
Smyth C J, Friedman-Kein A E and Salton M R J 1976 *Infect. Immunol.* **13** 1273–88
Stevens P, Alam S, Young L S and Chesebro K 1981 *J. Immunol. Meth.* **43** 199–207
Svendsen P J, 1973 *see* Weeke 1973
—— in 1979 *Electrophoresis, a Survey of Techniques and Applications* (*J. Chrom. Library 18(A)*) ed. Z Deyl, F M Everaets Z Prusik and P J Svendsen (Amsterdam: Elsevier)
Svendsen P J and Axelsen N H 1972 *J. Immunol. Meth.* 169–76
Svendsen P J and Rose C 1970 *Sci. Tools* **17** 13–7
Svendsen P J, Weeke B and Johansson B G 1983a *in* Axelsen 1983a pp 3–20
Söderholm J, Smyth C J and Wadström T 1975 *in* Axelsen 1975 pp 107–13
Sorensen S H, Norén O, Sjöström H and Danielsen E M 1982 *Eur. J. Biochem.* **126** 559–68
Takesue J 1975 *J. Biochem.* **77** 103–15
Tschesche H 1983 ed. *Modern Methods in Protein Chemistry* (Berlin: de Gruyter)
Tomita M, Kurakawa T, Onazaki K, Ichiki N, Osana T and Utika T 1971 *Experientia* **28** 84–5
Uriel J 1964 *Immunoelectrophoretic Analysis* ed. P Grabar and P Burtin (Amsterdam: Elsevier) pp 30–57
Wadström T 1983 *in* Axelsen 1983a pp 97–102
van Wezel A L, van der Marel P, Hazendonk T G, Boer-Bak V and Henneke M A C 1984 *5th Gen. meeting ESACT, Copenhagen 1982 'Develop. Biol. Standard'* vol. 55 (Basel: S Karger) pp 209–15
Wasserman R L and Carra J D 1977 *The Glycoconjugates* vol I ed. M I Horowitz and W Pigman (New York: Academic) pp 323–48
Verbruggen R 1975a *Biochem. J.* **151** 149–55
—— 1975b *Clin. Chem.* **21** (1) 5–43
Verbruggen R and Baeck A 1976 *Anal. Biochem.* **130** 144–55
Vesterberg O and Gramstrup-Christensen B 1984 *Electrophoresis 5* 282–285
Vestergaard B F 1975 *in* Axelsen 1975 pp 203–6
Weeke B 1968 *Scand. J. Clin. Lab. Invest* **22** 107–11
—— 1970 *Scand. J. Clin. Lab. Invest.* 25, 161–163
—— 1973 *in* Axelsen 1973 pp 203–206
Weeke B, Weeke E and Løwenstein H 1975 *in* Axelsen 1975 pp 149–154
Wieme R J 1959 *Studies on Agar Gel Electrophoresis* (Brussells: Arcia Uitgaven)
—— 1965 *Agar Gel Electrophoresis* (Amsterdam: Elsevier)
Williams C A 1971 *Methods in Immunology and Immunochemistry* vol II ed. C A Williams and M W Chase (New York: Academic Press) pp 237–73
Willoughby E W and Lambert A 1983 *Anal. Biochem.* **130** 353–8

Chapter 6

In situ Peptide Mapping of Proteins and Polypeptides Separated by Polyacrylamide Gel Electrophoresis

by
K Gooderham

6.1 Introduction

The extremely high resolution, simplicity and low cost have combined to make polyacrylamide gel electrophoresis a powerful and popular tool for protein analysis. However, although the separations which can be obtained by one-dimensional electrophoresis, let alone by two-dimensional electrophoresis, are very impressive, many important questions remain unanswered. For example two proteins may share the same mobility and yet be totally unrelated (see figure 6.1). Alternatively, related proteins may have markedly different mobilities, as for example in proteins from different tissues or species, or as a result of post-synthetic modifications, or proteolytic degradation. All of these difficulties can, theoretically at least, be overcome using traditional biochemical and chemical methods. In practice this is not so easily accomplished because the proteins remain trapped within the dense gel matrix and are very difficult to recover, especially in the relatively large quantities required for analysis.

In order to overcome these problems, Cleveland *et al* (1977) developed the technique of *in situ* peptide mapping. This method has the advantage of being a microanalytical technique, while generally being no more complex than standard gel electrophoresis. The principal steps in a typical peptide mapping experiment are outlined in figure 6.2. Proteins are separated by polyacrylamide gel electrophoresis in the usual way (figure

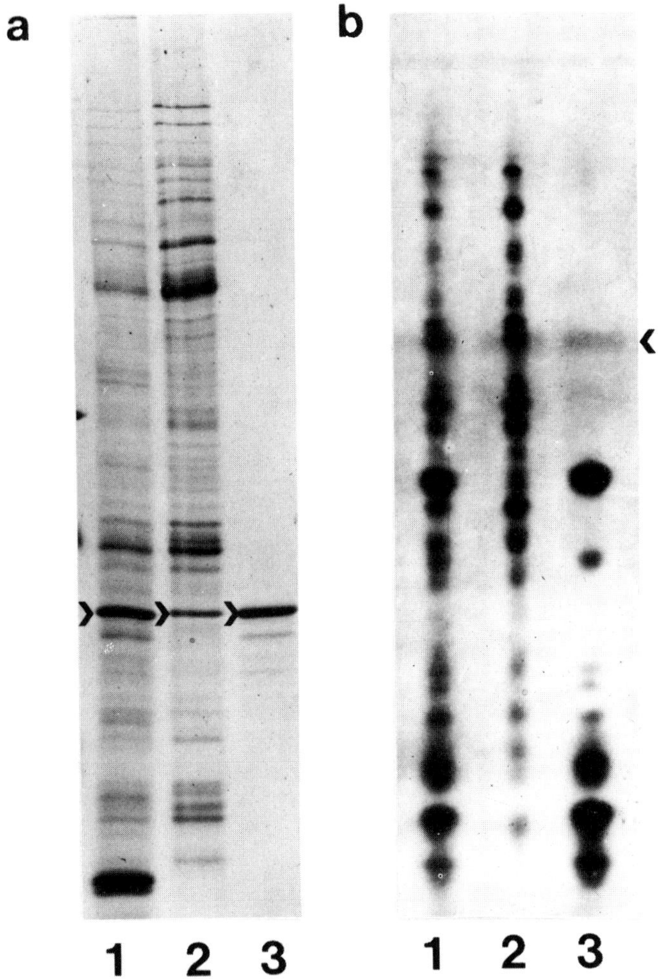

Figure 6.1 Peptide maps of nuclear lamin c and vimentin. Three different nuclear protein fractions panel ((*a*) 1–3) were separated by SDS-PAGE (8% acrylamide) and stained with Coomassie brilliant blue R-250. Proteins indicated by > in panel (*a*) were then cut out of the gel, equilibrated and rerun together with 50 pg of V8 protease on a second gel containing 15% acrylamide (panel (*b*) 1–3). The resulting peptide maps were then silver stained (Oakley *et al* 1981).

Comparison of the peptide maps in panel (*b*) shows that sample (*a*) 1 is a mixture of two proteins, lamin c and vimentin, while bands (*a*) 2 and (*a*) 3 represent lamin c and vimentin respectively. The position of the V8 protease in panel (*b*) is shown by <.

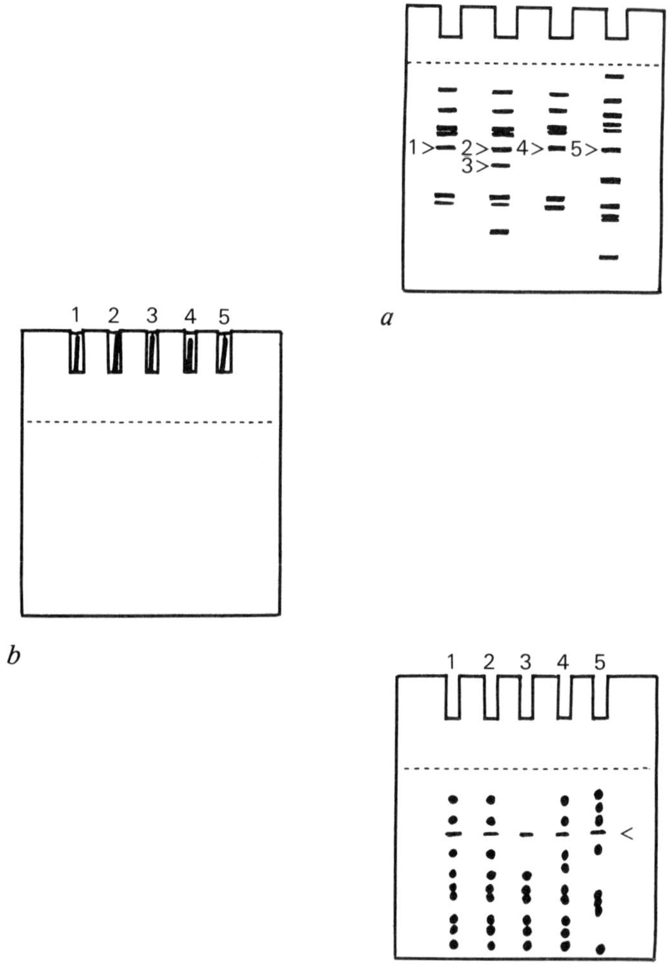

Figure 6.2 Diagrammatic representation of the key steps in an *in situ* peptide mapping experiment. (*a*) Protein samples are separated by gel electrophoresis and stained with Coomassie brilliant blue R-250. (*b*) Bands containing proteins of interest (> 1–5) are cut out of the gel, equilibrated and placed on a second gel together with a suitable protease. (*c*) The proteins and the protease are then run into the gel and the resulting peptides are visualised by silver staining. The position of the protease is indicated (<). Analysis of the peptide maps shows that proteins 1, 2 and 4 are homologous, while protein 3 is a specific degradation product of these proteins and protein 5 has no significant homology to any other protein.

6.2(*a*)). After the proteins have been visualised by staining, autoradiography or fluorography, the bands of interest are cut out of the gel and equilibrated in a suitable buffer. The gel pieces are then placed on top of a second gel and overlaid with a dense buffer solution containing protease (figure 6.2(*b*)). Next the protein and protease are slowly electrophoresed into the stacking gel where the protease digests the protein. The resulting peptides are resolved in the lower separating gel and subsequently detected by staining, autoradiography, etc (figure 6.2(*c*)) after which the peptide maps are compared and evaluated.

Since this method was originally developed, a number of modifications have been introduced. These include the use of chemical rather than proteolytic cleavages (see §6.4). In addition, this method is not restricted to using single bands as shown in figure 6.2, but complete longitudinal sections of the gel (Bordier and Crettol-Järvinen 1979, Tijssen and Kurstak 1983) as well as tube gels (Lam and Kasper 1980) containing large numbers of proteins can also be used.

Detailed laboratory schedules for performing peptide mapping experiments can be found elsewhere (Cleveland 1983, Fischer 1983, Gooderham 1984) and will therefore not be included here. Instead, the following sections will be devoted to a consideration of the various strategies which can be used at each stage of the experiment. Particular attention will be given to the different types of cleavage methods which can be used and the choice of primary and secondary gel systems and detection methods.

6.2 Primary Gel Systems

In theory almost any gel system e.g. SDS-PAGE, IEF, acid gels and 2D gels, may be used for the initial protein separation. However, in practice the most widely used gel system is the discontinuous SDS-PAGE system of Laemmli (1970).

In addition, a wide range of detection methods may be used to visualise the proteins before they are used in a peptide mapping experiment. In the original method (Cleveland *et al* 1977) only a very brief staining in Coomassie brilliant blue R250 was recommended. However, there is no real advantage in doing this and the gel can be completely stained in Coomassie brilliant blue without affecting the quality of the subsequent peptide maps (Spiker 1980). Alternatively, where the proteins are radioactively labelled, they may be dried down (with or without scintillant) and autoradiographed (Opperman *et al* 1981) or fluorographed (Rubin and Leonardi 1983) as appropriate. These complete detection systems have the advantage that a permanent record of the protein separation can be made before removing the bands of interest (see figure 6.1).

Irrespective of which gel system or detection method is used, it is important that only like should be compared with like, i.e. all of the samples should be taken from the same gel or where this is not possible they should have all been processed in the same way. This is particularly important for proteins which have peptide bonds between aspartic acid and proline residues, as these bonds will be hydrolysed by prolonged exposure to weak acids (see for example figure 1 in Gooderham 1984), or where the samples contain endogenous proteases.

6.3 Storage, Rehydration and Equilibration

Once the bands or gel strips containing the proteins have been cut out of the gel they can be stored at -20°C for at least six months. The samples may be stored in this way either before or after equilibration but in both cases any liquid should be drained off before freezing the samples. If this is not done, ice crystals will form in the gel pieces and may destroy them. Dried gels can also be kept for a considerable time, as long as several years at room temperature, before being used in peptide mapping experiments (Opperman *et al* 1981, Rubin and Leonardi 1983).

The same equilibration procedure can be used for both dried and wet gel pieces; the dried gel pieces being rehydrated and equilibrated in a single step. To equilibrate the samples each sample is placed in a disposable screw top plastic test tube containing 10 ml of equilibration buffer and gently agitated for 30 min. The most widely used equilibration buffer contains 0.125 mM Tris–HCl pH 6.8, 0.1% SDS, 0.1 M EDTA (Cleveland *et al* 1977). This buffer is designed to be used with samples which are to be run on a discontinuous SDS gel (Laemmli 1970). Where other gel systems are being used, the equilibration buffer must be modified to be compatible with the new gel system.

The length of the equilibration step can be varied and equilibration times of between 15 and 120 min are commonly used (Cleveland 1983; Gooderham 1984). Ideally the equilibration step should be as brief as possible since too long an equilibration will not improve the results and may in fact result in proteins being lost from the sample making the subsequent detection of the peptides more difficult. However, if too short an equilibration is used the gel may still contain chemicals which will interfere with the subsequent electrophoresis step. For example samples which contain traces of acetic acid will produce smeared and poorly resolved bands when run on the discontinuous SDS gel system of Laemmli (1970).

6.4 Protein Cleavage

6.4.1 Proteases

Both chemical and enzymatic cleavages are routinely used in *in situ* peptide mapping experiments but the enzymatic methods are by far the most popular and widely used. The popularity of this approach can be attributed to the method having being initially devised for use with proteases and also its simplicity. However, this method does suffer from a number of disadvantages, the most significant of which is that the protease also contributes to the final peptide map when general protein stains such as Coomassie brilliant blue or any of the various silver stains are used (see figure 6.1). This problem can be overcome in part by including a control track in which only the protease is run. However, in addition to suggesting false homologies, the background pattern produced by the protease may also obscure peptides and this can only be overcome by using radioactively labelled proteins. Also, while proteases are generally insensitive to quite wide variations in substrate concentration, only partial digests are obtained

Table 6.1 Proteases† used in *in situ* peptide mapping experiments

Enzyme	*Preferred cleavage site*	*Optimum concentration range (pg)*	*Selected references*
α-Chymotrypsin	Trp, Tyr, Phe	100–4000	Cleveland 1983, Dillmann *et al* 1984, Herman *et al* 1984
Elastase	Neutral amino acids	—	Cleveland *et al* 1977, Fischer 1983
Ficin	Lys, Arg, Leu, Gly	—	Cleveland *et al* 1977, Fischer 1983
Papain	Lys, Arg, Leu, Gly	0.5–1000	Calvert and Gratzer 1978, Tijssen and Kurstak 1983
Pepsin‡	Trp, Tyr, Phe, Leu, Asp, Glu		
Staphylococcus aureus V8 protease	Asp, Glu	10–1000	Cleveland 1983 Herman *et al* 1984
Streptomyces griseus protease	Broad specificity		Cleveland *et al* 1977
Subtilisin	Broad specificity	50–2000	Cleveland 1983 Fischer 1983
Trypsin	Lys, Arg	800–4000	Herman *et al* 1984 Lam and Kasper 1980

† Stock solutions at a concentration of 1–5 mg ml^{-1} in water may be conveniently stored for up to several months at −20°C.
‡ No references to the use of this peptide in *in situ* peptide mapping experiments have been found. However, this protease would be expected to be suitable for experiments which use a low pH mapping gel.

using this method and a number of different enzyme concentrations may have to be tested before the optimum digestion conditions are established.

A wide range of different proteases have been used in *in situ* peptide mapping experiments (see table 6.1). Of these the most widely used is *Staphylococcus aureus* V8 protease. This enzyme was first isolated by Drapeau *et al* (1972) and is commercially available through Miles Laboratories (Slough, UK). V8 protease has a pH optimum of 8 but still maintains considerable activity at pH 6.8 and is not activated by SDS. It is therefore ideally suited for peptide mapping experiments using the discontinuous SDS gel system of Laemmli (1970). Although V8 protease is generally the most suitable enzyme there are occasions when, due to a lack of suitable cleavage sites, one of the other enzymes listed in table 6.1 will produce better results. In addition, in many experiments it may be desirable to perform a second series of digests with a different protease in order to confirm the results obtained with the V8 protease (see §6.7).

Although most proteins will produce very good peptide maps using the standard method of Cleveland *et al* (1977), very large proteins, i.e. of 150 000 daltons (D) or greater, may not give such satisfactory results. The reason for this is that the large mass of the protein retards its migration through the stacking gel and the protease rapidly moves ahead of the protein leaving little or no opportunity for the digestion to occur. Dillmann *et al* (1984) working with myosin heavy chain (200 000 D) have recently devised a simple modification of the standard procedure which overcomes this problem. This was achieved by increasing the volume of protease from 10 μl to 30 μl, while decreasing the height of the stacking gel to 0.5 cm. The samples and protease were then slowly electrophoresed at 4 mA to the top of the stacking gel after which the current was increased to 15 mA for the remainder of the run. Using this technique excellent peptide maps were obtained using both V8 protease and chymotrypsin.

6.4.2 Chemical Cleavages

All of the major chemical cleavage methods have now been adopted for *in situ* peptide mapping experiments (see table 6.2). In contrast to the proteolytic cleavage techniques these methods are time consuming. The samples have to be cleaved prior to loading onto the peptide mapping gel because these methods are incompatible with most commonly used secondary gel systems. An additional disadvantage is that the gel pieces usually have to be equilibrated before and after the cleavage which is again time consuming and also involves a potential loss of proteins. However, unlike proteolytic digests, chemical cleavages do not contribute to the final peptide map. These methods also have the advantage of being relatively insensitive to substrate concentration and have well defined (though possibly rare) cleavage sites.

Table 6.2 Chemical cleavage methods for *in situ* peptide mapping experiments

Reagent	*Cleavage site*	*References*
BPNS-skatole	Trp—∣—*X*	Detke and Keller 1982
N-Chlorosuccinimide	Trp—∣—*X*	Lischwe and Ochs 1982
Cyanogen bromide	Met—∣—*X*	Lam and Kasper 1980
Formic acid	Asp—∣—Pro	Lam and Kasper 1980 Sonderegger *et al* 1982
Heat cleavage Tris–HCl pH 6.8	Asp—∣—Pro	Rittenhouse and Marcus 1984
Hydroxylamine	Asn—∣—Gly	Lam and Kasper 1980 Saris et al 1983

6.5 Secondary Gel Systems

As was the case for the primary gel system, almost any type of gel system can be used for the peptide mapping step, providing that it is compatible with the sample, and the protease if used. However, also like the primary gel systems the discontinuous SDS gel system of Laemmli (1970) is generally used. In peptide mapping experiments where proteases are used to generate the peptides, an extra long stacking gel, e.g. 5 cm rather than 2.5 cm long, is usually used. The increased length of the stacking gel is required to allow sufficient time for the protease to digest the protein, as well as to ensure the efficient 'stacking' of the peptides. Many workers also introduce a 30 min. break in the electrophoresis once the Bromophenol blue tracking dye has reached the bottom of the stacking gel to further ensure that the protease has enough time to digest the proteins. However, this step can usually be omitted provided that the samples are slowly electrophoresed through the stacking gel at a low constant current e.g. 20 mA, after which the current is increased to 40 mA for the remainder of the run (Gooderham 1984). In addition, we have also found it more convenient to make the peptide mapping gel slightly thicker than the primary gel, e.g. 1.5 mm instead of 1.0 mm thick, as well as to use deeper, narrower sample wells. These modifications allow the gel slices, which usually swell during the equilibration step, to fit easily into the sample wells and ensure that the gel pieces are held vertically in the wells (see figure 6.2(*b*)). Using this arrangement, the resulting peptides appear as concentrated spots rather than as diffuse bands (see figure 6.2(*c*)) increasing both the resolution and the sensitivity of the method.

For most applications a 15% separating gel will produce excellent results but this can be varied when particularly large or small peptides are being studied. Alternatively, Tijssen and Kurstak (1979) have used a 10–20% acrylamide gradient which gives better peptide separations, as well as producing sharper bands. For peptide mapping experiments with very small proteins (less than 10 000 D) one of the low molecular weight peptide gel systems (see for example West *et al* 1984) may produce better results.

6.6 Detection

Traditionally either Coomassie brilliant blue R250 or, in the case of radioactively labelled proteins, autoradiography and fluorography have been used to detect the peptides in the mapping gel. Unfortunately neither of these methods is totally satisfactory. The Coomassie based stains are not sufficiently sensitive for most applications while autoradiography and fluorography require radioactively labelled proteins. The recent development of high sensitivity silver stains (see Chapter 7) now offers an alternative simple method for overcoming these problems. Silver staining was first used by Spiker (1980) for *in situ* peptide mapping and it has now become the preferred detection method. *In situ* peptide maps prepared from 10 to 100 pg of protein can be visualised by silver staining compared with the 5 to 10 μg which are usually required for Coomassie staining. In addition to its high sensitivity, silver staining also has the advantage of not requiring any special sample preparation as is the case for radioactively labelled and fluorescently tagged (Tijssen and Kurstak 1979) proteins. Silver staining may also be used in conjunction with Coomassie staining, the peptides being first stained with Coomassie brilliant blue and then, if necessary, with silver. A large number of different silver staining methods are currently available and as yet no one method has been adopted for staining *in situ* peptide mapping gels. Each laboratory tends to have its own preferred method, for example we routinely use the ammoniacal silver nitrate stain of Oakley *et al* (1980). The sensitivity of this method is relatively low, i.e. only about 50 times more sensitive than Coomassie brilliant blue R250, but we have found it a simple and reproducible method to use (see figure 6.1). Other silver stains which have been used for staining peptide mapping gels include those of Merril *et al* (1981), Morrissey (1981), Sammons *et al* (1981) and Wray *et al* (1981). The last of these methods has the particular advantage of staining the peptides different colours which adds yet another dimension to the analysis of the peptide map.

6.7 Evaluation

A simple visual evaluation will usually provide all the information which is required from an *in situ* peptide mapping experiment. However, a few basic precautions must be taken when interpreting the results.

As previously mentioned, where proteolytic cleavages are used, the protease will also contribute to the final peptide map and these peptides will also be detected when either Coomassie or silver staining is used. In order to avoid any possible confusion, a control track containing only protease should be run on the peptide mapping gel. It is worth noting that V8 protease contributes remarkably little to the final peptide map (see figure 6.1).

Where molecular weight markers are to be run on the secondary gel they can be loaded in sample solution as normal. Alternatively, bands containing suitable marker proteins can be cut out of the gel from which the proteins for the peptide mapping experiment are taken. The latter method provides a more reliable estimation of molecular weights, as the marker proteins will also be retarded in their migration out of the gel slice. It is important to note that when using either enzymatic or chemical cleavage techniques, only partial cleavages are usually obtained. As a result the sum of the molecular weights of the individual peptides will be greater than the molecular weight of the parent protein.

Although both enzymatic and chemical cleavage techniques have well defined cleavage sites (see tables 6.1 and 6.2) the production of peptides using these methods should not be regarded as absolute proof that these sites are present. This is especially true for the chemical cleavage methods where rather harsh conditions are used, which may give rise to both specific and nonspecific cleavages. For example the methionine specific cleavage method using cyanogen bromide is performed in 75% formic acid which will cleave peptide bonds between aspartic acid and proline residues. Similarly acid hydrolysis may also occur during the staining and destaining of the primary gel and so generate additional peptides. All of these problems can be overcome by using suitable controls.

Where no apparent homology exists between two or more proteins this will generally be a good indication that they are unrelated. However, where two or more samples show an apparent homology a second series of peptide mapping experiments using a different cleavage method should be run in order to confirm the result. Either the original proteins or one or more of the peptides generated by the first mapping experiment can be used for this experiment. If the same degree of homology is seen in this second experiment there is probably sufficiently strong evidence for believing that the proteins are homologous. In experiments where a more rigorous analysis of the peptide maps is required the method of Calvert and Gratzer (1978) can be used. Using this method the distribution of the peptides is analysed statistically and the probability of two apparently similar or dissimilar patterns occurring by chance is calculated.

References

Bordier C and Crettol-Järvinen A 1979 *J. Biol. Chem.* **254** 2565–7
Calvert R and Gratzer W B 1978 *FEBS-Lett.* **86** 247–9
Cleveland D W, Fischer S G, Kirschner M W and Laemmli U K 1977 *J. Biol. Chem.* **252** 1102–6
Cleveland D W 1983 *Methods in Enzymology* **96** (New York: Academic) pp 222–9
Detke S and Keller J M 1982 *J. Biol. Chem.* **257** 3905–11
Dillmann W H, Barrieux A and Resse G S 1984 *J. Biol. Chem.* **259** 2035–8
Drapeau G R, Boily Y and Houmard J 1972 *J. Biol. Chem.* **247** 6720–6
Fischer S G 1983 *Methods in Enzymology* **100** (New York: Academic) pp 424–30
Gooderham K 1984 *A Laboratory Manual of Methods in Molecular Biology in Proteins* part I ed. J M Walker (New Jersey: Humana) pp 193–202
Herman H, Pytela R, Dalton J M and Wiche G 1984 *J. Biol. Chem.* **259** 612–7
Laemmli U K 1970 *Nature* **227** 680–5
Lam K S and Kasper C B 1980 *Anal. Biochem.* **108** 220–6
Lischwe M A and Ochs D 1982 *Anal. Biochem.* **127** 453–7
Merril C R, Goodman D, Sedman S A and Ebert M H 1981 *Science* **211** 1437–8
Morrissey J H 1981 *Anal. Biochem.* **117** 307–10
Oakley B R, Kirsch D R and Morris N R 1980 *Anal. Biochem.* **105** 361–3
Opperman H, Levinson A D and Varmus H E 1981 *Virology* **108** 47–70
Rittenhouse J and Marcus F 1984 *Anal. Biochem.* **138** 442–8
Rubin R W and Leonardi C L 1983 *Methods in Enzymology* **96** (London: Academic) pp 184–92
Sammons D W, Adams L D and Nishizawa E E 1981 *Electrophoresis* **2** 135–141
Saris C J M, van Eenbergen J, Jenks B G and Bloemers H P J 1983 *Anal. Biochem.* **132** 54–67
Sonderegger P, Jaussi R, Gehring H, Brunschweiler K and Christen P 1982 *Anal. Biochem.* **122** 298–301
Spiker S 1980 *J. Chromatography* **198** 169–71
Tijssen P and Kurstak E 1979 *Anal. Biochem.* **99** 97–104
Tijssen P and Kurstak E 1983 *Anal. Biochem.* **128** 26–35
West M H P, Wu R S and Bonner W M 1984 *Electrophoresis* **5** 133–8
Wray W, Boulikas T, Wray V P and Hancock R 1981 *Anal. Biochem.* **118** 197–203

Chapter 7

Protein Staining and Detection Methods

by

C R Merril, M G Harasewych and M G Harrington

7.1 Historical Introduction

Electrophoretic separation of proteins from complex mixtures has developed considerably since the initial observation in 1807 by the physicist Ferdinand Frederic Reuss that colloidal particles migrate in electrical fields (Gray 1951). Progress in applying electrophoresis to problems of protein purification would not have been possible, however, without the complementary development of methods to detect proteins. The earliest applications of electrophoresis relied on direct observation, and were limited to objects, such as cells, colloidal particles and bacteria, that could be detected by visible light, often with the aid of a microscope. Most protein molecules cannot be observed with visible light and are too small for direct observation. Despite their small size, their electrophoretic properties can be studied by adsorbing them onto microspheres (Porret 1816). The migrational properties of these protein-coated microspheres in electrophoretic fields were reviewed by Abramson (1934). However, Picton and Linder (1892) recognised that coloured proteins, such as haemoglobin, could be studied electrophoretically without quartz microspheres. Many naturally coloured proteins, including myoglobin, ferritin, and cytochrome C, were employed in early applications of electrophoresis (Davis and Cohn 1939).

Detection of non-coloured proteins by the specific absorption of ultraviolet light was first demonstrated by Tiselius in 1930. Tiselius (1937) was also successful in utilising this specific absorption of ultraviolet light to demonstrate the quantitative electrophoretic separation of ovalbumin, serum globulin fractions and Bence Jones protein. However, protein detection by ultraviolet absorption has never gained widespread acceptance, despite its early introduction, primarily because this technique requires a special light source, filters, and optical components that are transparent to ultraviolet light. Tiselius (1937) also detected schlieren, or shadows, created by boundaries between regions with different refractive indices due to different concentrations of proteins. Methods for visualising these refractive index gradients include schlieren optical and interferometric systems, both of which require complex optical systems.

The use of organic stains for the detection of proteins, along with the development of electrophoretic techniques, eliminated many of the complications inherent in the liquid moving boundary methods. The use of moist filter paper as an electrophoretic support medium or carrier for zonal electrophoretic separation (Koenig 1937) stimulated the adaptation of a number of histochemical stains for the detection of uncoloured proteins. Many of these had already been adapted for the visualisation of proteins separated by chromatography. These stains were employed after proteins were 'fixed' or made immobile, so that they would not be lost in the staining solutions. Heating the filter paper to 110°C after electrophoresis was one of the first methods of fixing proteins. Early general protein stains included Bromophenol blue (Durrum 1950) and Amido black (Grassman and Hannig 1952). Lipoproteins were preferentially stained by Oil red O (Durrum *et al* 1952), while glycoproteins were detected by a red colour that was produced upon their oxidation with periodic acid and subsequent reaction with fuchsin sulphurous acid (Schiff's reagent) (Koiw and Gronwell 1952).

Histochemical methods for detecting specific enzymes on the basis of their catalytic activity were first applied to electrophoretograms in the second half of this century (Hunter and Markert 1957). Although these techniques yield only qualitative results, their broad applicability in such diverse fields as clinical chemistry and population genetics has led to the development of staining systems for nearly a hundred different enzymes.

Radioactively labelled proteins may be visualised by autoradiographic or fluorographic detection methods. Autoradiographic methods were first introduced by Becquerel (1896) and Curie (1898) in their discovery of the phenomenon of radioactivity, while fluorography was first used in biological studies of metabolites involved in photosynthesis (Wilson 1958). Although it is often possible to incorporate large amounts of radioactive amino acids into the proteins of cells grown in culture, high specific activities are difficult to achieve in animal studies and unethical in research on humans. However, when proteins can be labelled with radioactive tracers to a high specific activity, they can be detected with sensitivities equal to, and often better than, those obtained with the most sensitive stains.

The acrylamide gel poses some limitations for protein detection due to the physical properties of its matrix. Towbin *et al* (1979) attempted to overcome these limitations by electrically transferring the proteins from the gel onto an alternate matrix, such as nitrocellulose paper. The paper may then be subjected to staining by various protein-binding agents, including immunological and chemical probes. The sensitivity of some detection methods, especially the immunostains, is much improved by this technique.

The increased resolution of electrophoretic techniques was augmented by the introduction of evermore sensitive organic stains, such as the Coomassie blue stains, which can detect as little as 0.5 μg of protein (Fasekas de St Groth *et al* 1963). Fluorescent stains, first introduced by Talbot and Yaphantis (1971), can now detect as little as 1 ng of protein (Barger *et al* 1976). The introduction of silver as a general protein stain for polyacrylamide gels (Merril *et al* 1979, Switzer *et al* 1979) increased the sensitivity of protein detection to as little as 0.1 ng (Merril *et al* 1984).

The quest for increased specificity in the detection of proteins has led to exploration of neutron activation and electron beam microanalysis as methods that might permit detection of trace elements in proteins separated by gel electrophoresis.

This chapter will critically review current gel-electrophoresis protein detection methods with the goals of helping the investigator to choose the best method for each application, and of indicating areas that might benefit from further development.

7.2 Post-Electrophoretic Organic Protein Stains

7.2.1 Coomassie Stains

The most commonly used organic stains for detection of proteins separated on polyacrylamide gels are the Coomassie blue stains, which were originally developed as acid wool dyes. A variety of wool dyes were named 'Coomassie dyes' to commemorate the 1896 British occupation of the Ashanti capital, Kumasi or 'Coomassie', in what is now Ghana. In a collaborative effort to find a highly sensitive protein stain that could be used in quantitative studies, a microbiologist of the Australian National University and textile chemists at the University of New South Wales School of Textile Technology (Australia) tested numerous dyes and dyeing techniques (Fazekas de St Groth *et al* 1963). Their studies showed that triphenylmethane Coomassie stains could detect proteins in the microgram range.

Coomassie brilliant blue R250 (the 'R' stands for a reddish hue while '250' is a dye strength indicator) was the first of the triphenylmethane stains to be introduced. It can detect as little as 0.5 $\mu g\,cm^{-1}$ of protein and gives a linear response up to 20 $\mu g\,cm^{-1}$. However, it should be noted that the slope of the response varied for each of the four proteins tested. Fazekas de St Groth *et al* (1963) found that this stain could be used on cellulose acetate, agar or starch gels. Meyer and Lamberts (1965) first utilised Coomassie brilliant blue R250 in polyacrylamide gels, using electrophoresis to remove excess stain, but a number of proteins displayed some mobility during this destaining (Marshall and Porath 1965; Polter, 1967).

Loss of proteins due to mobility after electrophoresis in the polyacrylamide gels is a problem that is common to all staining and protein detection methods. Chrambach and associates (1967) were able to reduce this problem after electrophoresis by fixing proteins in the gel with 12.5% trichloroacetic acid (TCA) for 30 minutes. They stained the gel by immersing it in a 1:20 dilution of an aqueous stock solution of 1% Coomassie brilliant blue R250 in 12.5% TCA for 1 h, followed by washing in 10% TCA, in which protein band intensity and contrast increased during the succeeding 48 h. The solubility of Coomassie brilliant blue R250 in 12.5% TCA permits the stain to penetrate deeply into the gel, maximising band intensity, but this property also produces a background stain that requires extensive washing.

7.2.2 Mechanism of Coomassie Brilliant Blue Staining

Coomassie blue staining requires an acidic medium for electrostatic attraction to form between the dye molecules and the amino groups of the proteins. This ionic attraction, together with van der Waal's forces, binds the dye–protein complex together. The binding, however, is fully reversible by dilution under appropriate conditions (Fazekas de St Groth *et al* 1963). The relatively high staining intensity of Coomassie blue stains compared to other organic dyes is apparently caused by secondary bonds formed between dye molecules. Additional dye may be bound by dye–dye interactions to dye molecules that are ionically bound to, or in hydrophobic association with, protein molecules (Wilson 1979).

7.2.3 Metachromatic Effects

Secondary dye-binding or dye–dye interactions may play a fundamental role in the metachromatic effects that have been observed with some polypeptides when stained with Coomassie brilliant blue R250 (Duhamel *et al* 1980, Miko and Schlaepfer 1977). Some peripheral nerve proteins stain to produce both red and blue bands with Coomassie Brilliant Blue R250 after separation by SDS-polyacrylamide electrophoresis. Calf skin collagen displays similar red-staining bands, as do most other collagen samples. Histone-1 protein produces a red colour with spectral properties similar to those observed in the collagen bands, while histone-2B displays only a faint red shift (Duhamel *et al* 1980). These shifts in spectral absorption, or metachromasy, are probably due to the orderly arrangement of dye molecules on a structurally similar portion of these red-staining proteins.

Metachromatic shifts are often observed when dye molecules are stacked together (Schubert and Hamerman 1956). Such dye stacking or close aggregation may affect the dye's electron resonance structure, resulting in an altered response to excitation by light. Duhamel *et al* (1980) have proposed that the closely spaced proline residues present in collagens, histone-1 and those in a small cluster in histone-2B, are important for this metachromatic shift of the Coomassie stain, rather than the repeating tripeptides that are present in the collagens but not in the histones. Furthermore, proline-rich salivary gland proteins often develop a reddish colouration with Coomassie blue stain (Wallach *et al* 1975).

Coomassie brilliant blue metachromatic shifts are affected by a number of variables. Bands containing more than 5 μg of a protein that normally produces a red hue have been shown initially to stain blue in the centre of the band. Only with continued destaining does the red become apparent. Red-staining proteins such as histone-1 or collagen may be converted to blue bands by placing the gels in solutions containing alcohol, SDS or TCA (Duhamel *et al* 1980). Metachromasy is also diminished by staining at elevated temperatures (Miko and Schlaepfer 1978). Effects of protein specificity and concentration, temperature, solvent, and the type of spectral shift observed with Coomassie blue are indicative of a metachromatic mechanism rather than of an artefact due to dye contaminants.

7.2.4 Coomassie Stain Variants

Variations of staining methods utilising Coomassie brilliant blue R250 have been reviewed by Righetti (1983). These vary mostly in the solutions used for fixation, usually containing TCA, sulphosalicylic acid, methanol or ethanol. Another modification, a high-temperature staining procedure (60 °C), was introduced to reduce the interference from ampholytes when staining isoelectric focussing gels (Vesterberg 1972).

Another Coomassie stain, Coomassie brilliant blue G250 ('G' indicates that this stain has a greenish hue), is only slightly soluble in 12% TCA. This diminished solubility permits its use as a colloidal dispersion. This stain is a dimethylated form of Coomassie brilliant blue R250 (figure 7.1). Diezel *et al* (1972) demonstrated that such a colloidally dispersed dye does not penetrate the gel, permitting rapid staining of proteins without an undesired background. The gel is usually fixed for 5 min in 12.5% TCA, after which an aqueous solution of Coomassie brilliant blue G250 (0.25% w/v) is added. Staining can be performed for as long as 30 min, after which the gels are stored in 5% acetic acid. Band intensification occurs in the 5% acetic acid because the protein-bound dye becomes soluble and diffuses into the gel to bind to interior regions of the protein band.

Another derived Coomassie stain, Coomassie Violet R150, has gained some favour for staining proteins on polyacrylamide gels by virtue of its ability to stain rapidly and its ease of destaining (Frater 1970, Radola 1980). Although Coomassie violet R150 is no longer produced by Imperial Chemical Industries Ltd, a very similar dye, Serva violet 49, is available. Serva violet 49 differs from Coomassie Violet 150 by the substitution of a diethylamine group for a dimethylamine group.

Imperial Chemical Industries Ltd no longer produces Coomassie blue R250 or G250. They still hold the trademark 'Coomassie', so that manufacturers who are currently producing Coomassie blue R250 and/or G250 have introduced their own trademark names, as reviewed by Wilson (1983).

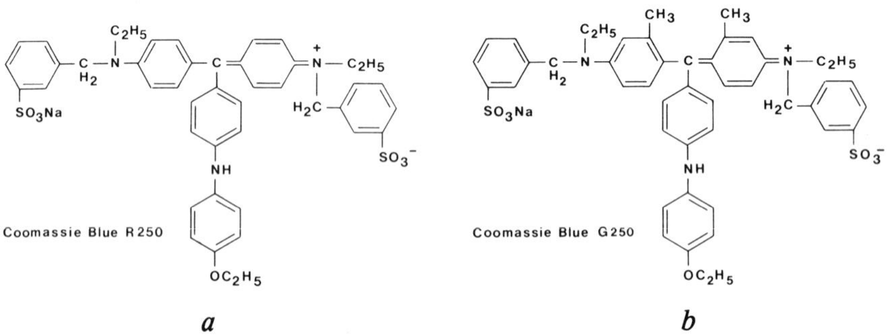

Figure 7.1 The Coomassie and triphenylmethane dyes, which permit detection of proteins separated on polyacrylamide gels by forming complexes with the proteins. Coomassie Blue G250 (*b*) is a dimethylated form of R250 (*a*).

7.2.5 Amido Black and Fast Green

Amido black (Acid black 1) and Fast green (Food green 3) are other organic stains that are commonly utilised for protein detection following polyacrylamide gel electrophoresis. This is so even though Coomassie blue R250 stains at three times the intensity of Fast green and six times the intensity of Amido black (Wilson 1983).

The staining intensities of these dyes are approximately proportional to their relative molar adsorption coefficients. Wilson (1979) estimated that 1 mg of protein will bind approximately 0.17 mg of Amido black, 0.23 mg of Fast green, 1.2 mg of Coomassie blue R250 and 1.4 mg of Coomassie blue G250. Since the variation in molecular weight between these dyes only ranges between 616.5 daltons (d) for Amido Black and 854.0 daltons for Coomassie Blue G250, the larger amount of bound Coomassie blue must be to an increased number of dye molecules bound per protein molecule. Many of these are believed to be bound by secondary mechanisms, such as dye–dye interactions or hydrophobic interactions between the dye molecules and the proteins.

7.2.6 Metachromatic effects

Secondary binding mechanisms also occur with Amido black and Fast green, although perhaps not at the levels observed with Coomassie blue R250, as these dyes display metachromic effects with certain proteins similar to the metachromasy observed with Coomassie blue R250. Amido black produces blue–green bands with certain histones rather than its characteristic blue–black colour, while Fast green produces a difference in the ratio of blue to green hues (McMaster-Kaye and Kaye 1974).

7.2.7 Quantitation using Organic Stains

Protein quantitation is possible with organic stains. Fazekas de St Groth *et al* (1963) demonstrated an accuracy of ± 10% in measuring the concentration of a protein in the range of 0.5–20 $\mu g/cm^{-1}$ using Coomassie Blue R250. It should be noted that, although individual proteins display linear relationships between absorbency and concentration, the slopes differ for each protein. Therefore, a standard curve must be produced for each protein assayed and quantitative comparisons limited to equivalent protein spots on two-dimensional electrophoretograms or equivalent protein bands on one-dimensional electrophoretograms. This is especially true for metachromatically staining proteins, which often display an additional complication, in that their bands fade more rapidly than the non-metachromatically staining proteins. In one study with Amido Black, certain metachromatic histone bands faded by 30–50%, while other protein bands in the same gel faded by only 5–15% (McMaster-Kaye and Kaye, 1974). This fading reflected a disproportionate loss of stain, rather than a loss of histone protein, since the bands themselves could be restained. In the case of smaller proteins such as insulin, decreases in band densities have been demonstrated to be due to the loss of protein during staining (Fazekas de St Groth *et al* 1963).

Another problem in utilising post-electrophoretic organic stains occurs because diffusion of the dye into the gel is slow. This results in 'ring-dyeing' if staining is terminated prematurely (Bosshard and Datyner 1977). In 'ring-dyeing', the stain concentration is less in the centre of a band or spot than at the edge because of the insufficient diffusion of dye molecules. Regional nonstoichiometric processes, as occur in 'ring-dyeing', require careful control of staining parameters to assure that they are not present in protein bands or spots that are to be quantitatively analysed.

7.3 Pre-electrophoretic Organic Stains

Use of pre-electrophoretic organic stains for protein detection usually involves the covalent binding of either a fluorescent or coloured moiety to proteins prior to electrophoresis. This group of stains have certain potential advantages: reactions with proteins can be performed stoichiometrically without the diffusion limitations imposed by staining within a gel matrix; the process of electrophoresis can be followed visually with 'stained' proteins; and there are no background problems to dye-trapping or reaction of the dye with the gel.

An often cited disadvantage of covalently bound pre-electrophoretic stains is that they usually alter the charge of the proteins (Bosshard and Datyner 1977). This objection is not of consequence for sodium dodecyl sulfate (SDS) electrophoresis, as the protein's mobility depends on molecular weight, and the dye molecules are usually too small to produce an appreciable effect. Furthermore, as long as the stains react with proteins in a stoichiometric manner, shifts in protein pattern should be highly reproducible, permitting construction of valid protein maps and protein identifications.

7.3.1 Nonfluorescent Stains

Remazol brilliant blue R, the first anionic dye used for prestaining proteins, was introduced by Griffith in 1972. A major disadvantage was its lower limit of sensitivity (3 μg of protein). Bosshard and Datyner (1977) introduced two anionic dyes, Drimarene brilliant blue K-BL and Uniblue A, both of which detect bands containing as little as 0.5 μg of protein (sensitivities comparable to the Coomassie blue stains). These anionic dyes react primarily with the amino groups of proteins as well as the hydroxyl groups of serine and tyrosine. A similar range and sensitivity was achieved with an unnamed cationic dye that binds to the protein's amino groups, causing less alteration in isoelectric focusing patterns than anionic dyes (Datyner and Finnimor 1973).

7.3.2 Fluorescent Stains

Fluorescent stains are the most sensitive pre-electrophoretic stains currently in use. The first fluorescent stain used to visualise proteins in gels was anilinonaphthalene sulphonate (Hartman and Udenfriend 1969), a post-electrophoretic stain. It is thought to bind the protein's hydrophobic sites, forming a fluorescent complex. The limit of sensitivity is about 20 μg of protein.

Pre-electrophoretic fluorescent staining was first introduced by Talbot and Yaphantis (1971) with the use of dansyl chloride to form fluorescent derivatives of proteins. This strain reacts with proteins in 1–2 min at 100°C, with sensitivity limit of 8–10 ng.

Reagents that were first designed to increase the detection limits of amino acid analysers (Stein *et al* 1973) have resulted in a new class of highly sensitive fluorescent stains. The first of these, fluorescamine, is a non-fluorescent compound. At room temperature and alkaline pH, it reacts with the primary amines of amino acids within 0.2–0.5 ms to yield a fluorescent derivative. Application of fluorescamine as a pre-electrophoretic protein stain for polyacrylamide gel electrophoresis permitted detection of as little as 6 ng of myoglobin (Pace *et al* 1974, Ragland *et al* 1974). A related compound, 2-methoxy-2,4-diphenyl-3(2H)-furanone (MDPF) has all of the attractive features of fluorescamine, including simplicity and speed. Furthermore, MDPF-labelled protein is 2.5 times as fluorescent as fluorescamine-labelled protein and does not fade as rapidly. As little as 1 ng of protein has been detected with MDPF, which has a linear response from 1–500 ng. As with most other protein stains, a plot of relative fluorescence versus protein concentration revealed a different slope for each of the four proteins studied (Barger *et al* 1976).

Although these fluorescent stains achieve greater sensitivity than other organic stains, they require ultraviolet light for visualisation, and direct quantitation requires fairly sophisticated equipment. These problems, coupled with the altered electrophoretic mobility during isoelectric focusing (a result of the altered protein charge(s) from the fluorescent derivatisation), have tended to inhibit utilisation of these fluorescent stains.

7.4 Silver Stains

Silver nitrate, the main ingredient in silver stains, was first described by an Arabian alchemist, Gâbir Dschâbir ibn Hajjam in the eighth century. Credit for the twelfth century observation that this compound has the ability to blacken when in contact with organic substances, including human skin, is usually given to Count Albert von Bollstädt (Eder 1932a). Modern scientific applications of this property of silver nitrate began with Krause's (1844) observation that silver nitrate solutions could be used to stain small pieces of fresh tissue. Golgi (1873) discovered that nervous tissue displayed net-like structures after 'hardening' in potassium dichromate and ammonia followed by staining with a 1% solution of silver nitrate. Use of this stain led Cajal to recognise that the net-like structures described by Golgi were individual neurons. These studies laid the foundations of modern neuroanatomy (Gibson 1963). Cajal (1903) also adapted photochemical methods to develop new histological silver stains that produced better definition of nerve fibres. Histological silver stains have, in general, become more complex as empirical alterations have been made in the procedures to limit the staining to specific cells or subcellular structures.

Introduction of silver as a general stain for proteins separated by polyacrylamide gel electrophoresis has permitted the detection of as little as 0.1 ng of protein. This silver stain (Merril *et al* 1979, Switzer *et al* 1979), an adaptation of a histological silver stain, was demonstrated to be a hundred-fold more sensitive than previous stains. Other histological silver stains were adapted, but these were limited to the detection of specific proteins, modifying subgroups, or to proteins electrophoresed in agarose. Frederick (1963) used a modified Gomori silver stain in studies of a phosphorylase. Hubbell *et al* (1979) applied a histological silver stain, originally developed to visualise nucleoli, to detect nucleolar proteins. Kerenyi and Gallyas (1972, 1973) adapted a histological silver stain to visualise cerebrospinal fluid proteins separated on agarose. This latter stain did not achieve widespread acceptance, perhaps because it did not work well in polyacrylamide, produced numerous staining artifacts, or was reported to be quantitatively irreproducible (Verheecke 1975). However, recent work by Peats (1983) has improved this stain's performance in agarose.

7.4.1 Mechanism of Silver Staining

The basic mechanism underlying all silver stains involves the reduction of ionic to metallic silver. Detection of proteins in the gel or membrane requires a difference in the oxidation–reduction potential between the sites occupied by proteins and adjacent sites of the gel or membrane. If a protein site has a higher reducing potential than the surrounding gel or matrix, then the protein will be positively stained. Conversely, if the protein site has a lower reducing potential than the surrounding gel or matrix, then the protein will appear to be negatively stained. These relative oxidation–reduction potentials can be altered by the chemistry of the staining procedure. As an example, proteins separated on polyacrylamide gels have been shown to stain negatively if the gel is soaked in the dark in silver nitrate followed by image development in an alkaline reducing solution (such as Kodak D76 photographic developer, which contains p-methylaminophenol sulphate as the reducing agent). By treating the gel with potassium dichromate prior to the silver nitrate incubation and following with development of the image in an alkaline reducing solution (utilising formaldehyde as the reducing agent), a positive image is produced (Merril and Goldman 1984). Positive images may also be obtained by substituting potassium ferricyanide (Merril *et al* 1981a), potassium permanganate (Ansorge 1983) or dithiothreitol (Morrisey 1981) for potassium dichromate in this stain.

Dichromate, permanganate and ferricyanide are thought to enhance the formation of a positive image by converting the protein's hydroxyl and sulphydryl groups to aldehydes and thiosulphates, thereby altering the oxidation–reduction potential of the protein. Although the formation or presence of aldehydes has often been suggested as essential for silver staining in certain histological stains, Thompson and Hunt (1966) noted that neither aldehyde-creating nor aldehyde-blocking reagents appreciably affected silver staining in some tissues. They suggested that, in some cases, tissue silver staining may depend on physical interface phenomena. For example, silver staining of fibrils appeared on electron microscopic observation to depend on whether the fibrils were in an ordered or random array.

Dithiothreitol, a reducing agent, also creates a positive image, perhaps by maintaining the proteins in a reduced state. However, other reducing agents, such as beta-mercaptoethanol, do not enhance positive image formation. Alternatively, all of the positive image enhancing compounds may form complexes with the proteins. These complexes may act as nucleation centres for silver reduction (Dunn and Burghes 1983).

Reduction of ionic to metallic silver is highly dependent on pH. In the photographic silver image development processes which utilise organic developers, a useful reducing activity is generally obtained only in alkaline solutions (James 1977). Metal-complex developers, vanadium ions and photons are exceptions in that they will reduce silver ions in acidic solutions.

In the six years since the introduction of silver staining as a general method for the detection of proteins separated by polyacrylamide electrophoresis, numerous staining protocols have been developed (reviewed by Dunn and Burghes 1983). These methods can be grouped into three categories:

(i) diamine or ammoniacal type silver stains;
(ii) adaptations of photographic chemical development processes;
(iii) photodevelopment stains.

(i) Diamine Silver Stains

The use of ammonium hydroxide to form silver diamine complexes offers a means of stabilising silver ions in an alkaline environment. Diamine silver stains were initially developed as histological stains for nerve fibres (Bielschowsky 1904). The concentration of silver ions in these stains is usually very low, as most of the silver is bound in diamine complexes (Nauta and Gygax 1951). If the concentration of silver ions is too low, diamine stains tend to be more sensitive for glycoproteins, which limits their use as general protein stains. This specificity can be minimised by maintaining a sufficient sodium to ammonium ion ratio in the diamine solution (Allen 1980). In some applications, emphasis of specificity has proven useful as in the adaptation of a diamine histological silver stain (Gambetti *et al* 1981). This stain, which was saturated with copper ions, detected neurofilament polypeptides in electrophoretic analyses of spinal cord homogenates. Copper is used in a number of diamine stains, and appears to increase sensitivity. Its mechanism may be similar to that of the biuret reaction (Merril *et al* 1979). In the diamine stains, the ammoniacal silver solution must be acidified, usually with citric acid, for image production to occur. Citric acid plays a dual role. It reduces the free ammonium ion concentration, liberating silver ions to a level where their reduction by formaldehyde to metallic silver is possible. The optimal concentration of citric acid results in a controlled rate of silver ion reduction, preventing non-selective deposition of silver.

(ii) Non-diamine Chemical Development Silver Stains

Non-diamine chemical development silver stains rely on the reaction of silver nitrate with protein sites in acidic conditions, followed by the selective reduction of ionic silver by formaldehyde in alkaline conditions. Sodium carbonate and/or hydroxide and other bases are used to maintain an alkaline pH during development. Formic acid, produced by the oxidation of formaldehyde, is buffered by the sodium carbonate (Ehrenfried 1952).

Alkaline conditions in the diamine silver stains may enhance silver reduction in the vicinity of protein molecules, producing a positive image. The negatively charged carboxyl groups could participate as electron donors in the initial reduction of ionic to metallic silver. Conversely, in non-diamine silver stains, silver ions are present in an acidic solution, in which the protein's amino groups tend to be positively charged. In this environment, electron donation through ionic interactions with silver would be less likely. This may be a factor in the requirement of compounds such as dichromate, permanganate, ferricyanide or dithiothreitol for the production of a positive image in the non-diamine silver stains. Many of these agents are used in photographic processes to achieve image reversal (Hillson 1977). Photographic emulsions prepared with ammonia are much more sensitive to image reversal (Mees 1952). This may be a factor in the positive image attained with the ammoniacal silver stains (Merril and Goldman 1984).

(iii) Photodevelopment Silver Stains

Photodevelopment stains utilise energy from photons of light to reduce ionic to metallic silver. Eder (1932b) noted that, as early as 1777, Scheele recognised that the blackening of ionic forms of silver by light was due to the formation of metallic silver. He demonstrated that silver chloride crystals exposed to light while under water produce hydrochloric acid and a black metallic silver precipitate (Eder 1932b). This ability of light to reduce ionic to metallic silver was adapted by William Fox Talbot, in 1839, as the basis of one of the first photographic processes. Talbot's process dominated photography from its introduction until 1862, when photodevelopment was replaced by 'chemical development' processes (Newhall 1983).

Photodevelopment silver stains provide a rapid and simple, yet sensitive method for detecting proteins separated by gel electrophoresis (Merril *et al* 1984, Merril and Harrington 1984). Most chemical development stains require a minimum of two solutions, in addition to the fixing solution. The presence of silver ions and an organic reducing agent in an alkaline solution usually results in an uncontrolled reduction of silver. However, since light can reduce silver in an acidic solution, a photodevelopment stain may utilise a fixation solution followed by a single staining solution.

A single-solution photodevelopment silver stain has two major advantages over chemical-development silver stains. First, pH gradient effects are eliminated. In chemical development, one solution, containing silver ions, diffuses out of the gel, while the solution containing the reducing agent diffuses into the gel. The interactions of these solutions create complex pH gradients within the gel. A single-solution photodevelopment stain reduces such diffusion effects, minimising staining artefacts caused by variations in gel thickness or the use of plastic gel 'backings'. Proteins on ultra-thin supporting membranes such as cellulose nitrate stain poorly with the 'chemical stains' because they retain very little silver nitrate when transferred into alkaline solution for image development. Because the photodevelopment stain contains the silver ions in the image-developing solution, proteins may be visualised even when bound to thin membranes.

In the photodevelopment silver stain, light photons are utilised to liberate electrons. The mechanism is probably best described by the Gurney-Mott photochemical theory: silver forms when a liberated photoelectron combines with a silver ion. It is generally accepted that a single silver atom will rapidly undergo oxidation, reverting to a silver ion, unless additional silver atoms are formed nearby. Once a critical number of silver atoms are formed in a local region, they become an autocatalytic centre for the reduction of additional silver ions. The rate of silver ion reduction after the photocatalytic formation of stable metallic silver depends on the electron availability or the local oxidation–reduction potential (Hamilton 1977).

Prior to silver staining by either the chemical or photodevelopment methods, proteins must be 'fixed'. Fixatives play a dual role in both methods. They retard diffusion of the protein from the gel or membrane, and they elute substances that might interfere with staining (such as reducing agents, detergents, and ampholytes) from the matrix. Electrophoretic systems may, in some cases, require more than one fixation to clear the gel of these substances. In the photodevelopment stain, the fixative also impregnates the gel with chloride ions.

The need for the presence of both silver nitrate and silver chloride in the photodevelopment process was addressed by the 19th century photochemist, Hermann Vogel. He reasoned that silver chloride was more sensitive to reduction by light than silver nitrate. Since the silver chloride was fixed in position because of its insolubility, the density of the image would be limited unless free silver ions (supplied by the soluble silver nitrate) were present to diffuse into image developing regions (Reilly 1980). However, the mechanism of chloride enhancement to detect proteins separated on gels must be more complex than Vogel's interpretation of the effects of chloride in photography. When a gel that has been treated with a fixative, containing a small amount (0.2% w/v) of sodium chloride, is transferred into a silver nitrate solution, a fine bluish-white precipitate of silver chloride forms in the gel. This image can be visualised either with transillumination or incident light. Proteins appear as clear regions, while the rest of the gel contains a fine bluish-white precipitate. This effect is especially noticeable if the protein concentration is greater than 1 μg. With transillumination by a light source of sufficient intensity, the clear regions containing most proteins will darken (Merril *et al* 1984). One interpretation of the appearance of these clear areas at the sites occupied by the proteins was that proteins exclude silver ions. This interpretation cannot, however, be correct, since previously clear regions develop the characteristic black or brown images formed by the reduction of silver ions on exposure to light (Merril *et al* 1984). Further evidence that the silver ions are not excluded from gel regions containing electrophoresed proteins was obtained by placing a polyacrylamide gel into a radioactive silver nitrate solution, using ^{110}Ag. At very low concentrations of silver nitrate (10^{-9} M) sufficient silver was bound to the proteins to visualise their positions by autoradiography. At higher silver concentrations, such as those used in silver staining (0.1 M), the distribution of silver was fairly uniform. In no case was there evidence for the exclusion of silver ions (Merril *et al* 1984). The evidence for the presence of chloride ions in the clear zones is more indirect. If chloride ions are not present in the gel, or not used in the fixer, there is a significant loss of the stain's sensitivity.

One explanation for the lack of a visible silver chloride precipitate in gel regions that contain relatively high concentrations of proteins may be an altered solute structure in these gel regions which affects the interaction of silver and chloride ions. Biological molecules often reorganise solvent molecules in the neighbourhood of certain reactive groups. X-ray diffraction studies of proteins have demonstrated water oxygen atoms that have been immobilised by the proteins, such that they are visible in a specific position in the resulting electron density maps (Wolfenden 1983). Effects of proteins on their surrounding solvents may also be augmented by the physical organisation of the proteins themselves. Examples of histological silver staining appear to exhibit physical interface phenomena, including the previously discussed electron microscopic observations which suggest that the degree of silver staining differences depends on the arrangement of fibrils in a random array or in organised bundles (Thompson and Hunt 1966).

7.4.2 Colour Effects

Most proteins have a monochromatic brown or black colour when stained with silver. However, in a study of cerebrospinal fluid proteins, Goldman *et al* (1980) noted that certain lipoproteins tend to stain blue while some glycoproteins appear yellow, brown or red. This colour effect is most likely an analogue of the Herschel effect in photography (Merril and Goldman 1984). Herschel (1840) discovered that if he projected the spectrum of visible light obtained by passing sunlight through a prism onto a silver chloride-impregnated paper, the colours of the spectrum appeared on the paper, particularly a 'full and fiery red' at the focal point of the red light. While images produced by reducing ionic to metallic silver are usually black or gray, coloured images may be obtained if the particles of metallic silver are small in comparison to the wavelength of light. The colour produced depends on three variables: the size of the silver particles, the refractive index of the gel, and the distribution of the silver particles within the gel. In general, small grains (less than 0.2 μm in diameter) transmit reddish or yellow-red light, while grains above 0.3 μm give bluish colours, and larger grains produce black images (Mees 1952).

Modifications of the silver staining procedures, such as lowering the concentration of reducing agent in the image development solution, prolonging the development time, adding alkali, or elevating the temperature during staining will often enhance colour formation. Some silver stain protocols have been developed to produce colours that may aid in identification of certain proteins (Nielsen and Brown 1984, Sammons *et al* 1981, 1984). Production of colour with silver stain depends on many variables. Nielsen and Brown (1984) have shown that charged amino acid side groups play a major role in colour formation. However, variations in protein concentration and conditions of image development may produce colour shifts, confusing identification. Furthermore, colour-producing silver stains tend to become saturated at low protein levels and often produce negatively stained bands or spots. These factors tend to make quantitative analysis more difficult.

7.4.3 Specificity

In some applications, silver stains can demonstrate considerable specificity. For example, Hubbell *et al* (1979) used an histological stain that is highly specific for nucleolar proteins, while Gambetti *et al* (1981) adapted a silver stain specifically for neurofilament polypeptides. Other silver stains detect not only proteins but also DNA (Somerville and Wang 1981, Boulikas and Hancock 1981, Goldman and Merril 1982, Merril *et al* 1982), lipopolysaccharides (Tsai and Frasch 1982), and polysaccharides (Dubray and Bezard 1982).

The utilisation of a silver stain followed by Coomassie brilliant blue R250 has enhanced the differentiation of sialoglycoproteins. In a study of erythrocyte membrane proteins, sialoglycoproteins and lipids stained yellow with silver, while the other membrane proteins were counterstained with Coomassie blue (Dzandu *et al* 1984).

Some silver stains do not detect proteins such as calmodulin or troponin

C. However, pretreatment with gluteraldehyde often permits positive silver staining of these proteins (Schleicher and Watterson, 1983). Histones also often fail to stain with silver, but fixation with formaldehyde coupled with simultaneous prestaining with Coomassie blue partly alleviates this problem. However, even with this fixation procedure, the method is ten-fold less sensitive for histones than for neutral proteins (Irie and Sezaki 1983). Another example of differential sensitivity was demonstrated in a study utilising four different stains with salivary proteins. Different protein bands were visualised with each of the stains, and this selectivity may be related to the proline content of these salivary proteins (Friedman 1982).

7.4.4 Silver Quenching of Autoradiography or Fluorography

Quenching of ^{14}C-labelled proteins is barely perceptible with most of the non-diamine silver stains. Even with the most intense diamine stain, ^{14}C-labelled proteins can be detected by autoradiography with only a 50% decrease in image density. However, the detection of ^{3}H-labelled proteins is severely quenched by all silver stains. Destaining of the silver stained gel with photographic reducing agents can permit detection of approximately half of the fluorographic density of ^{3}H-labelled proteins, providing that the initial silver staining involved a non-diamine silver stain. Many diamine stains continue to quench, even after treatment with photographic reducing agents, so that fluorographic detection of ^{3}H-labelled proteins is not feasible. This impediment to ^{3}H detection with diamine stains is likely to be caused by a greater amount of residual silver deposited in the gels by the diamine stains, which block the weak β emissions. Residual silver has been demonstrated in gels that have been cleared by photographic reducing agents. A faint silver image of the protein can be observed after drying the gel with heat. Silver has also been demonstrated in these 'cleared' gels by electron beam analysis (Van Keuren *et al* 1981).

7.4.5 Sensitivity and Quantitation

Silver stains offer the most sensitive non-radioactive method for detecting most proteins separated by gel electrophoresis. Proteins labelled *in vivo* with high specific activity ^{14}C and separated by two-dimensional electrophoresis produced an image after 56 h of autoradiography that was comparable to the image obtained with silver staining (Merril *et al* 1979). Some increase in sensitivity could have been obtained by utilising fluorography or by increasing the exposure time of the autoradiography. However, both of these techniques generally result in some loss of resolution: in autoradiography, a ten-fold increase in exposure results in a doubling of the spot size (O'Farrell 1975).

Silver stains are more sensitive than the most sensitive post-electrophoretic organic stain, Coomassie blue (table 7.1), with chemical development silver stains being, in general, more sensitive than the photodevelopment silver stains. However, the photodevelopment stain is faster than most chemical development stains, producing an image within 10 to 15 min of gel electrophoresis (Merril *et al* 1984). Unfortunately, photodevelopment often produces negatively stained regions, even within positively stained bands. Many of the chemical development silver stains which have been modified to enhance colour may also display this effect. The presence of a collection of negative, positive, and mixed negative and positive regions in photodeveloped or colour-enhanced silver-stained gels makes quantitative analysis difficult (Merril and Harrington 1984, Merril *et al* 1984).

Table 7.1 Increase in the limits of sensitivity of silver stains as compared with Coomassie brilliant blue R250 stains

	Sensitivity increase (n-fold)	
Proteins†	*Photodevelopment silver stain*‡	*Chemical development silver stain*‡
Bovine serum albumin	5	10
α-lactalbumin §	> 10	> 20
Soybean trypsin inhibitor	10	100
Ovalbumin	10	200
Carbonic anhydrase §	> 20	> 200

† Proteins were Pharmacia-purified molecular weight markers. They were separated by SDS gel electrophoresis in 10% polyacrylamide.
‡ Coomassie blue and silver stains were performed as described (Merril *et al* 1984). Proteins were determined to be beyond the limits of detection when they were either no longer visible by eye, or by computer enhancement of the gel image.
§ Protein bands could not be detected with the Coomassie blue stain at the highest concentration of α-lactalbumin or carbonic anhydrase (15 and 10 ng respectively).

Attaining the highest sensitivity with silver stains requires care in selecting reagents, particularly the water which is used to make solutions (the water's conductivity should be less than 1 μmho cm). Additional sensitivity can often be obtained by recycling the electrophoretogram through a silver nitrate solution followed by a sodium carbonate solution containing formaldehyde. Such recycling permits visualisation of trace proteins that otherwise might not have been detected (Merril *et al* 1982). Many image intensification methods that have been developed for photography may also be used to intensify silver-stained gels. Such intensifiers may add more silver to the existing deposits, as in the recycling procedure, or they may combine other dense metals with the silver in the image. Some intensifying procedures exhibit an increase in density that is proportional to the amount of silver in the gel prior to intensification, while others increase contrast by combining primarily with the heaviest original silver deposits (superproportional) or preferentially build up the less dense regions (subproportional). Intensification with silver is usually proportional, while copper, mercury, copper iodide and mercuric chloride intensifiers are superproportional. Uranium, mercuric iodide and chromium tend to be subproportional intensifiers (Wall and Jordan 1976).

Destaining may also be proportional, subproportional or superproportional, with some of the silver stains being more resistant to destaining than others. The diamine stains are most resistant, while the photodevelopment stains are most sensitive to destaining. The most generally used photographic reducing agent that may be used as a destainer is Farmer's reducer (introduced by Howard Farmer in 1884). Farmer's reducer is a subproportional reducer and, if allowed, will remove all the silver from the lighter regions of the gel. Farmer's reducer utilises ferricyanide as a silver solvent and thiosulphate to complex and solubilise silver salts (Wall and Jordan 1976). The first reducer used to destain silver-stained polyacrylamide gels was a copper reducer employing ammonium hydroxide, copper sulphate and sodium thiosulphate (Switzer *et al* 1979). It is also a subproportional reducer.

If quantitation is intended, only proportional processes such as intensification by recycling in silver, or destaining in a proportional reducer should be employed. Even so, given the fine balance required to achieve proportional destaining or intensification, caution should be employed in quantitatively analysing gels which have been manipulated by these techniques.

Chemical development silver stains have been shown to be linear over a forty-fold range in concentration for most proteins (Merril *et al* 1982, 1984, Merril and Harrington 1984, Switzer *et al* 1979). However, protein concentrations greater than 2 ng mm^{-2} generally cause saturation of silver images, resulting in non-linearity above that concentration. This range of linearity can be extended by photographing gels during image development, before bands or spots become non-proportional or saturated. Saturation often results in band or spot centres which are less intensely stained than the regions near the edges. This effect is similar to the 'ring dyeing' noted with the post-electrophoretic organic stains.

An often quoted report by Poehling and Neuhoff (1981) states that 'Silver does not stoichiometrically stain proteins, unlike Coomassie blue'. However, their silver stain data actually is linear over a thirty-fold range in protein concentration, while their Coomassie blue data demonstrated linearity over a twenty-fold range (Dunn and Burghes 1983, Merril *et al* 1982). Plots of the optical density of silver-stained proteins against their concentration produce different slopes for each of the proteins studied. These protein-specific staining slopes have also been observed with most of the organic dyes including Coomassie blue (Fazekas de St Groth *et al* 1963), and even the commonly used Lowry protein assay (Lowry *et al* 1951).

Protein-specific staining slopes are indicative of the basic mechanism of most protein detection and assay methods. There are three fundamental mechanisms underlying most protein stains. The first depends on specific reactive groups, such as aromatic structures, which will produce staining slopes that are dependent on the number and accessibility of such specific groups in each protein. Since most current silver and organic protein stains produce protein-specific slopes, this appears to be the primary mechanism underlying most current silver and organic protein stains. The second mechanism would apply to stains based exclusively on a fundamental subunit, such as the peptide bond. One would then expect similar specific staining slopes for all proteins when the staining densities for each of the protein bands or spots are plotted against the mass of each of the proteins in their respective bands or spots. This argument assumes that the proteins are not modified, and that the average molecular weight of the fundamental subunit in each of the proteins is similar. The third mechanism is applicable to a stain that reacts with a single element, such as either the terminal amino group, or the terminal carboxyl group in each protein, and produces a similar staining slope for all proteins when the staining densities for each protein's bands or spots are plotted against the number of molecules of each protein in their respective bands or spots.

As most currently employed protein stains appear to depend on specific reactive groups, specific staining slopes may be utilised as identification parameters for proteins. Specific staining slopes also require that quantitative inter-gel comparative studies limit comparisons to homologous protein bands or spots on each gel. For example, the actin spot on one gel can be compared with an actin spot on another gel, but not with a transferrin spot. These limitations to homologous comparisons are also applicable to most of the organic stains, including the Coomassie blue stains (Fazekas de St Groth *et al* 1963).

Quantitative inter-gel comparisons require the presence of reference proteins for the normalisation of spot or band staining densities. One scheme for such normalisation utilises 'constitutive' proteins, a subset of proteins contained in each gel that have constant density ratios to each other in all of the gels examined. The sum of the densities of the constitutive proteins in an arbitrarily designated 'standard gel' are compared with the sums of the densities of the constitutive proteins in all other gels, and a specific normalisation factor is thus determined for each gel. These normalisation factors are then utilised to correct the densities of all the proteins on each gel to those of the standard gel. This scheme corrects for variations in staining, in image digitisation (Merril and Harrington 1984) and initial protein loading; a variation of initial protein loading of up to ten-fold may, however, be tolerated (Merril *et al* 1982).

7.5 Enzyme Stains

During the last few decades, techniques have been developed for the detection of specific proteins on the basis of their enzymatic reactions. The occurrence of multiple molecular forms of proteins with similar enzymatic activity was shown by Neilands (1952) in studies on lactate dehydrogenase. Early work on techniques to identify serum enzymes employed paper electrophoresis (Baker and Pellegrino 1954, Sayre and Hill 1957) and relied wholly, or in part, on elution and spectrophotometric methods for detection.

Hunter and Markert (1957) are credited with being the first to use histochemical staining methods to demonstrate enzymes directly in electrophoretic gels. Such methods rely on the enzyme's catalytic activity actually surviving electrophoresis, and on the ability to couple one of the catalytic reaction products to a second, colour-producing reaction. The earliest methods detected enzymes that catalysed simple reactions, the products of which could be assayed colorimetrically. One example is the use of phenolphthalein phosphate as a substrate for acid phosphatase. Phenolphthalein, one of the phosphatase reaction products, is pink at a pH above 9.0. In another example, amylase is detected by incubating the electrophoretic gel in contact with a starch plate, followed by staining of the starch with iodine. Enzymes with amylase activity appear as clear areas on a purple background.

The majority of enzyme staining methods in use today involve electron transfer dyes such as methyl thiazolyl tetrazolium (MTT), which is reduced by electron donors to form a dark blue insoluble formazan. This reaction, catalysed by phenazine methosulphate (PMS), can be used to detect enzymes whose activities lead to the production of the reduced forms of the coenzymes NAD or NADP (nicotinamide adenine dinucleotide, nicotinamide adenine dinucleotide phosphate (figure 7.2(*a*)). This method can be expanded to stain for any enzyme that can be coupled by one or more intermediate reactions to the reduction of NAD or NADP (figure 7.2(*b*)). As these detection methods rely on enzymes retaining their catalytic activities, electrophoretic parameters such as the pH and composition of the buffers must also be optimised for specific enzymes of interest. These limitations are not absolute, as Scheele *et al* (1980) were able to assay the enzymatic activity of 15 human pancreatic proteins separated by two-dimensional electrophoresis under denaturing conditions by excising and renaturing the enzymes.

Applications of isoenzyme techniques are qualitative, but in some cases measure relative quantities of enzyme. Quantitative analyses are generally performed spectrophotometrically.

To date, methods have been developed for nearly a hundred specific enzymes, many with modifications tailored to several different electrophoretic buffer systems. Details of specific stains are listed in such review works as Brewer (1970), Harris and Hopkinson (1976) and Shaw and Prasad (1970).

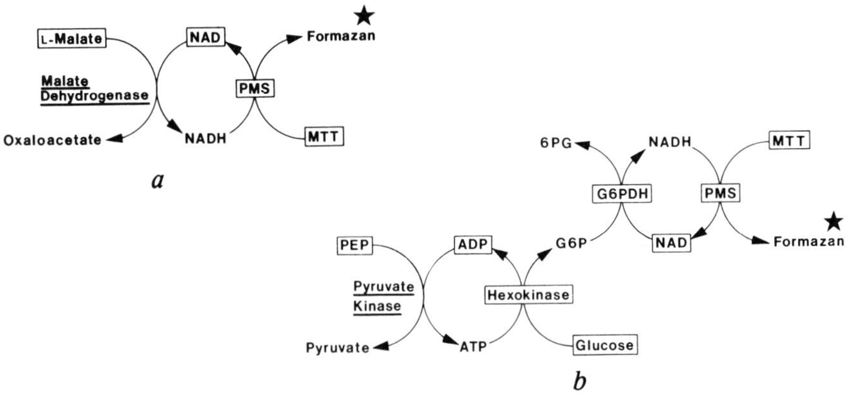

Figure 7.2 Reaction mechanisms of enzyme staining systems that utilise electron transfer dyes. The required ingredients of the staining mixtures are boxed. Enzymes that will produce a stain are underlined. NADH is the reduced form of NAD.

7.6 Detection of Radioactive Proteins

Detection of radioactive proteins separated by gel electrophoresis is most commonly performed by autoradiography. The development of these detection methods is closely interwoven with the discovery of radioactivity. The first autoradiographs were made by exposing photographic plates to crystals of radioactive uranium salts (Saint-Victor 1867). Becquerel (1896) studied this phenomenon by placing various crystals next to photographic plates covered with dark paper. He wished to determine whether, after exposure to sunlight, these crystals would emit rays, similar to those discovered by Roentgen, that could penetrate the paper and expose the film. Becquerel found that crystals of uranium salts emitted such rays, whether or not they were exposed to sunlight. Marie Curie discovered that similar rays were emitted by other elements and named this phenomenon radioactivity (Weinberg 1983).

Application of autoradiography to biological problems began with the use of photographic film to locate radioactive polonium in histological tissue sections (Lacassagne and Lattes 1925). The technique was also used to identify ^{14}C-labelled amino acids separated by chromatography (Fink *et al* 1947), while 'scintillation autographs' or fluorographs were first utilised as a detection method to visualise chromatographically separated compounds labelled with ^{3}H (Wilson 1958).

7.6.1 Radiolabelling of Proteins

Methods of radiolabelling proteins vary from *in vivo* incorporation of the radiolabel to *in vitro* labelling of different chemical groups. *In vivo* labelling most commonly utilises cell or tissue culture techniques. The most frequently selected isotopes for electrophoretic studies are ^{3}H, ^{14}C, ^{35}S, ^{32}P and ^{125}I. Generally, a radiolabelled amino acid that is fairly ubiquitous (e.g. methionine or leucine) is used to supplement an unlabelled culture medium that is deficient in that amino acid. Such radiolabelling can be used as an aid in protein detection, or, by labelling specific cell phases and/or physiological events it can be used to study protein turnover. Radiolabelling can also be utilised to label specific proteins, such as the use of D-6-(^{3}H)-glucosamine as a precursor for glycoproteins (Haley and Gouras 1984).

Analysis of complex protein mixtures can be facilitated by the use of multiple radio-isotopes, each with different characteristics, such as differing energies of emission or half lives. A number of investigators have labelled one protein preparation with one isotope, another protein preparation with a different isotope, and then electrophoresed the mixture. After autoradiography, the two protein preparations may be distinguished by the specific energies of their radiolabel (Choo *et al* 1980, McConkey 1976, Walton *et al* 1979), or by their different half lives.

In vivo radiolabelling has permitted amino acid compositions of proteins to be determined. Latter *et al* (1984) used 20 different radiolabelled amino acids, each to label a different culture from the same original cell line. Quantitation of autoradiographs of two-dimensional electrophoretograms run from each of the 20 cultures provided sufficient information to determine the amino acid composition of some proteins. This technique requires the presence of a protein of known composition, and the assumptions that *in vivo* labelling is uniform and interconversion between amino acids minimal.

In vitro radiolabelling may be performed either before or after electrophoretic separation. If radiolabelling is performed prior to electrophoresis, alteration in the migrational characteristics of the protein may occur, such as those due to alterations in charge produced by iodinated p-hydroxyphenylpropionic acid, N-hydroxysuccinimide ester (Bolton–Hunter reagent, Bolton and Hunter 1973, Hunter and Greenwood 1962), radiolabelled lectins (Burridge 1978), and 1,3,4,6-tetrachloro-3,6-diphenylglycouril (Iodogen, Tuszynski *et al* 1983). Other procedures, such as reductive methylation with boro-(^{3}H)-hydride (Means and Feeney 1968), do not appear to effect migration (Kuhn and Wilt 1980). Recently introduced radioactive post-electrophoretic stains include ^{59}Fe-ferrous bathophenathroline sulphonate, which allows detection of as little as 10 ng of protein (Zapolski *et al* 1982). Neutron activated Ponceau red S and Serva blue R have recently been introduced as post-electrophoretic stains also (Zapolski *et al* 1984).

Some *in vitro* techniques have inherent problems of high background staining. It has been shown that this problem could result from iodination of the contaminants in acrylamide (Elder *et al* 1977), or it may arise from neutron activated contaminants as noted in the stains of Zapolski *et al* (1984).

7.6.2 Autoradiography

Radiolabelled proteins separated on electrophoretograms are most frequently detected by autoradiography, a process that uses photographic film as the detector. Ionising radiation, including β, γ or X-rays, can create a latent image centre of metallic silver from the silver halide crystals in the film, which is then made visible by photographic development.

The main factor determining autoradiographic exposure time is the activity and quantity of radio-isotope present. This can be estimated from the quantity of label in the proteins that are subjected to electrophoresis. Many additional factors influence optimal autoradiographic detection. These include temperature, whether the autoradiograms are made from hydrated or dehydrated gels, the quenching potential of prior protein stains, the type of film and the use of intensification techniques in the photographic process.

In general, a photographic film's sensitivity to ionising radiation increases linearly with temperature between -100 and $60^{\circ}C$ and is independent of the energy of the radiation or the temperature before or after exposure (Bass 1950, Mees 1952, Morgan 1944, Reckie 1939).

Films may be exposed to hydrated gels but then a barrier must be placed between the film and the gel to protect the film. Damage can result from the fusion of the hydrated gel with the photographic emulsion, or by moisture-induced destabilisation of the latent image centres (Mees 1952). A barrier, however thin, will significantly impair or prevent weak β emitters such as ^{3}H from exposing the film. By dehydrating gels, the path lengths from the emitting isotope to the film are shortened and the barrier or membrane required in the use of hydrated gels can be eliminated.

Quenching of the radiolabel due to staining before autoradiography may diminish sensitivity. This has only been fully studied for silver stains, where it is a significant problem for weak β emitters (^{3}H, ^{14}C and ^{35}S). Tritium is most easily quenched, and the thickness of the acrylamide matrix and silver staining procedures have been shown to effect this quenching (Van Keuran *et al* 1981). Destaining prior to autoradiography may lessen quenching, as discussed in the section on silver staining.

Choice of photographic film depends on the isotope(s) studied. For autoradiography of low and medium energy β emitters, higher speed X-ray films are most suitable (Randerath 1970). For higher energy β or γ emitters, a double emulsion film minimises exposure time, while single emulsion film leads to better image resolution. To enable alignment of proteins on a stained gel with their images on an autoradiograph (or fluorograph) to take place, it is convenient to mark the periphery of the gel with radioactive ink (Bonner 1983).

A number of intensifying techniques that are applicable to autoradiography were originally developed for astronomical research. Askins (1978) reported image intensification when faint photographic images were treated with ^{35}S-thiourea, which labels the silver grains in the photographic image with ^{35}S. This radiolabelled photograph is then used to expose a fresh film, producing a second autoradiograph with an 80–90% increase in image density. Amplification of faint images on photographic film may also be achieved by making contact prints (Malin 1978). Since the photographic grains of faint images are concentrated in the upper layers of emulsion, while background fog is uniformly distributed throughout the film, the image from the upper layer of the original photograph is preferentially recorded in the contact print, resulting in an increased signal-to-noise ratio. Other intensification methods include bathing a pre-exposed film in 8.3×10^{-4} M silver nitrate (Carroll and Hubbard 1932, Clark 1946, Jenkins and Farnell 1978), or baking the pre-exposed film in a hydrogen atmosphere for 2 h at 66°C (Smith 1982 a, b, 1983). Both of these methods produce a 15- to 35-fold increase in sensitivity.

7.6.3 Fluorography

Bonner and Laskey (1974) first applied fluorography to improve the detection of ^{3}H-, ^{14}C- and ^{35}S-labelled proteins separated by polyacrylamide gel electrophoresis. The mechanism of the fluorographic process involves conversion of some of the energy from the radio-emission to light by the addition of a fluor or scintillant to the gel. Photographic film is used to record this fluorographic image. If low energy β emitting isotopes, such as ^{3}H, ^{14}C, ^{33}P or ^{35}S are used as the radiolabel, the fluor must be dispersed throughout the gel, since the path length of β particles from these isotopes is often too short to penetrate through the gel to a fluor on the gel surface (Bonner 1983). The optimal amount of fluor to be incorporated in the gel must be empirically determined (Bonner and Laskey 1974). Furthermore, a film with a spectral sensitivity that complements the emission spectra of the fluor should be chosen. Most fluors and intensification screens emit ultraviolet and blue light, and films sensitive to this region are appropriate. However, some intensification screens emit primarily green light, requiring the use of orthochromatic films.

The use of low temperature during fluorography often results in an increase in sensitivity. This is contrary to the effect of temperature in autoradiography, in which sensitivity increases with temperature from -100 to 60°C (Mees 1952). The mechanism for this paradoxical response is unclear and two theories have been proposed. Luthi and Waser (1965) have suggested that energy from a fluor may be radiated partly as light and partly as thermal radiation from intramolecular vibrations. They suggested that, at low temperature, there is less intramolecular vibration thus making more energy available for light production. A personal communication from Prydz was cited by Randerath (1970) as demonstrating that the increase in light production by fluors at lower temperatures is minimal, or absent as in the case of anthracene, the fluor used by Luthi and Waser. Another proposed mechanism elaborated by Randerath (1970) is based on the difference between the interaction of light and elementary particles with the silver halide crystals in the film emulsion. Elementary particles cause a darkening of the film which is proportional to the product of the intensity of the particle emissions and the exposure time, irrespective of temperature. Light, however, has this relationship only at medium light intensity. At low light intensity, this relationship is lost (reciprocity failure). Although the basis for this reciprocity failure is not completely understood, it has been suggested that at low light intensity, one photon of light causes one silver ion to be reduced to metallic silver. Since it takes four silver atoms to form a stable latent image centre (Hamilton and Logel 1974), any unstable latent image centres formed by less than four light photons will decay rapidly. By lowering the temperature, the lifetime of an unstable latent image centre is increased to 4.8 ms (at 77° K), as compared with a lifetime of 1.7 μs at room temperature (293 K) (Kellog 1977). This increased lifetime at low temperature permits the accumulation of sufficient photons from the radioactively induced fluorescence to form a stable latent image centre containing at least 4 silver atoms. Most investigators maintain their fluorographs at -70°C to -80°C during exposure, obtaining a twelve-fold increase in sensitivity for ^{3}H and nine-fold increase for ^{14}C and ^{35}S. West and Gilman (1977) attributed some of this increase in sensitivity at low temperature to a decrease in desensitisation of the latent image by oxygen and moisture.

Pre-exposure of the photographic film to light that is sufficient to produce a uniform fog of 0.15 optical density results in a greater range of linearity and a two- to three-fold increase in the relative densities of faint spots (Bonner and Stedman 1978, Laskey and Mills 1975). As sensitised film has a reduced threshold for the creation of stable latent image centres, extra caution should be taken to avoid exposure to contaminating light, such as from 'safelights' in the darkroom. Endogenous radioactive emissions from intensification screens may also blacken a pre-exposed film during a long exposure (Laskey 1980).

7.6.4 Intensification Screens

Fluorography has also been useful for the detection of higher energy emissions, such as from ^{32}P and ^{125}I. Much of the emission from these isotopes passes directly through the photographic film. More efficient detection of these high energy emissions can be achieved by placing the film between the gel and an image intensification screen (Laskey and Mills 1977). The strong emission that passes through the film excites the fluor or phosphor in the screen, and a fluorographic image is then superimposed on an autoradiographic image. Swanstrom and Shanks (1978) suggested that a second screen arranged: sample, screen, film, screen, improves the sensitivity of detection of ^{32}P, but it leads to loss of resolution (Bonner 1983). Two screens do not increase detection sensitivity for ^{125}I.

7.6.5 Dual Isotope Techniques

Dual isotope detection systems can be used to advantage by exposing film(s) to an electrophoretogram containing different isotopes under different conditions. A fluorograph will detect all proteins containing ^{3}H, ^{14}C or ^{35}S, while an autoradiograph will record the emission from only the ^{14}C or ^{35}S. Overlay of the fluorograph on the autoradiograph allows interpretation, by subtraction, of the differential labelling of the proteins (McConkey 1976, Walton *et al* 1979). Another method (Lecoq *et al* 1982) uses a combination of the isotopes ^{35}S and ^{75}Se. First, fluorography is performed to detect the emissions from both isotopes. The γ emitter (^{75}Se) is then separately recorded by blocking β emissions with an exposed X-ray film that is interposed between the sample and the autoradiographic film. Another dual isotope technique uses isotopes with differing half lives. By exposing two films, one before and one after the decay of the shorter lived isotope, a record of differential labelling is obtained from the mixture. Depending on the isotopes selected, the time of decay may make this procedure impractical.

Kronenberg (1979) used colour negative film to differentiate radioactive emissions according to the colour of the radiographic images that were produced. He used a film containing three photographic emulsion layers; the first layer produces a yellow image, the second a magenta image, and the third layer a cyan image. Tritium exposes only the first layer, distinguishing it from ^{14}C and ^{35}S which expose both first and second layers, and ^{32}P or ^{125}I, which expose all three layers. This method has the advantage that ^{3}H can be distinguished from the other emitters on the same film.

7.6.6 Quantitation

Quantitation presents three concerns that are peculiar to autoradiographic and radiolabelling processes. The curve relating silver deposition in the autoradiograph to increments of radioactivity is sigmoidal. Quantitation is only possible if the linear portion of the curve is utilised, or if correction factors are used for the nonlinear portions of the curve (Bossinger *et al* 1979). This problem is partly obviated by the use of pre-exposed film (utilising a light flash of 1 ms duration to hypersensitise the film) when the film response is linear from an optical density of 0.15 (the background fog level achieved by pre-exposure) up to an optical density of 1.5 (Laskey and Mills 1975).

In vivo labelling of proteins with specific amino acids results in proteins that contain different molar ratios of radio-isotopes, depending upon the content of the labelled amino acid or subgroup in each protein. These molar ratios result in a protein-specific relationship between the density of each band or spot visualised in the autoradiograph or fluorograph, and the actual amount of protein. Quantitative comparisons are thus limited to homologous proteins, just as with most organic dyes and silver detection methods.

Autoradiographic or fluorographic optical densities can be converted to units of disintegrations per minute (DPM) for each isotope employed, but only when standards containing the specific isotope are assayed along with the sample (Bossinger *et al* 1979). The minimum amounts of radioactivity detected using fluorography with intensification screens at $-70^{\circ}C$ and pre-exposed film, were 300 DPM ^{3}H and 30 DPM ^{14}C in a 1 cm × 1 mm band in a gel (Laskey and Mills 1975).

7.6.7 Non-autoradiographic Detection and Quantitation of Radiolabelled Proteins

Radiolabelled proteins, located on a gel either by autoradiography or by chemical staining, may be excised from the gel and their emission quantitated by using liquid scintillation spectrometry. This method gives results comparable to the quantitation of autoradiographs (Van Keuran *et al* 1981). Although feasible (Lieberman *et al* 1980), this technique is fairly laborious for the routine analysis of the many proteins separated by two-dimensional electrophoresis.

Single- and multi-wire proportional counters have been used to quantitate β emissions in a chamber of ionising gas (Goulianos *et al* 1980, Scott *et al* 1982). This system is more than a thousand-fold faster for ^{3}H detection than conventional autoradiography, but resolution is limited to 2 mm. Other techniques for rapid electronic recording of β emissions include the detection of light photons from fluorography using an electronic image intensification system (Burbeck 1983, Davidson and Case 1982). These methods offer faster detection than autoradiography, but need further development to increase both their resolution and their ability to differentiate samples that are radiolabelled with more than one isotope.

7.7 Detection of Protein-bound Trace Elements

Of the more than 70 elements that have been detected in human tissues, most occur at very low levels (less than 1 $mmol\,kg^{-1}$) and are considered to be trace elements. Many are bound to, or carried by, specific proteins. Although the majority of trace element research has been limited to the determination of the total concentrations of specific elements in tissues or body fluids, there is a growing body of knowledge on the metabolism of trace elements and their interactions with proteins. Such studies have been restricted by the quantitative limitations of current analytical methods.

Early studies involved the injection of radiolabelled phosphorus (Maurer and Schild 1951) into rats and mice, followed by paper electrophoresis of serum. The paper electrophoretograms were cut into 5 mm strips and protein fractions containing phosphorus identified by using a Geiger-Müller tube to detect radiation. Wallenius (1952) performed a similar study with ^{59}Fe, using autoradiography to locate the iron-containing proteins.

Subsequent workers utilised a variety of preparative techniques to purify proteins from large samples (2–100 ml) of biological fluids prior to analysis for trace metals. Himmelhock *et al* (1966) used ion-exchange chromatography for separation and spark emission spectroscopy for quantitation of trace metals. Gel filtration was used in conjunction with neutron activation analysis (NAA) by Evans and Fritze (1968) in studies of copper binding proteins in serum, while Schmelzer and Behne (1975) employed isoelectric focusing to separate protein fractions of serum for quantitation of trace elements by NAA. Norheim and Steinnes (1975) were able to detect as little as 0.3 ng of arsenic and 2 ng of copper in liver fractions using gel filtration, followed by NAA.

In-gel detection of metalloproteins was pioneered by Drysdale (1970) who used the Prussian blue reaction to identify iron in ferritin following isoelectric focusing of serum proteins in polyacrylamide gels. Latner (1973) used 2,4-dinitroso-1,3-naphthalenediol to stain for the iron of transferrin in a two-dimensional electrophoretic gel. Both methods were qualitative and required high protein concentrations.

Polyacrylamide gel electrophoresis was used to separate liver (Ribas 1983) and brain (Ribas Ozonas 1983) proteins of rats injected with ^{109}Cd. These gels were sliced into 1 mm bands and γ activity was counted. Legrum *et al* (1983) subjected strips of a polyacrylamide gel on which microsomal proteins were separated to NAA. They were able to detect as little as 0.01 ng of cobalt per mg of gel.

Because of the low levels of trace elements in biological samples, studies of proteins that contain trace elements have relied on their purification from relatively large samples. Another approach has involved the introduction of radiolabelled trace elements into the biological system. Efforts are currently underway to develop techniques for in-gel mapping and quantitation of specific elements in electrophoretically separated metalloproteins and phosphorylated proteins. Two approaches that have proven most promising are electron microprobe analysis and autoradiography of neutron activated gels (Harasewych *et al* 1985a, b).

Electron microprobe analyses of a polyacrylamide electrophoretogram with phosvitin, a protein containing nearly 10% phosphorus, and haemoglobin, with 0.34% iron, is illustrated in Figure 7.3. The distribution of sulphur is thought to represent sulphur in the ammonium persulphate, added to catalyse polymerisation of the acrylamide, migrating in the electrophoretic field. Of the two phosphorus peaks, the larger more anodal peak migrates at the same rate as Bromophenol blue and is indicative of phosphorus that is not protein bound. It is not known whether it is due to impurities or to the dissociation of phosvitin. The smaller, broader peak corresponds to the electrophoretic mobility of phosvitin. The broadening and displacement seen in the microprobe profile are thought to be caused by the diffusion of protein during drying of the unfixed gel. The distribution of iron appears to be in two peaks, both near the lower limit of detection (56 pg μm^{-1}). Neutron activation of polyacrylamide electrophoretograms containing the same two proteins were performed and permitted detection of as little as 10 ng of phosphorus by autoradiography. Although autoradiographic images corresponding in electrophoretic mobility to haemoglobin were detected at haemoglobin levels of 640 μg, gamma spectrometry was unable to confirm the presence of iron, and it is thought that these images were due to the activation of sulphur atoms in haemoglobin into ^{32}P (Harasewych *et al* 1985b).

Analysis of protein-bound trace elements in electrophoretograms is limited by certain methodological considerations. Denaturation of proteins during SDS electrophoresis, fixation or staining results in the loss of non-covalently bound trace elements (Harasewych *et al* 1985a, b). Steifel *et al* (1980) demonstrated that the pH at which the samples are electrophoresed is also a major factor affecting the lability of trace elements, with recovery of beryllium ranging from 15 to 80% depending upon pH only. Careful consideration must also be given to all components of the samples, gels and buffers to avoid unacceptable background. Sodium dodecyl sulphate (SDS) for example, cannot be used in gels that will be neutron activated, as a portion of its ^{32}S is converted to ^{32}P, making it impossible to distinguish activated ^{32}P of phosphorylated proteins from the uniformly high background.

The lower limit of detection for trace elements is comparable to that for proteins on a molar basis, but the molecular weight of most proteins exceeds that of their trace elements by several orders of magnitude. Therefore, to study the distribution of proteins and their trace elements on

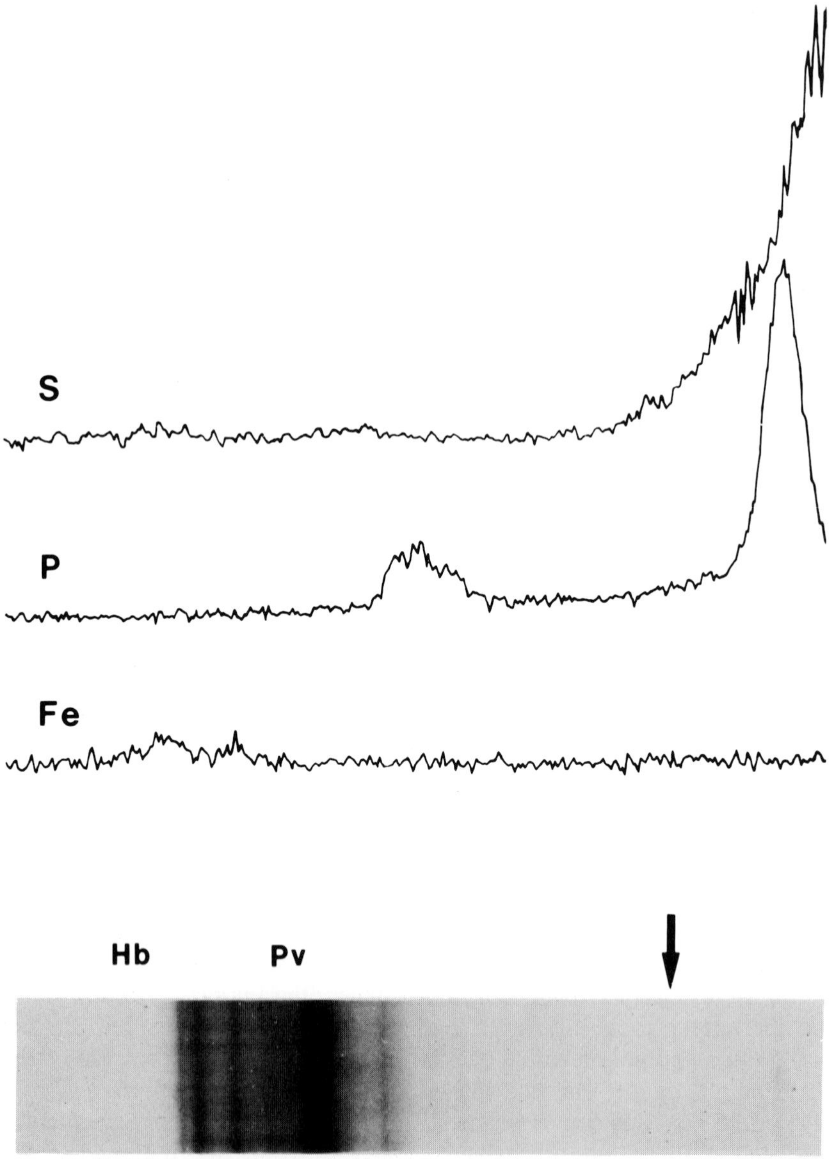

Figure 7.3 Electron microprobe scans for the elements sulphur (S), phosphorus (P) and iron (Fe) of a one-dimensional polyacrylamide gel on which haemoglobin (Hb) and phosvitin (Pv) have been separated. The bottom portion of the figure illustrates the silver-stained electrophoretogram. The arrow indicates the position of the Bromophenol blue front.

the same electrophoretic gel, trace element detection methods must be several orders of magnitude more sensitive than those for proteins. This still remains to be achieved.

7.8 Detection of Proteins on a Support Matrix after Electroblotting

After separation of proteins in a gel matrix such as acrylamide, staining is partly limited by features of this matrix. Its thickness and non-uniformity affect permeability to stains, while difficulty of handling and limits in durability restrict the use of multiple consecutive stains. These limitations are all lessened by the electrotransfer of proteins from the gel to a thin matrix, most commonly nitrocellulose paper, as first described by Towbin *et al* (1979). This electrotransfer technique has come to be known as electroblotting or Western blotting. This rapidly expanding field was reviewed by Gershoni and Pallade (1983), with additional technical information provided in reviews by Tsang *et al* (1983) and Symington (1984).

Protein transfer from gel to paper can be achieved by diffusion or solvent flow, but by far the most rapid, sensitive and commonly used technique involves electrotransfer. Under optimal conditions, over 90% of the protein is transferred in this manner (Burnette 1981). Many variations in transfer buffer, electrical running conditions and the receiving support matrix have been selected to optimise the transfer of specific proteins. Proteins are generally transferred immediately after electrophoresis, since even contact for only a few seconds with fixing solution inhibits transfer. Phelps (1984), however, has described a method of transferring proteins from fixed gels after Coomassie blue staining, by SDS treatment of the gel.

Nitrocellulose paper is the most commonly utilised matrix for electroblots, and most staining methods have been developed for this material. Nitrocellulose with a pore size of 0.45 μm is used most frequently, but a smaller matrix of 0.20 μm is recommended to avoid loss of small proteins, which may pass directly through the larger pore sizes (Burnette 1981). Alternative matrices include diazo-modified or cyanogen bromide activated cellulose papers, and various nylon sheets. The nylon membranes have greater strength and protein binding capacity than nitrocellulose paper, which leads to greater sensitivity with immunostains, but the chemical stains tend to work poorly on nylon because of nonspecific background staining (Gershoni and Palade 1983). Whatever stain is used on an immunoelectroblot, background staining is reduced or avoided by blocking non-specific binding. This is usually achieved by incubation of the electroblot with a protein such as bovine serum albumin, or with a detergent such as polyoxyethylene sorbitan monolaurate (Tween 20).

All protein stains that are used directly on gels can be employed on electroblots. Stains usually have a greater sensitivity on the electroblot than on the gel, because the thinner matrix of the electroblot allows greater access to the protein. There is also the ability to re-probe the same blot with sequential detection reactions, after erasure of each preceding stain. Erasure methods depend on the stain; for example, pH alteration removes antigen/antibody complexes, alternatively, probes may be removed by denaturation or dissociation from the matrix. It is feasible to probe the same blot up to ten times. This versatility is emphasised by the ability to perform such staining over a period of at least a year.

Chemical stains, including Coomassie blue (Burnette 1981), Amido black (Towbin *et al* 1979), Fast green (Reinhart and Malamud 1982) and silver have all been used on nitrocellulose paper. India ink stains proteins more sensitively on nitrocellulose than Amido black, Coomassie blue and Fast green, detecting some proteins at levels of 80 ng (Hancock and Tsang 1983). Silver stains have proven capable of detecting ng quantities of proteins that have been electroblotted onto nitrocellulose (Brada and Roth 1984, Merril *et al* 1984, Yuen *et al* 1982). Furthermore, after silver staining, the electroblotted proteins still retain their antigenic properties which subsequently permit the use of immunostains (Yuen *et al* 1982).

Immunostains have been extensively used on nitrocellulose paper with both poly- and monoclonal antibodies. High sensitivity may be achieved by utilising dual antibody techniques. Xu and McFarlin (1984) demonstrated that 10 ng of an IgG oligoclonal band from the cerebrospinal fluid of patients with multiple sclerosis could be detected on nitrocellulose paper by immunostaining. One ng of protein can be detected from an electroblot of a two-dimensional electrophoretogram, using a 1:40000 dilution of primary antibody, with immunoperoxidase staining of the antigen/antibody complex, amplified by a secondary antibody with an avidin/biotin/peroxidase complex developed by Hsu *et al* (1981). This is illustrated in figure 7.4, where spinal fluid proteins from a patient with multiple sclerosis are separated by two-dimensional electrophoresis, and four kappa immunoglobulin light chains are detected equally by both silver and immunostaining. Other dual antibody techniques employ secondary antibodies that are radiolabelled, enzyme-linked, and fluorescein- or rhodamine-conjugated. These methods allow detection of 1 pg to 1 ng of protein (Gershoni and Pallade 1983). Nearly all the applications have superior speed and sensitivity on nitrocellulose when compared with acrylamide gels (Symington 1984). Electrotransfer of ^{14}C- or ^{35}S-labelled proteins permits more efficient autoradiography, but does not improve the efficiency of fluorography (Erickson *et al* 1982, Symington *et al* 1981).

Proteins on electroblots have also been reacted with many types of ligands, which are then visualised with an immuno- or radio-labelled detection system. Some ligands, such as dinitrophenol (Wojtkowiak *et al* 1983) and pyridoxyl 5′-phosphate (Kittler *et al* 1984), are fairly general and bind to all proteins with sensitivities in the ng range. Other ligands are more

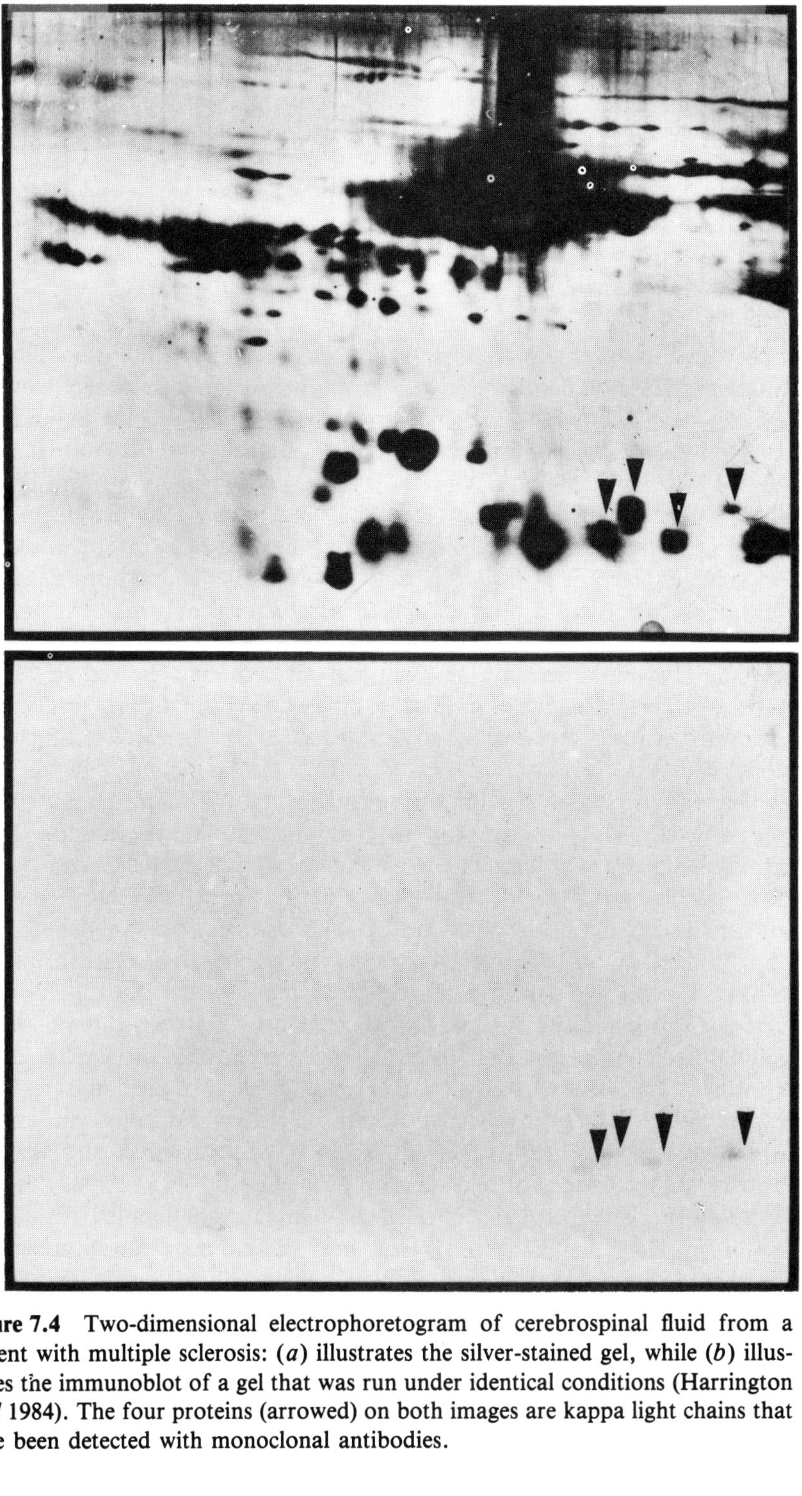

a

b

Figure 7.4 Two-dimensional electrophoretogram of cerebrospinal fluid from a patient with multiple sclerosis: (*a*) illustrates the silver-stained gel, while (*b*) illustrates the immunoblot of a gel that was run under identical conditions (Harrington *et al* 1984). The four proteins (arrowed) on both images are kappa light chains that have been detected with monoclonal antibodies.

specific and these include binding of electroblotted proteins to lectins, calmodulin, receptors, protein A, collagen or other proteins, DNA, RNA, or whole cells (Anderson *et al* 1982, Gershoni and Pallade 1983, Symington 1984, Towbin *et al* 1979).

7.9 Conclusion

The introduction and development of electrophoretic techniques has produced ever more powerful means of resolving proteins from complex mixtures. Development of these separation techniques has been paralleled by the development of protein detection methods with ever increasing sensitivities. Techniques have progressed from direct observation of protein-coated microspheres and coloured proteins, to the detection of proteins by their absorption of ultraviolet light, the observation of schlieren patterns, labelling with radionuclides, and, more recently, by direct staining with organic, fluorescent and silver stains. Methods have also been developed that permit the identification of specific types of proteins. Proteins may be stained on the basis of specific side groups (e.g. glycoproteins), or for specific enzymatic activity. Immunological methods are used to identify those proteins with specific antigenic sites, while radiolabelling or elemental analysis permit proteins to be distinguished on the basis of trace element content.

During this technical evolution, sensitivity has increased from the mg to the pg level. This range of sensitivity, coupled with high resolution separation techniques, now permits clinical studies of proteins in fluids that were not possible even a decade ago. As an example, over 300 proteins can now by visualised from less than 100 μl of cerebrospinal fluid using silver staining, instead of the 30 proteins that were visualised with organic stains. Similar sensitivities can be achieved with proteins labelled to high specific activities with radioactive tracers, however, this is often not possible in animal and human studies for economic and, in the latter case, ethical considerations. Electrotransfer of proteins from the gel matrix to a thin membrane can be performed with high efficiency and sensitivities of the order of 1 pg have been achieved. These electrotransferred proteins may also be subjected to multiple staining procedures.

Many of the protein detection techniques may be employed quantitatively, provided that their methodological limitations are respected. Most protein stains and autoradiographic methods exhibit protein-specific quantitative responses. Optical density/concentration relationships are usually linear over a thirty- to forty-fold range in concentration. Protein-specific slopes are indicative of the dependence of these detection methods on the content of specific groups within each protein. These protein-specific staining slopes may be utilised to differentiate proteins and emphasise the need to limit quantitative comparisons to homologous proteins. So long as inter-gel studies are confined to comparisons of homologous proteins and observations are made within the linear range of the detection procedures, valid quantitative results may be obtained.

Future advances in the quest for increased sensitivity may utilise laser-induced fluorescence, which has already detected as little as 28 attograms (1 attogram = 10^{-18} g) of the dye Rhodamine 6G (Dovochi *et al* 1983). Photoacoustic spectroscopy has already been utilised to detect and characterise specific proteins in both stained and unstained polyacrylamide electrophoretograms (Möller *et al* 1983). As the fundamental mechanisms of the protein stains are revealed, it may also be possible to achieve even greater sensitivities and specifications with these reagents. Continued efforts in the development of protein separation and detection methods may someday permit the visualisation of all proteins synthesised in every tissue or present in every body fluid.

Acknowledgments

We are grateful to Dr Julia H Onuferko, Research Division, W R Grace and Co. for assistance with, and helpful discussion of, electron microprobe techniques, and to Dr Mark L Moore, Armed Forces Radiobiology Research Institute, for making facilities for neutron activation available. This work was supported by the Intramural Program of the NIMH and the Donald Yarborough Fund.

References

Abramson H A 1934 *Electrokinetic Phenomena and their Applications to Biology and Medicine* Chemical Catalog Co. Inc. (New York) pp 17–104

Allen R C 1980 *Electrophoresis* **1** 32–7

Anderson N L, Giometti C S, Gemmell M A, Nance S L and Anderson N G 1982 *Clin. Chem.* **28** 1084–92

Ansorge W (1983) *Electrophoresis '82* ed. D Stathakos (Berlin: de Gruyter) pp 235–242

Askins B S 1978 *Science* **199** 684–6

Baker R W and Pellegrino C 1954 *Scand. J. Clin. Invest.* **6** 94–101

Barger B O, White F C, Pace J L, Kemper D L and Ragland W L 1976 *Anal. Biochem.* **70** 327–35

Bass H 1950 *AEC Report NYO-1522* Aug 3 pp 1–17

Becquerel A H 1896 *C. R. Acad. Sci., Paris* **122** 420–1

Bielschowsky M 1904 *J. Psychol. Neurol.* **3** 169–89

Bolton A E and Hunter W M 1973 *Biochem. J.* **133** 529

Bonner W M 1983 *Methods in Enzymology* Vol. 96 ed. S Fleischer and B Fleischer (New York: Academic) pp 215–21

Bonner W M and Laskey R A 1974 *Eur. J. Biochem.* **46** 83–8

Bonner W M and Stedman J D 1978 *Anal. Biochem.* **89** 247–56

Bosshard H F and Datyner A 1977 *Anal. Biochem.* **82** 327–33
Bossinger J, Miller M J, Vo K-P, Geiduschek E P and Xuong N H 1979 *J. Biol. Chem.* **254** 7986–8
Boulikas T and Hancock R 1981 *J. Biochem. Biophys. Methods* **5** 219–22
Brada D and Roth J 1984 *Anal. Biochem.* **142** 79–83
Brewer G J 1970 *An Introduction to Isozyme Techniques* (New York: Academic)
Burbeck S *Electrophoresis* **4** 127–33
Burnette W N 1981 *Anal. Biochem.* **112** 195–203
Burridge K 1978 *Methods in Enzymology* vol. 50 ed. V Ginsburg (New York: Academic) pp 54–64
Cajal S R 1903 *Trab. Lab. Invest. Biol. Univ. Madrid* **2** 129–222
Carroll B H and Hubbard D 1932 *The Photographic Emulsion* Focal (London: Focal) p 241
Choo K H, Cotton G H and Danks D M 1980 *Anal. Biochem.* **103** 33–8
Chrambach A, Reisfeld R A, Wyckoff M and Zaccari J 1967 *Anal. Biochem.* **20** 150–4
Clark W 1946 *Photography by Infrared* 2nd ed (New York: Wiley) p 94
Curie M 1898 *C. R. Acad. Sci.* Paris **143** 332–41
Datyner A and Finnimore E D 1973 *Anal. Biochem.* **55** 479–91
Davidson J B and Case A L 1982 *Science* **215** 1398–1400
Davis B D and Cohn E J 1939 *Ann. NY Acad. Sci.* **39** 209–12
de Olmos A (1969) *J. Brain Behav. Evol.* **2** 213–37
Diezel W, Kopperschläger G and Hofman E 1972 *Anal. Biochem.* **48** 617–20
Dovichi N J, Martin J C, Jett J H and Keller R A 1983 *Science* **219** 845–7
Drysdale J W 1970 *Biochem. Biophys. Acta* **207** 256–8
Dubray G and Bezard G 1982 *Anal. Biochem.* **119** 325–9
Duhamel R C , Meezan E and Brendel K 1980 *Biochem. Biophys. Acta* **626** 432–42
Dunn M J and Burghes A H M 1983 *Electrophoresis* **4** 173–89
Durrum E L 1950 *J. Am. Chem. Soc.* **72** 2943–8
Durrum E L, Paul M H and Smith E R B 1952 *Science* **116** 428–30
Dzandu, J K, Deh M H, Barratt D L and Wise G E 1984 *Proc. Natl. Acad. Sci. USA* **81** 1733–7
Eder J M 1932a *History of Photography* (transl. E Eptean 1945) (Columbia University Press: New York) pp 22–4
—— 1932b in Eder J M 1932a pp 96–9
Ehrenfried G 1952 *Photgr. Sci. Tech.* **18B** 2–6
Elder J H, Pickett R A, Hampton J and Lerner R A 1977 *J. Biol. Chem.* **252** 6510–15
Erickson P F, Minier L N and Lasher R S 1982 *J. Immunol. Methods* **51** 241–9
Evans D J R and Fritze K 1969 *Anal. Chem. Acta* **44** 1–7
Fazekas de St Groth S, Webster R G and Datyner A 1963 *Biochem. Biophys. Acta* **71** 377–91
Fink R M, Dent C E and Fink K 1947 *Nature* **160** 80–3
Frater R 1970 *J. Chromatography* **50** 469–74
Frederick J F 1963 *Phytochemistry* **2** 413–15
Friedman R D 1982 *Anal. Biochem.* **126** 346–9
Gambetti P, Autilio-Gambetti L and Papasozomenos S C H 1981 *Science* **213** 1521–2

Gershoni J M and Palade G E 1983 *Anal. Biochem.* 131 1–15
Gibson W C 1963 *Creative Minds in Medicine* (Springfield Ill.: Thomas) pp 53–70
Goldman D and Merril C R 1982 *Electrophoresis* **3** 24–6
Goldman D, Merril C R and Ebert M H 1980 *Clin. Chem.* **26** 1317–1322
Golgi C 1873 *Gazz. Med. Ital. Lombarda* **33** 244–6
Goulianos K, Smith K K and White S N 1980 *Anal. Biochem.* **103** 64–9
Grassman W and Hannig K 1952 *Z. Physiol. Chem.* **290** 1–27
Gray G W 1951 *Sci. Am.* **185** 45–9
Griffith I P 1972 *Anal. Biochem.* **46** 402–12
Haley J E and Gouras P 1984 *Clin. Chem.* **30** 1906–13
Hamilton J F and Logel P C 1974 *Photogr. Sci. Eng.* **18** 507–510
Hamilton S F 1977 *The Theory of the Photographic Process* ed. T H James (New York: Macmillan) pp 105–32
Hancock K and Tsang V C W 1983 *Anal. Biochem.* **133** 157–62
Harasewych M G, Onuferko J H and Merril C R 1985a *Submitted for publication.*
Harasewych M G, Merril C R and Moore M L 1985b *Submitted for publication*
Harrington M G, Merril C R, Goldman D, Xu X and McFarlin D 1984 *Electrophoresis* **5** 236–45
Harris H and Hopkinson D A 1976 *Handbook of Enzyme Electrophoresis in Human Genetics* (Amsterdam: North Holland)
Hartman B K and Udenfriend S 1969 *Anal. Biochem.* **30** 391–4
Herschel J F W 1840 *Phil. Trans. R. Soc. (London)* **131** 1–59
Hillson P J 1977 *The Theory of the Photographic Process*, 4th edn ed. T H James (New York: Macmillan) pp 170–182.
Himmelhoch S R, Sober H A, Vallee B L, Peterson E A and Fuwa K 1966 *Biochemistry* **5** 2523–30
Hsu S-M, Raine L and Fanger H 1981 *Am. J. Clin. Path.* **75** 816–21
Hubbell H R, Rothblum L I, Hsu T C 1979 *Cell. Biol. Int. Rep.* **3** 615–22
Hunter R L and Markert C L 1957 *Science* **125** 1294–5
Hunter W M and Greenwood F C 1962 *Nature* **194** 495
Irie S and Sezaki M 1983 *Anal. Biochem.* **134** 471–8
James T H 1977 *The Theory of the Photographic Process* 4th edn ed. T H James (New York: Macmillan) pp 404–32
Jenkins R L and Farnell G C 1978 *AAS Photo Bull.* **17** 3
Kellogg L M 1977 *The Theory of the Photographic Process* 4th edn ed. T H James (New York: Macmillan) p 145
Kerenyi L and Gallyas F 1972 *Clin. Chem. Acta* **38** 465–7
—— 1973 *Clin. Chem. Acta* **47** 425–36
Kittler J M, Meisler N T, Viceps-Madore D, Cidlowski J A and Thomassi J W 1984 *Anal. Biochem.* **137** 210–16
Knecht D A and Dimond R L 1984 *Anal. Biochem.* **136** 180–4
Koenig P 1937 *Act. E. Tab. Terceiro Congr. Sud-Americano Chim (Rio de Janeira e Sao Paulo)* **2** 334–7
Koiw E and Gronwell A 1952 *Scand. J. Clin. Lab. Invest.* **4** 244–6
Krause C 1844 *Handwörterbuch der Physiologie* ed. R Wagner (Braunschweig: Vieweg and Sohn) pp 108–86
Kronenberg L H 1979 *Anal. Biochem.* **93** 189–95
Kuhn O and Wilt F H 1980 *Anal. Biochem.* **105** 274–80

Lacassagne A and C Lattes 1925 *J. Radiol. et Elect.* **9** 1–15
Latner A L 1973 *Annals NY Acad. Sci.* **209** 281–98
Laskey R A 1980 *Methods in Enzymology* vol. 65 ed. L Grossman and K Muldave (New York: Academic) pp 363–71
Laskey R A and Mills A D 1975 *Eur. J. Biochem.* **56** 335–41
—— 1977 *FEBS Lett.* **82** 314–316
Latter G I, Burbeck S, Fleming J and Leavitt J 1984 *Clin. Chem.* **30** 1925–32
Lecocq R, Hepburn A and Lamy F 1982 *Anal. Biochem.* **127** 293–9
Legrum W, Brüller H J and Birebent G 1983 *Trace Element Analytical Chemistry in Medicine and Biology* vol. 2 ed. P Brätter and P Schramel (Berlin: de Gruyter) pp 227–33
Liebermann D, Hoffman-Liebermann B and Sachs L 1980 *Devel. Biol.* **79** 46–63
Lowry O H, Rosenbrough N J, Farr A L and Randall R J 1951 *J. Biol. Chem.* **193** 265–75
Luthi U and Waser P G 1965 *Nature* **205** 1190–2
Malin D F 1978 *Nature* **276** 591–3
Marshall W E and Porath J 1965 *J. Biol. Chem.* **240** 209–17
Maurer W and Schild K T 1951 *Klin. Wchnschr* **29** 514–5
McConkey E M 1976 *Anal. Biochem.* **96** 39–44
McMaster-Kaye R and Kaye J S 1974 *Anal. Biochem.* **61** 120–32
Means G E and Feeney R E 1968 *Biochemistry* **7** 2192–201
Mees K C E 1952 *The Theory of the Photographic Process* 1st edn (New York: MacMillan) pp 563–83.
Merril C R, Dunau M L and Goldman D 1981a *Anal. Biochem.* **110** 201–7
Merril C R and Goldman D 1984 *Two-Dimensional Gel Electrophoresis of Proteins* ed. J E Celis and R Bravo (New York: Academic) pp 93–109
Merril C R, Goldman D, Sedman S A and Ebert M H 1981b *Science* **211** 1437–8
Merril C R, Goldman D and Van Keuren M L 1982 *Electrophoresis* **3** 17–23
Merril C R and Harrington M G 1984 *Clin. Chem.* **30** 1938–42
Merril C R, Harrington M and Alley V 1984 *Electrophoresis* **5** 289–97
Merril C R, Switzer R C and Van Keuren M L 1979 *Proc. Natl. Acad. Sci. USA* **76** 4335–9
Meyer T S and Lamberts B L 1965 *Biochem. Biophys. Acta* **107** 144–5
Miko S and Schlaepfer W W 1978 *Anal. Biochem.* **88** 566–72
Möller U, Kost H, Schneider S and Coufal H 1983 *Electrophoresis* **4** 148–52
Morgan R H 1944 *Radiology* **43** 256–62
Morrisey J H 1981 *Anal. Biochem.* **117** 307–10
Nauta W J H and Gygax P A 1951 *Stain Technology* **26** 5–11
Neilands J B 1952 *Science* **115** 1431–2
Newhall B 1983 *Latent Image, the Discovery of Photography* University of New Mexico Press pp 8–17 and 117–18
Nielsen B L and Brown L R 1984 *Anal. Biochem.* **141** 311–15
Norheim G and Steinnes E 1975 *Analyt. Chem.* **47**(9) 1688–90
O'Farrell P M 1975 *J. Biol. Chem.* **250** 4007–21
Pace J L, Kemper D L and Ragland W L 1974 *Biochem. Biophys. Res. Commun.* **57** 482–8
Peats S 1983 *Biotechniques* **1** 154–6
Phelps D S 1984 *Anal. Biochem.* **141** 409–12

Picton H and Linder S E 1892 *Jour. Chem. Soc.* **61** 148–172
Poehling H M and Neuhoff V 1981 *Electrophoresis* **2** 141–7
Polter C 1967 *Z. Naturforsch* **22b** 340–7
Porret R 1816 *Annals of Philosophy* July 78–83
Radola B J 1980 *Electrophoresis* **1** 43–56
Ragland W L, Pace J L, Kemper D L 1974 *Anal. Biochem.* **59** 24–33
Randerath K 1970 *Anal. Biochem.* **34** 188–205
Reckie J 1939 *Proc. Phys. Soc.* **51** 683–90
Reilly J M 1980 *The Albumen and Salted Paper Book* Rochester, NY Light Impressions Corp. p 4
Reinhart M P and Malmud D 1982 *Anal. Biochem.* **123** 229–235
Ribas B 1983 *Trace Element Analytical Chemistry in Medicine and Biology* vol. 2 ed. P Brätter and P Schramel (Berlin: de Gruyter) pp 181–97
Ribas Ozonas B 1983 *Trace Element Analytical Chemistry in Medicine and Biology* vol. 2 ed. P Brätter and P Schramel (Berlin: de Gruyter) pp 585–97
Righetti P G 1983 *Laboratory Techniques in Biochemistry and Molecular Biology* vol. 11 ed. T S Work and E Work (Amsterdam: Elsevier) pp 148–259
de Saint Victor N de 1867 *C. R. Acad. Sci. (Paris)* **65** 505–7
Sammons D W, Adams L D and Nishizawa E E 1981 *Electrophoresis* **2** 135–41
Sammons D W, Adams L D, Vidmar T J, Hatfield A, Jones D H, Chuba P J and Crooks S W 1984 *Two-dimensional Gel Electrophoresis of Proteins* ed. J E Celis and R Bravo (New York: Academic) pp 112–27
Sayre F W and Hill B R 1957 *Proc. Soc. Exp. Biol. Med.* **96** 695–7
Scheele G, Pash J and Bieger W 1980 *Anal. Biochem.* **112** 303–13
Schleicher M and Watterson D M 1983 *Anal. Biochem.* **131** 312–7
Schmelzer W and Behne D 1975 *Progress in Isoelectric Focusing and Isotachophoresis* ed. P G Righetti (Amsterdam: North Holland) pp 257–64
Schubert M and Hamerman D 1956 *J. Histochem. Cytochem* **4** 159–89
Scott B J, Bateman J E and Bradwell A R 1982 *Anal. Biochem.* **123** 1–10
Shaw C R and Prasad R 1970 *Biochem. Genet.* **4** 297–320
Smith A G 1982a *AAS Photo-Bull* **30** 6–9
—— 1982b *AAS Photo-Bull.* **31** 9–12
—— 1983 *AAS Photo-Bull.* **34** 3–5
Somerville L L and Wang K 1981 *Biochem. Biophys. Res. Commun.* **10** 53–8
Stein S, Böhlen P, Stone J, Dairman W and Udenfriend S 1973 *Arch. Biochem. Biophys.* **155** 203–12
Stiefel T, Schulze K and Tölg G 1980 *Trace Element Analytical Chemistry in Medicine and Biology* ed P Brätter and P Schramel (Berlin: de Gruyter) pp 427–38
Swanstrom R and Shanks P R 1978 *Anal. Biochem.* **86** 184
Switzer R C, Merril C R and Shifrin S 1979 *Anal. Biochem.* **98** 231–7
Symington J 1984 *Two-dimensional Gel Electrophoresis of Proteins* ed. J E Celis, R Bravo (New York: Academic) pp 127–68
Symington J, Green M and Brakmann K 1981 *Proc. Natl. Acad. Sci. USA* **78** 177–81
Talbot D N and Yaphantis D A 1971 *Anal. Biochem.* **44** 246–53
Thompson S W and Hunt R D 1966 *Selected Histochemical and Histopathological Methods* (Springfield ILL: Thomas) pp 798–802

Tiselius A 1930 *Nova Acta Reg. Soc. Sci. Upsal.* **IV 7** No. 4.
—— 1937 *Trans. Faraday Soc.* **33** 524–31
Towbin H, Staehelin T and Gordon J 1979 *Proc. Natl. Acad. Sci. USA* **76** 4350–4
Tsai C M and Frasch C E 1982 *Anal. Biochem.* **119** 115–9
Tsang V C W, Peralta J M and Simons A R 1983 *Methods in Enzymology* vol. 92 ed J J Langone and H Van Vunakis (New York: Academic) pp 377–91
Tuszynski G P, Knight L C, Kornecki E and Srivastaua S 1983 *Anal. Biochem.* **130** 166–70
Van Keuran M L, Goldman D and Merril C R 1981 *Anal. Biochem.* **116** 248–55
Verheecke P 1975 *J. Neurol.* **209** 59–63
Vesterberg O 1972 *Biochem. Biophys. Acta* **257** 11–19
Wall E J and Jordan F I 1976 *Photographic Facts and Formulas* ed. J S Carrol (New York: American Photographic) pp 168–89
Wallach D, Kirshner N and Schramm M 1975 *Biochem. Biophys. Acta.* **375** 87–105
Wallenius G 1952 *Scand. J. Clin. Lab. Investigation* **4** 24–9
Walton K E, Styer D and Gruenstien E Z 1979 *J. Biol. Chem.* **254** 7951–60
Weinberg S 1983 *The Discovery of Subatomic Particles* (New York: Freeman) pp 111–14
West W and Gilman P B 1977 *The theory of the Photographic Process* 4th edn ed. T H James (New York: MacMillan) p 283
Wilson A T 1958 *Nature* **182** 524
Wilson C M 1979 *Anal. Biochem.* **96** 236–78
—— 1983 *Methods in Enzymology* vol. 91 ed. C H W Hirs and S N Timasheff (New York: Academic) pp 236–47
Wojtkowiak Z, Briggs R C and Hnilica L 1983 *Anal. Biochem.* **129** 486–9
Wolfenden R 1983 *Science* **222** 1087–93
Xu X and McFarlin D E 1984 *Neurology* **34** 769–74
Yuen K C L, Johnson T K, Denell R E and Consigli R A 1982 *Anal. Biochem.* **126** 398–402
Zapolski E J, Gersten D M and Ledley R S 1982 *Anal. Biochem.* **123** 325–8
Zapolski E J, Podrasky F, Ledley R S and Gersten D M 1984 *Electrophoresis* **5** 354–7

Chapter 8

Quantifying Patterns from Two-dimensional PAGE

by
S P Spragg, R Amess, M I Jones and R Ramasamy

8.1 Introduction

The two underlying reasons for analysing mixtures of protein using PAGE are to assess the number and relative quantities of components present and to transmit this information to other workers. In order to fulfil these requirements it is necessary to provide more than just a photograph of the pattern of the gel experiment. It is for these reasons that scanners were produced for single dimensional gels where a graphical representation of bands against distance is produced. Areas and hence quantities of material, can be calculated from these graphs as well as relative positions of peaks. With the widespread use of two-dimensional gel electrophoresis, a similar reduction of a gel pattern from the wet gel to a quantitative map presents new problems. It is largely the technical aspects of these problems that are discussed in this chapter. The basic experimental procedures have been reviewed by Dunn and Burghes (1983).

One can list the requirements from an analysis as follows:

(i) Convert the arbitrary cartesian coordinates of the spots in a wet gel into an acceptable and reproducible projection;
(ii) measure relative quantities of all or selected components in the wet gel;
(iii) provide a transmittable set of data for universal distribution;
(iv) subtract a common pattern from the set of patterns in order to emphasise the differences between samples.

In the case of one-dimensional gels most, if not all, of these requirements can be satisfied by established commercial scanners but this has yet to be achieved for two-dimensional gels. Thus, it is relevant to consider the basic requirements for a scanner which records data from either wet gels or films as well as filtering the records to remove noise, etc. These analyses will then satisfy requirements (i), (ii) and (iii) above but not (iv). Comparing sets of data assumes that the patterns are reproducible. However the procedures used for comparing large sets of data are best run on commercial sorting systems which normally will not form part of a scanner. Before discussing details of scanners it is worth remembering that the requirements for the fast recording of coordinates, including absorbances which produce contrast between the features in a pattern, are different from those where the measured absorbances are to be integrated in order to give volumes within boundaries. The second case imposes more rigorous specifications on the dynamic range of the optical transducer than the first and it is this second case that is normally expected in quantitative two-dimensional PAGE.

A further feature to note is that although two-dimensional scanners can be used to scan one-dimensional gels the reverse is not true. Furthermore, although clever algorithms can correct for imperfections between patterns, this procedure can never replace the care taken in ensuring reproducibility both during the preparation of samples and the running of the gels. In other words the old maxim 'garbage in, refined garbage out' is as true for gel analyses as for any other scientific pursuit.

8.2 Basic Outlines

Most one-dimensional scanners provide inclusive routes to requirements (i) and (ii) (above), since the result of a scan is a two-dimensional map bearing a spatial coordinate in one dimension and absorbance in the other. A similar processing of two-dimensional gels would require a three-dimensional map in which two axes carry the spatial coordinates and the third the absorbance. This means that representation can only be achieved through computerised displays.

It is possible to compare gels for qualitative differences (i.e. changes in the position of components caused by charge and/or size) without specialised scanners by laying the gels on top of one another over a lighted surface and then aligning common components. This approach is made easier by using optical comparators in which the images of two separately illuminated gels are superimposed and the lights are switched on and off out of phase (Anderson *et al* 1979, Spragg *et al* 1983). With these systems one of the gels can be moved in order to superimpose the common components in the image plane, then as the lights flash on and off the altered components appear to jump from side to side or flash on and off themselves. With the aid of this simple comparator one can readily note changes even in complex patterns.

However if the whole pattern cannot be superimposed then comparisons have to be made in segments by moving one gel relative to the other. In order to store the differences resulting from many comparisons as well as use internal standards to correct for minor variations between patterns, it is useful to couple this optical comparator to a microcomputer system (Spragg *et al* 1983). Using this system it is possible to store the coordinates of qualitative differences between a standard gel and many others. The results from these measurements are an idealised map which is linked through a code to a photograph of the standard gel (figure 8.1).

These comparators are relatively cheap to produce in addition to utilising the knowledge of the experimentalist to the maximum. They are not suitable, however, for situations where many gels are to be processed nor where it is necessary to measure quantitative differences between components. For these situations a true photometric scanner is required. Before discussing instrumental designs of such a scanner it is necessary to remind oneself of the relationship assumed in photometry and which is used to evaluate concentrations.

The relationship between light absorbed and concentration is described by the Beer–Lambert equation

$$I_0/I = \exp(zcl) \tag{8.1}$$

where I_0 and I are the light intensities before and after passing through the sample respectively; l is the path length of solution; c the concentration and z, the extinction coefficient, which has units of inverse $c \times l$. In most situations where many different components are present in one map (although they may all be of the same general type, e.g. peptides) z is not a universal coefficient but varies from substance to substance. Thus a separate calibration will be required for each, and all must be made under the correct conditions. Naturally this calibration is not easy to do since there are so many unknowns. Hence, $\ln(I_0/I)$ is usually measured (this is the absorbance, A, if a common logarithm is used in the transformation) at a fixed pathlength and a comparison of the integrated absorbance is made between the components. In this comparison it is useful to normalise with respect to a selected common component thought to be invariant between gels as far as concentration is concerned.

In assessing the statistical significance of this procedure it is important to note that the detector in the scanner measures intensity, leaving the logarithm of the ratio to be calculated afterwards. Hence, the error in A is inversely proportional to the intensity. Thus, lowering the intensities increases the errors in A. A similar proportional relationship holds when dividing one absorbance by another, which means that it is wise to choose a component present in moderate concentrations for the reference

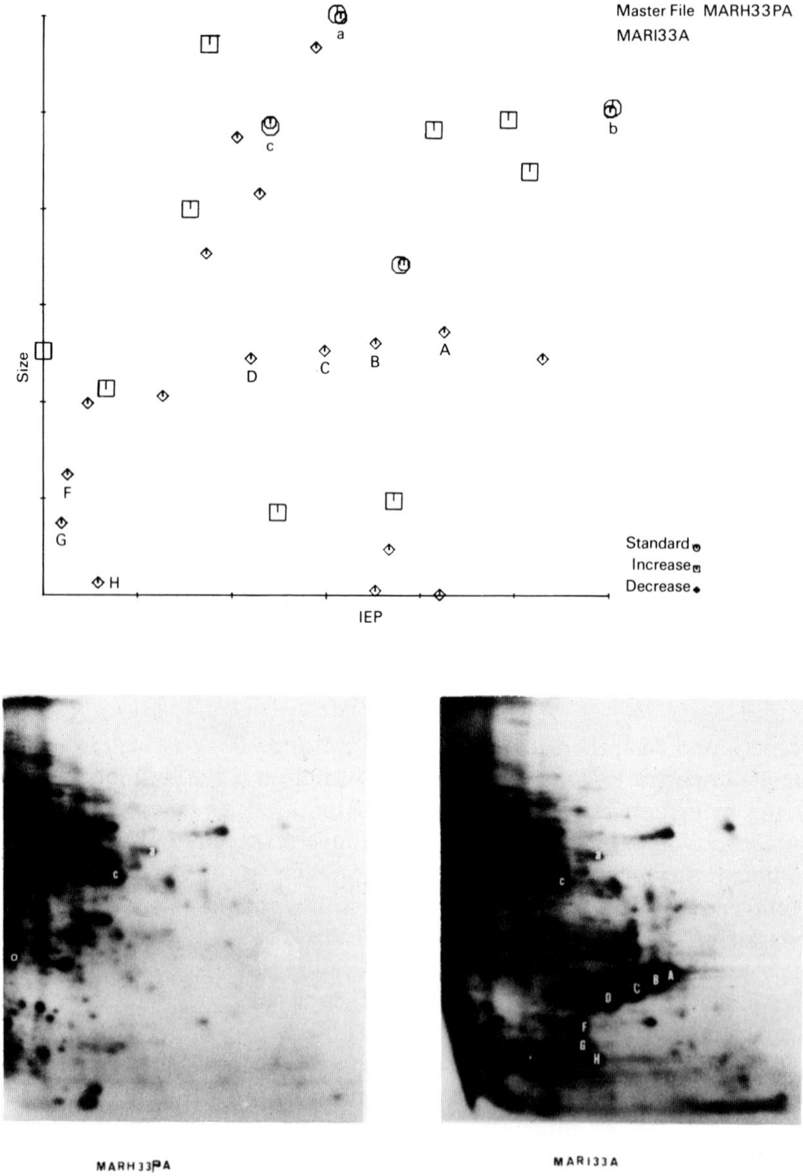

Figure 8.1 An example of an idealised map produced by a comparater (Spragg *et al* 1983). The samples were plant protoplasts run by reverse non-equilibrium two-dimensional PAGE. One sample was infected with a virus and the other acted as the control. The map shows the differences between the two patterns plotted on a background of a few common components (open circles): these components were used to align the patterns. Some points have been labelled in order to relate the map to the two photographs of the original autoradiographs.

substance (optimum absorbance is between 0.3 and 0.4 but as the absorbance for the reference substance approaches zero so the error approaches infinity). It is also important in this comparison to use as high a light intensity as is acceptable to the detector (see later discussion on detectors).

In addition to the influence that random errors will have on the design of the scanner, consideration must be given to the effect that flaring of the light around the spot in the gel has on the absorbance. In order to detect an absorbance either the gel or the light must move and in most one-dimensional scanners the gel is mechanically pulled over a stationary light source. The width of the light beam is determined by a slit after the source, hence, when an absorber in the gel completely covers the slit, then maximum absorbance occurs. If the absorber is smaller than the slit then light is detected which does not pass through the absorber and it can be seen from equation (8.1) that small amounts of unabsorbed light destroy the linearity between $\ln(I_0/I)$ and concentration. This process is equivalent to the scattering or refraction of the light beam around a spot in the gel. It is for this reason that one-dimensional scanners designed to measure bands or large spots are generally unsuitable for measuring absorbance from two-dimensional gels where the components are resolved into small discrete spots. If a one-dimensional scanner must be used then it is essential to ensure the aperture sizes are less than the diameter of the spot measured. Further, since a spot is spread in two dimensions, one pass by the scanner may not be sufficient to give a reliable estimate (e.g. it may not always include the maximum absorbance). Therefore it is necessary to make several passes across the aperture at different positions and sum the total absorbance (i.e. $d\sum cz$, where d is the interval for the measurement in one axis).

It should be mentioned that for high absorbances ($A > 1$), stray light from any source (including flare) limits the accuracy of the measurements. The presence of stray light equivalent to 1% of I_0 means that a true absorbance of 1 becomes 0.96, while a true absorbance of 2 becomes 1.7! If it is not possible to reduce the level of stray light to <1% then measurements of high absorbance are not reliable. A realistic limit for most instruments would be $A < 2$.

Since the quantity of material in a spot can be calculated only by integrating a series of measurements across a spot, the accuracy achieved in measuring the spatial dimensions and absorbances with two-dimensional scanners is thus dependent on the size of the aperture or spot of light. These limits can be defined if one can calculate the expected range of the spot diameters. To do this it is reasonable to assume that diffusion is the main cause of the spreading of a spot. If a substance of relative mass 10^4 is allowed to diffuse in two dimensions for about 8 h then the spot would reach about 5 mm in diameter (this is the diameter between positions where for a substance of relative mass 10^5 this diameter would be 2 mm (in a 2 h experiment the diameters would be about 3 mm and 1 mm respectively).

These values define the range of sizes expected for spots present in a gel having a scanned area greater than 10^4 mm^2 (100 × 100 mm^2 slab of gel).

If the step interval of the scan had a precision of 1 part in 256 (eight-bit d/a converter, see later for discussions of these converters) then this leads to an uncertainty of about 0.4 m in both dimensions. The sampling theorem shows that if ideal measurements which are free from noise are available then at least two measurements are required in each direction in order to calculate the volume. Measurements are never free from noise and depending on the relative level of this noise then upwards from five measurements per direction may be required to produce an adequate estimation of absorbance. In other words for a small spot containing a constituent of relative mass 10^5 the aperture of the spot of light should be between 0.2 mm and 0.3 mm. This is similar to the uncertainty found for the spatial coordinates generated by an eight-bit d/a converter, thus the two specifications are compatible. It seems therefore, that for a slab of the dimensions 100 mm × 100 mm, an accuracy of 1 part in 256 in the coordinates is sufficient. It shows also that an aperture of 0.2 mm is sufficient for most estimates of absorbances even with small gel-spots. For slabs larger than 100 mm × 100 mm more accuracy would be required in measuring the *x*- and *y*-coordinates but the size of the aperture remains unaffected.

8.3 Detectors

Today there are three types of detector used for photometry: photomultipliers (PM), vidicon cameras (VT) and charge coupled devices (CCD, essentially solid state vidicon cameras made in the form of arrays of phototransistors). The PM differs in many ways from the other two but since it is the most sensitive light detector of the three it is useful to use the PM as a reference for comparing other photon detectors.

The first and most obvious difference between the detectors is that whereas a VT and CCD can accumulate light signals, the PM can give only single responses. Thus, with a PM the current is constant with time for a constant luminosity, while with VT and CCD the signal level increases with increasing time of exposure (i.e. they integrate the light). The effects these differences have on the operations of these detectors is large but possibly the most important are the limits this imposes on the upper speeds for collecting data. A reasonably priced PM will operate in excess of 10 MHz (i.e. it will separate photons arriving one per 100 ns) and if operated correctly has a wide dynamic range for linear relationships between light intensities received at the photocathode and output from the anode (>1000). In contrast the maximum reading rates for most VT and CCD are about 1 KHz and here the dynamic range is <50.

More obvious differences occur in the physical operations where the photocathode of a PM is homogeneous and has no spatial separation of the light signals while the photosensitive part of the VT and square CCDs (256 × 256 elements in a 100 mm square) can be regarded as a two-dimensional grid of detectors. This major difference means the speed limits are not defined only by upper signal frequencies of the detectors. If a PM is the detector for a two-dimensional scanner, either the light or the gel must move in two dimensions relative to the PM. With a VT and square CCD no movement is required and lenses only must be included to match the gel dimensions to those of the array. If a linear and not a square CCD is used (up to about 2000 elements per inch), then the gel or light must be moved in one direction. Thus the scanner is beginning to look more like those using a PM detector. These physical considerations suggest that the VT or square CCD would seem the obvious choice of detector for scanners since the high bandwidth of the PM seems to be offset by the need to mechanically scan the gel. However, because of the limited spatial resolution of a square CCD it is usual to employ a linear CCD array which means the relative dynamic light ranges become more important. This is so much larger for the PM that its merits should be examined further.

Practically, the photocathode of a PM is less efficient than a CCD in converting photons to electrons but whereas the PM has a gain of about 10^6 after the photocathode, no such gain is built into a CCD. Because the CCD is essentially a phototransistor, the signal/noise from a PM is considerably better than from a CCD at room temperature. These factors mean that for low lighting conditions the PM is the best detector. Furthermore, the response of all the pixels to the same light intensities for VT and CCD are not uniform over the area of the detector, although they do remain constant for relatively long periods. Hence, these detectors must be calibrated to give a set of spatial coefficients which must be used for normalising the raw readings.

All modern spectrophotometers employ a split-beam system in which I_0 and I are measured but this procedure is not easily achieved with VT and CCD. It is possible, however, to make a 'split-beam' scanner with PMs in which one PM monitors the incident light (I_0) while a second measures the light after passing through the gel (I) then by combining these signals via a log–amplifier it is possible to obtain an absorbance directly and uninfluenced by variations in the incident light or detector performance. A simple double beam system using only two photomultipliers is not perfect and minor misalignments of optical components after splitting the beam do produce a base pattern which is constant and must be subtracted from scans before reaching an unbiased recording.

There is one further factor in favour of PM detectors involving the wide dynamic linear range of a PM. This facility means the linearity between absorbance and concentration continues into high absorbances (>3). In fact it is possible to produce an offset absorbance in order to maintain precision in absorbance despite dense backgrounds by inserting a neutral filter between the incident light and reference PM (Spragg *et al* 1985).

These comments on the two groups of detectors show that accurate photometry using low levels of illumination is achieved best with a PM. The future, however, may be with CCD since the sensitivity, speed and uniformity of elements in an array will improve with time. In the meantime many users favour the VT (for example Tracy and Young 1984) and accept the limits this imposes on integrated absorbances when using commercially produced television cameras.

8.4 Scanning Method and Speeds of Collection

If a vidicon camera or two-dimensional CCD is used as the detector there is no need to move either gel or light. The reduction of object size to match the detector will be about 1 : 6 if one is to match 25 mm detectors to commonly used gels. To do this lenses of moderate quality are required (commercial camera lenses are suitable). If a linear CCD is used then the gel or light must be moved in one direction, while with a PM the movement must be in two directions. It would seem that moving the gel would be simplest since this involves only mechanical stages driven by suitable motors. Obviously using mechanical stages will limit the speed of scanning and since speed is important when handling many gels it must play a major part in deciding which scanning system to use. If one starts with the detector, then it is possible to define maximum scanning speeds.

With a CCD the signal level is determined both by the instantaneous intensity of light falling on the CCD (so defining the signal/noise ratio for the detector) and the integration time allowed for the detector. Limits can be calculated for the required light intensities by using typical specifications of commercially available CCDs and integrating the light for 10 ms. This shows that if optical filtering is not necessary to select wavelengths then a 5 W strip light, 20 cm long, will give sufficient light. Either the power of the light or the integration time must be increased beyond 10 ms if filtered light is required. Optical filtering is unnecessary for simple photometry of gels stained with silver or autoradiographs of gels. Increasing the integration time for more than 10 ms is not acceptable since there is a dark output current from the CCD which also integrates with time. Manufacturer's data suggests that this current will be about 10% of the signal after integration for more than 10^{-1} s. Thus, for successful use of CCDs, reasonably high light intensities are desirable and it is safer to limit the integration time to <10 ms (this time must be compared with the further 10 ms or more required to read 2000 pixels into a computer). This limits the scanning speeds to about 0.1 mm per 20 ms or 5 mm^{-1} which is a very acceptable rate of scan (about one 100 mm gel can be scanned per 20 s). This produces about 2×10^6 pixels, each having 8 bits of intensity.

A similar scanning speed is possible with linear CCDs and a mechanical stage providing space is left to allow the stage to accelerate up to this speed before the operational part of the scan reaches the detector. The motors driving the stage should be stepping motors driven within their limits so they can be accurately pulsed without 'slippage'. The gel could be moved continuously if this increases the scanning speed, however, here the limit on speed is set by the need to allow a small overlap between each line of readings. A small amount of blurring of the image will occur with continuous feed but if required this can be corrected using Fourier transforms (Gonzalez and Wintz 1977).

Most VT detectors in commercial TV cameras scan at about 20 frames per second and require reasonably high light intensities to give reasonable performance. The dynamic light range is about 30 (i.e. absorbance 0 to 1.4 is recordable but not necessarily linearly with intensity) providing the light is intense (more than the 5 W strip light quoted above).

With a PM detector the usable light intensity can be reduced over a thousand-fold compared with CCD/VT and still give the same signal/noise ratio. Further, because of its large bandwidth, a PM can be read at 10 MHz, so 10^6 pixels could be read in 0.1 second if conversion from analogue to digital data is made at this rate. However, with a PM the reading rate is limited by this conversion time so the minimum economic time is 10 s. Since this is similar to the time that is possible with a CCD, the PM is a better detector if any spatial movement needs to be included. A scanner can be made which uses a cathode ray tube to produce a spot of light whose position and intensity are controlled electrically. This system gives scanning rates equal to those achieved by VT yet includes a split-beam system to correct for variations in the incident light and so maintain linearity between absorbance and concentration over a wide range of absorbances. Flying spot scanners built using cathode ray tubes are used in nuclear physics for measuring films from spark and bubble chamber experiments and have recently been redesigned to scan a two-dimensional gel (Spragg *et al* 1985). Tubes are produced which scan wide areas (200×200 mm^2) through small angles and give precisions of better than 5 μm in the positioning of the spot. One disadvantage of using a CRT is the spectral range of the light is determined by a fixed phosphor and so changing the wavelength is not easy. There are, however, many different phosphors available of which P4 (white) has the most useful range of wavelengths. Its persistence is 470 μs (the time for its intensity to decay from 100% to 1%) which means the speed of scanning must be about one pixel per 0.5 ms; for a 100×100 mm^2 gel to be scanned using a 0.2 mm spot this then takes 150 s. A considerably faster phosphor is available (A5, persistence about 0.1 μs) which gives an upper rate of scan of 1 gel per 10 ms (i.e. television scanning rates) using a fast analogue/digital (A/D) converter. This phosphor has a peak wavelength at 540 nm and would be suitable for scanning Coomassie stained gels as well as silver stained gels and autoradiographs.

This brief survey shows that the technology for making a fast scanner is well established. Commercial scanners are now available and are generally based on VT detectors. Several types of accurate film scanners exist using PM detectors. These employ mechanical systems (rotating films or moving flat beds) which make them unacceptably slow in comparison with flying spot scanners. The need is for a system having about 1% precision in measured absorbance within the range 0 to 2 and the scanning speed should be more than one gel per minute. These limits are readily achieved using a cathode ray tube (CRT) scanner with a PM detector or two-dimensional CCD detectors. Thus the future should see the production of several models which can scan stained wet gels as well as films at acceptable speeds.

8.5 Collecting Data

It must be assumed that scanners will be controlled by a digital processor. The collecting system can be either a dedicated unit which combines a processor with memory and which possibly acts as a slave to a larger computer, or it may be a computer system which is sufficiently large that it can collect data as well as process it. In both cases the voltage from the detector must be converted to its logarithm using a reference voltage (which in flying spot scanners is from the unabsorbed light) and so provide analogue measurements of absorbance (equation 8.1). This logarithmic voltage must then be digitised using an analogue to digital converter (A/D). These units can provide accuracies in conversion equivalent to 8, 10, 12 or 16 bits per maximum allowable voltage (i.e. the voltage is divided into a maximum of 256, 1024, 4096 or 65536 increments) but for gel scanners an 8-bit converter is sufficient. These are economic to interface to modern microprocessors and even cheap versions can convert a 10 volt signal in 25 μs while more expensive versions respond in 750 ns so making themselves suitable for video scanners. Their eight output data lines and two control lines can be linked directly through an interface circuit (e.g. Intel 6522) to the computer.

Often a further circuit will be required to convert digital words from the computer into analogue voltages (D/A; 8-bit is sufficient for $100 \times 100\ mm^2$ scan, see earlier) in order to control the *x*- and *y*-coordinates of the scanner. These scanning voltages can be provided either by the main computer or through a logical interface which only needs the increment on top of the starting and finishing positions in order to complete the scan without further reference to programmes in the main computer.

The net result must be a unit which has the following features

(i) an analogue multiplexer which, via the program, can switch between two analogue channels into the single output supplying the A/D. This makes it possible to monitor several voltages (such as the absorbance and the reference PM) with the same A/D;
(ii) three D/A channels of which two are for x and y-coordinates while the third controls the light intensity in response to signals collected through a reference detector;
(iii) data lines directly connected to a store from the A/D converter and which can be in the computer or special memory.

Interfaces incorporating these aspects make it possible to collect one pixel every μs at the fastest rate and one every 25 μs using slower and more economical converters. These rates are either equal to or faster than most access times of microprocessors for reading or sending out data so the choice of the computer determines the maximum speed of reading data. 8- and 16-bit microprocessors require at least 5 μs to accept from or dispatch to interfaces, so even allowing for inefficiency in programmes a byte could be read every 30 μs.

It should be remembered that 256 grey levels of intensity will be recorded even though the dynamic range of VT or CCD does not warrant this accuracy. Practically this range cannot be displayed and generally only eight levels are possible with a simple colour visual display unit (these are visually more definitive than a monochromic display) so each level spans 32 units of absorbance. This means discrete gel spots of low concentration or varying concentration but still within one of the intervals will not appear as separate spots if they are clustered. The programmer thus needs to provide a non-linear scale for grading the displayed pixels. The art of computer graphics is well defined and many excellent texts contain suitable coded routines for displaying isometric pictures etc, so routines are not given here.

8.6 Programming Languages

The choice of which language to use for programming a scanner is not always clear. An overriding determinant must be the speed needed to implement the algorithms and here the rule of thumb for relative speeds is

Assembly (1) > FORTRAN, ALGOL (10) > BASIC (100)

where approximate relative times to execute the same routines are given in brackets. However, if one is considering ease of writing and correcting programmes then the order is reversed. An example can be drawn from a system employing a 6502 8-bit processor where it took 50 ms to read a byte from an A/D converter using BASIC but only about 50 μs for the same routine written in assembly language, here the factor was 1000 not 100! It took a morning to get the Assembly routine going, assuming little or no knowledge of the language at the beginning, while less than half an hour was required before numbers were appearing with the BASIC programme. Thus, it is best to code the collection routines first in BASIC in order to get the strategy correct then they should be converted to Assembly programmes for routine operations. Obviously for all languages the routine operations such as writing to disks, consoles, etc should be via the operating system and it is only the control and reading of non-standard devices that needs Assembly codes.

In choosing a system it is wise to get one that allows intermixing of Assembly routines with high level languages within a compiled programme. Great improvements in the efficiency of programmes can be made if special routines are written for executing functions and simple arithmetic using Assembly languages rather than FORTRAN etc. Many of these precautions are well known and have been applied to scientific computing for years so many optimised procedures exist for these purposes and are collected in the older literature (see for example Spragg 1973).

One additional hardware unit required for processing images from gels is an interactive colour display. At first it may appear that a colour television will suffice. However, because the modulated UHF signals which control the television have limited bandwidths the results are discouraging. A colour monitor requiring the three primary colours (hence, the unit is called a RGB monitor) plus synchronisation is necessary and although a vector display is a great advantage, this is not essential. At a minimum the graphics should be capable of a resolution of 256×512 pixels and have the ability to write (in small letters) in the graphic area. Many small personal computers connected to larger computers fill this requirement since the arithmetic and storage can be carried out in the large machine leaving the small unit to control the display. Whatever system is used the computer memory should be equal to running compiled programmes using 256×256 bytes of data. This means that a 16-bit micro- or minicomputer is the minimum system for these operations. Smaller processors will give tantalising glimpses of what is possible but will not provide a usable system for processing images. Economic systems utilising an Intel 8086/8087 processor or Motorola 68000 processors are available having at least 10^6 bytes of memory and these could provide economic controllers for scanners. Normally this local processor would be connected by cable to a mainframe which can act as the backing store.

8.7 Numerical Analyses

Assuming the data are collected, stored and can be displayed then the next stage is to reduce the noise (if any) on the data before removing background imperfections. The first operation is easier and less subjective than the second because noise by definition is random having defined frequency characteristics. This can be treated using statistical or signal processing procedures but the background is less analytical.

To filter the noise using a digital filter means changing the raw data from the time domain in which it was collected to the frequency domain by applying a Fourier transform (see appendix for details and Gonzalez and Wintz 1977, for detailed coverage on this and other operations used in image analysis). Unless one has access to special processors or optimised transform routines written in assembly languages (thus not transportable between different types of computer) the Fourier procedure for removing noise is slow. There is a bonus in using it since once the image is transformed into the frequency domain many manipulations can be made on the transform which are not possible with the original data. For example, a high pass filter can differentiate the data in order to find edges or blurred images can be restored.

It is still possible, however, to filter noise from data in the time domain using simple orthogonal filters based on Legendre polynomials (Milne 1949, and given in the appendix) but this process will smear the edges of some features (a deficiency not found with frequency filtering). Nevertheless, filtering in the time domain is relatively fast compared with that in the frequency domain and since the edges of spots in gels are not sharp this is an acceptable procedure.

A procedure commonly used to find edges is based on Sobell operators. This is an average procedure and this type of operator has been used in gel analyses to compensate for backgrounds etc (Brown and Ezer 1982, Taylor *et al* 1980). In this method the pixels in a 3×3 matrix are used to produce either an average at the centre, or, in the case of finding edges, the derivatives (see Gonzalez and Wintz 1977, for further details). The 3×3 window is moved over the image and the central calculated values replaces the original value. A similar but simpler way to detect limits for background correction is to calculate the point-derivatives and then use simple rules to find peaks, plateaus, etc (e.g. a plateau occurs after a positive gradient, while a peak has a positive followed by negative, etc.). Usually it is better to use intensities around a point (i.e. rows and columns) to calculate a derivative because ideally a vector should be used in these manipulations rather than the sum of two differences in the two directions (as normally calculated). Calculations made on data from gels in which only rows and not rows and columns were used do not appear to produce any obvious polarisation (Fox 1982) but this cannot be guaranteed.

It is also possible to use derivatives calculated using Sobell operators to outline edges but here the choice of what is a significant level for a derivative is subjective and needs guidance from the operator. The technique is to include a test for a threshold for the derivative which when exceeded means that the operational pixel should be replaced by a black pixel. Continuing this process over the whole pattern will give an image containing black outlined edges of peaks. Most of these features are included within existing systems for analysis gel patterns (Fox 1982, Lemkin *et al* 1982, Miller *et al* 1982, Taylor *et al* 1980) and the present discussions only outline the underlying concepts. Naturally the idea is to reduce the original gel or photograph to a pattern without noise or variable background by computer 'massaging' but this is no substitute for clean 'wet' experiments.

After cleaning the original data the next step is to label the spots and produce estimated 'volumes' or reduced masses of the individual components. This means locating peaks and summing the absorbances within the total area of the peak. At its simplest this can be done through the interactive displays in which markers are placed at the four quadrants around the peak. Then travelling along rows and columns between the limits of the markers the absorbances are summed and after multiplying by the square of the step interval used for the image the result is the mass (or volume). It is important to subtract backgrounds from the pixels before integrating otherwise the masses will be very inaccurate. For example if a background was equal to 10% of the maximum absorbance at the spot then the integrated value would be in error by more than 10% and could reach 50% depending upon the number of pixels summed and the spread of the spot. This factor probably constitutes the major source of inaccuracy when comparing quantities in different spots within a gel or between gels. Hence, reducing background levels should form the major activity in any analytical system. For more automatic systems (Lemkin *et al* 1982, Miller *et al* 1982, Taylor *et al* 1980) the background is reduced by a Sobell operation and the volumes calculated indirectly by using least squares procedures to fit empirical functions to the corrected data. These models contain parameters which equate with masses or volumes of the spots (see below).

8.8 Reducing Data to Models

Following the smoothing and filtering there remains a clean set of data but the number of pixels stored remains the same as when recorded. Much of this data covers blank regions and they only separate the spots, hence, its information is low. It can be discarded provided the *x*- and *y*-coordinates for the pixels can be unambiguously recorded. This coding can be by a variety of defined procedures but all are dependent on having more than 16 bits per computer word. Thus, by using logical rather than numerical procedures much of the total area in a record can be discarded without degrading the pattern. A more satisfactory way of reducing the number of stored values is to fit each spot to a mathematical function which contains estimated parameters for describing *x*- and *y*-coordinates of the peak centre, intensity at the centre and widths (x and y) of the peak. One proven solution is to fit a two-dimensional Gaussian equation to each peak since it seems logical to assume that spread in both directions is caused by diffusion which produces Gaussian shapes. Thus, once the pattern has been mapped as peaks (using Sobell or other systems to locate the peaks) then the intensities around each peak maximum can be fitted using non-linear least squares criteria (see appendix) (Taylor *et al* 1980). This fitting procedure may take some time to transform a whole map containing many peaks but this can be reduced if array processors are used in the computation. The time is reduced if tables of exponentials are included in the programme and values interpolated from the lists as required since this is faster than computing an exponential each time it is required. The overall result of the fitting is an idealised map in which each spot is stored as five parameters. From these it is still possible to reconstruct the whole map but now there will be no irregularities or background. The questionable part of this procedure occurs when streaks are found in the patterns. These represent proteins which for a variety of reasons have not resolved into spots. In general these streaks contain intense regions so the algorithms used to cope with streaks concentrate on these regions and treat them as separate spots. If a Fourier procedure is used in the analysis then special procedures can be applied to treat streaks using the same approach as for restoring a blurred photograph (see Gonzalez and Wintz 1977). It is not possible to quote recovery figures after condensation of patterns using these models because the overall conservation of mass appears not to have been calculated. The total mass of protein in a gel would be computed by summing all the intensities after filtering for a complete gel and this would be compared with a similar value computed using the fitted Gaussians. One would hope for better than 90% matching between the two but it is not known whether this has yet been realised.

Fitting a model reduces a set containing 3×10^6 pixels and collected from 1000 spots down to about 5×10^3 parameters. Naturally the Gaussian is not the only model that could be fitted and other models could be tried. There are sound physical arguments to support a Gaussian profile for peaks formed by diffusion of substances present in dilute solutions. However, this is not necessarily true for cases where very high concentrations of charged species occur (as in the centres of spots). In these cases the diffusion coefficient increases with increasing concentration and so the centre of the peak diffuses faster than the more dilute edges. The result is a shape having Gaussian shaped sides but with a flattened top. Thus, since it seems that the profile of a spot cannot be an exact Gaussian shape it seems reasonable to enquire whether other simple functions could replace the Gaussian function. For example an ellipsoid (again having five determinable parameters) could describe the spots. Here linear least squares routines could be used to fit the functions to the data and this is a comparatively fast procedure.

The criterion for fitting would be to conserve the mass for each peak so this would be the optimised model for the whole map.

Whatever model is fitted the objective in analysing single gels is to produce acceptable estimates of the mass in the spots as well as store simplified records of the experiments. Under these circumstances the actual shape described by the mathematical model is aesthetic rather than scientific providing it gives an accurate representation of the volumes and positions of spots.

So far the discussion has been concerned with individual gels but if gels are to be compared either within a laboratory or with results from other laboratories then distortions caused by inhomogeneous experimental effects should be removed. This can be achieved by taking a known pattern and stretching the other patterns so that known common spots superimpose. The extent of the stretching can be varied within the pattern providing many spots over the whole area are known to superimpose. If however, only a few in a limited area are known to be common throughout the patterns then only this area can be accurately corrected, the rest will have to use averaged values for correcting the coordinates. It is not too important whether these spatial corrections are done before or after fitting a model because it seems likely that comparisons between gels will be the most important reason for carrying out the experiments and it is unlikely a standard pattern will emerge until well into the project. Thus, a set of routines for comparing gels must contain a reliable routine for stretching the coordinates and must also allow interactions between the experimentalist and computer to be included.

8.9 Presentation of Data

The gels are themselves the primary data for the experiments and hence, in any publication there must be at least one photograph. This precludes the measurement of intensities and if this is required then a photograph and a photometric record are required. If the latter is carried out correctly (i.e. proven sensitivity and accuracy) then the photograph is superfluous. Comparisons between experiments can be made by transmitting coordinates between laboratories once a complete set of rules are agreed to guide the differing experimental regimes. These rules should include standards within each gel which can be readily identified from the other spots using properties such as abnormally high or low pI compared with native peptides plus a reasonable range of relative masses. A further possibility for synthetic standards is to covalently label them with a dye which is permanent and identifiable in the scan (e.g. having spectral characteristics which differ from the stained or labelled native peptides). These standards would mean that usable comparisons between maps produced by different laboratories would be possible.

These considerations naturally lead to the question of labelling spots of an unknown biochemical function—a problem the Andersons have considered at length (Anderson and Anderson 1982). If a cell contains 2000 or more subunits then it is unlikely that more than 200 will be assigned biochemical activities, the rest have unknown biological significance but must be labelled in a gel. Again it is not important at this stage how the labelling of spots is done as long as a labelled map is produced and the labelling system adhered to in subsequent publications.

Computer methods which can be used to compare maps can only be successful if the normalisation of the axes are successfully made through reliable standards. The data may be different in detail from that found in commercial operations but the algorithms used to sort commercial data can be applied to gels. Not too much discussion is necessary to show that these activities can be carried out only by using large computer systems with plenty of fast backing store. If only a microcomputer system were applied to the sorting operation then only limited comparisons could be made.

Appendices

A1 Smoothing Procedures

Noise is random, spikey and therefore high frequency so in order to reduce noise either an averaging procedure which includes nearest neighbours must be applied to the time domain data, or the image must be converted to the frequency domain before removing high frequency components. Smoothing in the time domain is faster to compute as well as using less memory than transforming to a frequency domain, but it has the disadvantage of reducing the sharp edges of features in the image. The harmonic analysis using a Fourier series is relatively slow in computation (even when using fast Fourier transforms, FFT), it requires large memories and yet after transformation it is possible to remove noise and other unwanted features without disturbing the geometrical features in the image.

(i) *Time domain smoothing* Since the data will be collected at equal intervals in the x- and y-coordinates, smoothing by a least squares analysis using Legendre orthogonal polynomials is possible using integer arithmetic. Milne (1949) lists suitable polynomials for this procedure. Experience has shown that a five point cubic equation is a satisfactory equation for smoothing data from gels

$$Y'(2) = (-3Y_0 + 12Y_1 + 17Y_2 + 12Y_3 - 3Y_4). \qquad (8.2)$$

Here $Y'(2)$ replaces the measured $Y(2)$ after smoothing over the five points. The whole process consists of moving this five point window along first the rows and then the columns of the two-dimensional array and is equivalent

```
C          SMOOTHING                                SUBROUTINE       1
C          USING ORTHOGONAL POLYNOMIALS FOR EQUAL INTERVALS
C          SMOOTHING IS BY 3RD. POWER AND 5 POINT
           SUBROUTINE SMOOTH(IY)
           INTEGER*2 IARRAY,ICON
           COMMON /A/IARRAY(128,128),IVAL
           DIMENSION ICON(5),JDIF(128)
           DATA ICON/-3,12,17,12,-3/
C          FIRST READ IN THE ARRAY
           CALL FREAD(IXINC,IYINC,IXC,IYC,NDATA,MAXOD)
C          SMOOTH ALONG ROWS FIRST
C          AT BEGINNING & END OF LINE WHERE DATA REMAINING
C          IS < 5, IT ASSUMES VALUES OUTSIDE THE RANGE
C          ARE ZERO,HENCE THE TESTS FOR J
101        DO 3 I=1,IYC
           DO 4 J=1,IXC
           IF(J.NE.1)GO TO 5
           NTEMP=17*IARRAY(I,J)+12*IARRAY(I,J+1)-3*IARRAY(I,J+2)
           GO TO 6
5          IF(J.GT.2)GO TO 7
           NTEMP=IARRAY(I,J-1)*12+IARRAY(I,J)*17
           NTEMP=NTEMP+IARRAY(I,J+1)*12-IARRAY(I,J+2)*3
           GO TO 6
7          IF(J.LT.IXC-1)GO TO 9
           NTEMP=ICON(2)*IARRAY(I,J-1)+ICON(1)*IARRAY(I,J-2)
           NTEMP=NTEMP+ICON(3)*IARRAY(I,J)
           IF(J.EQ.IXC)GO TO 6
           NTEMP=NTEMP+IARRAY(I,J+1)*ICON(4)
           GO TO 6
9          K=-2
           NTEMP=0
           DO 8 M=1,5
           NTEMP=NTEMP+IARRAY(I,J+K)*ICON(M)
           K=K+1
8          CONTINUE
6          IARRAY(I,J)=NTEMP/35
4          CONTINUE
3          CONTINUE
C          REPEAT ALONG COLUMNS
102        DO 10 J=1,IXC
           DO 11 I=1,IYC
           IF(I.NE.1)GO TO 12
           NTEMP=17*IARRAY(I,J)+12*IARRAY(I+1,J)
           NTEMP=NTEMP-IARRAY(I+2,J)*3
           GO TO 13
12         IF(I.GT.2)GO TO 14
           NTEMP=12*IARRAY(I-1,J)+17*IARRAY(I,J)
           NTEMP=NTEMP+IARRAY(I+1,J)*12-IARRAY(I+2,J)*3
           GO TO 13
14         IF(I.LT.IYC-1)GO TO 15
           NTEMP=ICON(2)*IARRAY(I-1,J)+ICON(1)*IARRAY(I-2,J)
           NTEMP=NTEMP+ICON(3)*IARRAY(I.J)
           IF(I.EQ.IYC)GO TO 13
           NTEMP=NTEMP+IARRAY(I+1,J)*ICON(4)
           GO TO 13
15         K=-2
           NTEMP=0
           DO 16 M=1,5
           NTEMP=NTEMP+IARRAY(I+K,J)*ICON(M)
           K=K+1
16         CONTINUE
13         IARRAY(I,J)=NTEMP/35
           IF(IVAL.GT.IARRAY(I.J))GO TO 11
           IVAL=IARRAY(I,J)
11         CONTINUE
10         CONTINUE                                                  2
C          CONTINUE PROGRAM FOLLOWING 2-DIM. SMOOTH
```

to a cross-window used in Sobell functions. The procedure is successful in removing 'pepper and salt' noise without causing undue distortion of the image. A suitable subroutine coded in FORTRAN is given in figure 8.2 while figure 8.3 illustrates the results of applying this routine.

(ii) *Harmonic analysis* This procedure works by fitting Fourier series to data to give estimates of the amplitudes of frequencies arbitrarily assigned to the data. This version of least squares analysis is often employed in image analysis (Gonzales and Wintz 1977, but see Milne 1949, for a derivation of the least squares approach). Many texts describe the mathematics of the transformation and modern books show how the FFT can be implemented. The FFT requires data collected at equal intervals and whose total number of pixels is a power of 2 (i.e. 8, 16, 32, 64, etc). Once the image has been transformed to the frequency domain then high frequencies can be removed using simple filters in which high frequencies (noise) are removed by low pass filters and differentiation is achieved by high pass filters. The effects of different types of filters on the pattern is beyond the scope of this review but to illustrate the procedure the effect of a simple wedge filter is shown in figure 8.4. The FORTRAN programme used to carry out these operations is given in figure 8.5 and is written to allow segmentation of the image into units containing 64×64 complex numbers which in turn are taken from an image containing 128×128 pixels. The arithmetical operation for the Fourier transform requires real and complex numbers but the measured pixels are held in bytes. In each segment the transformed, filtered and inverted arrays are stored back in the original bytes before starting the next segment. Segmentation is not an ideal procedure because it means the highest frequency of the transform is 32 (a sine or cosine requires two points to describe each frequency so the highest frequency of the transform is $N/2$, where $N =$ number of data points, here 64 per pass), yet for an array of 128×128 it should be 64. Thus, when filtering high frequencies (coded to be <32, figure 8.5, but the program includes an option for no filtering) some polarisation occurs in the inverted image. The routine assumes that the contributions from frequencies up to a given value (called variable RC in the program) are preserved intact while those from a lower frequency (also read in before starting the transform and called variable RL) to 0 are proportionally reduced before inverting the array (the filtering involves the calculation of the vector). The results of high frequency and low frequency filters are shown in the image shown in figure 8.4. In general Fourier

---

Figure 8.2 A subroutine in FORTRAN for smoothing along rows and columns using five point Legendre cubic equation (equation (8.2)). The routine assumes the data are held in IARRAY ( ) and the smoothed points replace the original data. This routine can take up to two minutes to smooth an array containing 128×128 absorbances when using a microcomputer system.

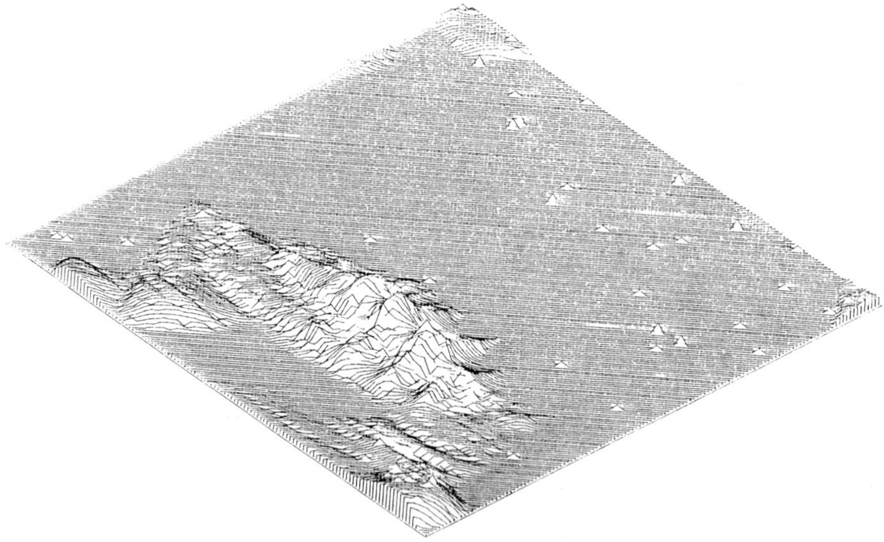

a

b

Figure 8.3 An example of the effect smoothing has on a set of data: (*a*) unsmoothed data collected from a region of a pattern from rat's liver microsomes; (*b*) the same data smoothed by the Legendre routine given in figure 8.2.

Figure 8.4 An illustration of the filtering effected by a wedge filter after Fourier transforming the data and inverting back using the routine given in figure 8.5. The original data was that shown in figure 8.3(*a*) and the limits used for the filter were: high frequency cut off (RC, figure 8.5) was 27, and low frequency was 5 (RL). The filtering reduced the maximum absorbance from about 1 in the original to less than 0.5 after filtering. The 'whiskery' effect on the top left edge is caused by numerical approximations affecting the data at this edge.

filtering highlights the spots but it can be used to remove artefacts (e.g. lens aberrations or blurring by motion of the gel) using band-reject filters.

A2 Fitting Gaussian Profiles to Spots

For an image having m spots, the intensity $(I, (x, y))$ can be described as (Taylor *et al* 1980)

$$I(x, y) = \sum_{k=1}^{m} F_K(x, y). \qquad (8.3)$$

Here $F(x, y)$ represents the two-dimensional distribution for spot k, and this function can be factorised into two parts, the amplitude (a_k) and the two component argument of the exponential (g_k and h_k, g_k being the function of x-coordinate while h_k is for the y-coordinate). The new expression becomes,

$$F_k(x, y) = a_k \exp\{-[r_k^2(x-\bar{x}_k)^2 + q_k^2(y-\bar{y}_k)^2]\} \qquad (8.4)$$

which contains the five parameters a_k, x_k, y_k, r_k and q_k. Thus, there will be m equations each carrying five parameters which must be fitted to the total image (this assumes the total number of spots, m, is known before starting the fitting).

Many computer systems have libraries which contain corrector routines for non-linear least squares fitting for known functions. If several routines are available then choose one which allows the user to include analytical first derivatives of the functions (in this case an exponential). Many routines calculate these derivatives numerically and this significantly increases the computational time. The derivatives are required in order to calculate the corrections for the approximate values of the five parameters for each spot and given at the start or calculated during each iteration. The whole procedure resembles the calculation of roots using a Newton–Raphson procedure in which the least squares evaluation Q (equation (8.5)) is the function to be evaluated

$$Q = \sum_{k=1}^{m} [F_k(x, y)(\text{expt}) - F_k(x, y)(\text{theo})]^2 \qquad (8.5)$$

(the terms expt and theo respectively refer to values found experimentally and after evaluating the expression in equation 8.4). By taking derivatives of equation (8.5), expanding in a Taylor series and retaining only first order derivatives, a relationship between the derivative of Q and the difference between the guessed and estimated parameters, (A_i) can be found for each Gaussian equation (see McCalla 1967 for a full treatment of this procedure). This linearises the non-linear equations to produce a matrix of the form

$$\boldsymbol{AB} = \boldsymbol{C} \qquad (8.6)$$

in which each row is

$$\boldsymbol{A} = \prod_{p=1}^{5m} \sum_{i=1}^{n} \frac{\partial F_i}{\partial C_l} \cdot \frac{\partial F_i}{\partial C_p} \qquad (8.7)$$

(N = number pixels) while there will be $p = 1$ to $5m$ columns. Each element of the column matrix $\boldsymbol{B}$ is the correction term for each parameter (a_k, k etc equation (8.4)) calculated for m spots (so giving three parameters plus two means for each spot). The elements of the column matrix $\boldsymbol{C}$ are

$$\boldsymbol{C} = -\sum_{i=1}^{n} \frac{\partial F_i}{\partial C_l} \cdot R_i. \qquad (8.8)$$

R_i are the individual differences calculated by $f(x_i, y_i, A_1, A_2, \ldots, A_5) - P_i$ (P_i = pixel intensity at x_i, y_i, while A_1, A_2, etc are the values for the five parameters in the last cycle). The object of each inversion is to determine C_l in order that a better estimate of A_i can be calculated for the zth cycle from A_i, $(z) = A_i\,(z-1) + \delta C_l$. The iterations are continued until either $\boldsymbol{C}$

```
SSS/SUPERSOFT 8086/88/87 Fortran IV Ver 1.04 cpm                PAGE 0001

C       FOURIER TRANSFORM CONTROL PROG
C       INCLUDES FILTERING AND READING IN BLOCKS
C       MAX DIMESIONS 64X64 FOR EACH PASS
C         ORIGINAL DATA SERIALLY IN IDU( )
C         ACCEPTS UP TO 10 SETS OF DATA TRANSFORMING
C         EACH SEPARATELY AND WRITING BACK FILTERED
C         DATA TO ORIGINAL FILENAMES APPENDED .FFT
        DIMENSION ARRAY(64,65),R(64),MCH(11)
        DIMENSION IDU(16600),ITRA(4),NCH(11),ICH(11),IAP(4)
          DIMENSION LIST(10,11),NREC(10),MTRAN(10,11)
        COMMON /FR1/ARRAY/CO/R/LIN/IDU,ICH,JREC
        COMPLEX ARRAY,R,CONJG
        CHARACTER*2 LCH,LPRI,LSTO,JCH
        INTEGER*1 IDU,IDIR,ICH,LCHA,IAP,ITRA,NCH
          INTEGER*1 LIST,MTRAN
        DATA IAP/".FFT"/,IAR/64/,ITRA/".TRA"/
          DATA LAP/3/
C         AFTER GETTING EACH FILENAME IT TESTS
C         TO SEE IF VALID BEFORE INCLUDING IN
C         LIST(  ).
          NFILE=0
          WRITE(1,999)
  999     FORMAT(1X,"ENTER NAMES OF FILES:AFTER LAST ENTER END")
          WRITE(1,990)
  990     FORMAT(1X,"ASSUMES ALL FILES ON SAME DISK")
          WRITE(1,889)
  889     FORMAT(1X,"INPUT HOME(=0) OR AWAY(=1) DISK"$
          READ(1,2002)ID
  20    WRITE(1,1000)
  1000  FORMAT(1X,"FILE NAME TO READ "$
        READ(1,2010)ICH

          IF(ICH(1).EQ.69)GO TO 21
C         E=69 IN ASCII
          IER=1
C         IER=1 SO CHECK FOR FILE WITH APPEND .FFT(IN IAP)
          CALL TESTF(ICH,IAP,IER,ID)
C         IF IER=-1 THEN FAULT SO GET NAME AGAIN
          IF(IER.EQ.-1)GO TO 20
          INCREMENT NFILE
C         KHAR IS SYSTEM FUNCTION TO HELP FIND
C         FIND NO CHARACTERS STORED IN FILENAME
C         THEN PUTS INTO LIST( )
          DO 22 I=1,11
          LCHA=KHAR(ICH,I)
          IF(LCHA.LT.44.OR.LCHA.EQ.46)GO TO 23
  22      CONTINUE
  23      DO 24 J=1,I
          LIST(NFILE,J)=ICH(J)
  24      CONTINUE
  2010  FORMAT(11A1)
C         SOME SYSTEMS NEED TO KNOW NO. RECORDS IN FILE
  2001  FORMAT(1A0)
  239   WRITE(1,1021)
  1021  FORMAT(1X,"NO RECORDS STORED ?" $
        READ(1,2002)NREC(NFILE)
          GO TO 20
C         END OF CLERICAL PART ENTERING FILENAMES
C         NORMALLY NOT POSSIBLE TO PRINT OUT ALL DATA
```

Figure 8.5 A program for (*a*) organising the data to be Fourier transformed, filtered and inverted. The filter is a simple wedge (shown in the inset diagram).

```
SSS/SUPERSOFT 8086/88/87 Fortran IV Ver 1.04 cpm                 PAGE 0002

2002  FORMAT(I0)
21        WRITE(1,1005)
1005    FORMAT(1X,"FILTER TRANSFORM? Y/N"$
        READ(1,2001)JCH
        IF(JCH.EQ."N")GO TO 33
C       SINCE TRANSFORM IS ON 64 OR LESS DATA,THE LIMIT ON THE HIGH
C       FREQUENCY CODE IS 32
        WRITE(1,1002)
1002    FORMAT(1X," NORMAL UPPER LIMIT IS 32,BUT LESS IF < 64X64")
        WRITE(1,1003)
C       USUALLY NEEDS TO INVERT BUT IN SOME SITUATIONS THIS
C       IS NOT NECESSARY
1003    FORMAT(1X,"INPUT UPPER LIMIT  OF HI.FREQ. FILTER"$
        READ(1,2003)RC
C       LO-FREQ. DIFFERENTIATES SO ERR ON HIGH SIDE,5 HAS BEEN
C       FOUND TO BE USEFUL FOR MANY GELS BUT IT REDUCES THE
C       HEIGHT OF PEAKS.TO OMIT LO-FREQ. FITERING ENTER -1
226     WRITE(1,1004)
1004    FORMAT(1X,"INPUT UPPER LIMIT FOR LO-FREQ(NORMALLY 5)= "$
        READ(1,2003)RL
2003    FORMAT(F0.0)
        IF(RL.GT.RC)GO TO 226
33      WRITE(1,1026)
1026    FORMAT(1X,"INVERT TRANSFORM?,Y/N "$
        READ(1,2001)LCH
C       HERE STARTS THE CONTROL OF DATA FROM THE
C       FILES (NO.=NFILE) AND THIER FOURIER TRANSFORM,ETC.
        DO 30 NUM=1,NFILE
        DO 31 I=1,11
        ICH(I)=LIST(NUM,I)
        LCHA=KHAR(ICH,I)
        IF(LCHA.LT.44.OR.LCHA.EQ.46)GO TO 32
31      CONTINUE
32      JREC=NREC(NUM)
C       ABOVE SETS UP THIS FILE NAME & RECORD
C       NBLOK TESTS FORROWS=COLUMNS=2**NP
      CALL NBLOK(NRD,IBLK,ICH,JREC,IER,NP)
      IF(NP.LE.6)GO TO 237
      NP=6
237   IF(NP.GT.1)GO TO 238
      WRITE(1,1006)
1006  FORMAT(1X,"ERROR IN DATA,2**N N<2 ")
      STOP
C     NRD=NO BLOCKS 64X64.IN THIS FILE
C     IBLK=0 IF THERE ARE NRD*64 BLOCKS IN THE FILE
C     IBL;K NE. 0 IF THE FILE NEEDS PADDING TO
C     MAKE UP TO NEXT POWER OF 2-ONLY NEEDED
C     IF NRD >= 1
C
C     FIRST READ IN ALL THE DATA
238   CALL FFRD1(IXINC,IYINC,IXC,IYC,NDATA,MAXOD)
C     ALL DATA IN LINEAR IDU(   )
C     SEE IF PADDING REQUIRED
        NAXOD=-100
1031  FORMAT(1X,10I4)
223   IF(IBLK.EQ.0)GO TO 2
      WRITE(1,3000)
3000  FORMAT(1X,"BLOCK FILLING ")
      N=NDATA
```

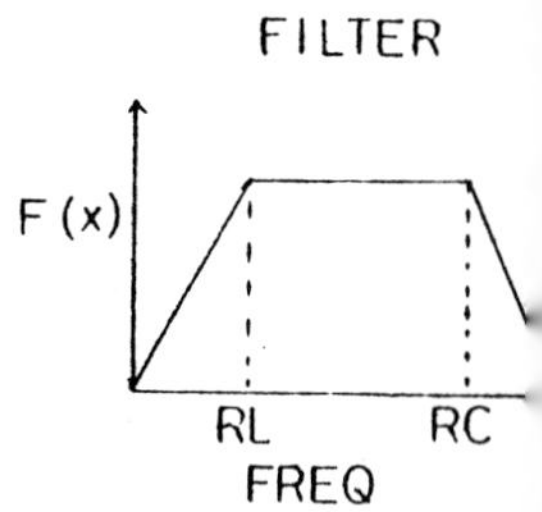

```
      NDATA=NDATA+IBLK
      DO 1 I=N,NDATA
      IDU(I)=0
1     CONTINUE
C     FILLED BETWEED NDATA & NRD*(64*64)
C     NOW POSSIBLE TO LOOP IN 64X64 BLOCKS
2     NX=IXC/IAR
      IF(NX.GT.0)GO TO 10
      NX=1
10    NY=NX
      LS=1
C     CHECK IF NDATA>64X64
C       NDATA=TOTAL AMOUNT OF DATA ORIGINALLY IN IDU
      IF(NDATA.GT.4096)GO TO 11
      LE=NDATA
      NJ=IXC
      LI=1
      NI=IXC-1
      GO TO 255
11    LE=(4064*NX)
      LI=IXC/IAR
      NJ=NX*IAR
      NI=IAR-1
C       STORES THE FIRST PART OF ORIGINAL
C       DATA SINCE THESE ARE CONSTANTS ASSOCIATED WITH
C       THE COLLECTION OF THE DATA
255   DO 201 JLOOP=1,NY
C     START OF MAJOR LOOP
      DO 202 NLOOP=1,LI
C     START OF X-LOOP,FIXED Y
        WRITE(1,1040)JLOOP,NLOOP,LS,LE
1040  FORMAT(1X,"Y-LOOP NO ",I6," X-LOOP NO ",I6,/,1X,
     1"START COUNT ",I6," END COUNT ",I6)
      IY=0
      DO 203 KLOOP=LS,LE,NJ
C     STEPPING LOOP
      IY=IY+1
      IX=0
      DO 210 LOOP=KLOOP,KLOOP+NI
C       CONVERTS FROM UNSIGNED BYTES IN IDU TO REAL POSITIVE
C     PICKS UP BLOCKS OF DATA FOR PROCESSING
      IX=IX+1
      IF(IDU(LOOP+9).GE.0)GO TO 206
      ICON=IDU(LOOP+9).AND.127
      ICON=ICON+128
      GO TO 207
206   ICON=IDU(LOOP+9)
207   ARRAY(IY,IX)=(0.0,0.0)
      ARRAY(IY,IX)=FLOAT(ICON)
210   CONTINUE
203   CONTINUE
C     ENDS THE TWO DIM. LOOPING & STEPPING
      NCOL=NI+1
C     CHECK NCOL=NLINE
      IDIF=IXC-IYC
      IF(IDIF)100,101,101
C     NEG THEN IYC>IXC
100   NCOL=NI+1
101   LINE=NCOL
```

```
        N2=ISHIFT(NCOL,-1)
        WRITE(1,1020)
 1020   FORMAT(1X,"TIME DOMAIN DATA")
        NF=1
 C        THE TRANSFORMATION STARTS HERE,ROWS FIRST
        DO 104 I=1,NCOL
        DO 300 MF=1,LINE
        R(MF)=ARRAY(I,MF)
 300    CONTINUE
 3003   FORMAT(1X,2E14.4)
 C        FFT IS THE TRANSFORMATION SUB. TREATING LINEAR COMPLEX ARRAYS
        CALL FFT(NP,N2,NCOL)
        DO 301 MF=1,LINE
        ARRAY(I,MF)=R(MF)
 301    CONTINUE
 1001   FORMAT(1X,I6)
 104    CONTINUE
 C      NOW ROTATE & REPEAT
        DO 105 LL=1,LINE
        DO 302 MF=1,NCOL
        R(MF)=ARRAY(MF,LL)
 302    CONTINUE
        CALL FFT(NP,N2,LINE)
 C        REPLACES ORIGINAL DATA BY THE TRANSFORM
        DO 303 MF=1,NCOL
        ARRAY(MF,LL)=R(MF)
 303    CONTINUE
 105    CONTINUE
 C        NOT NORMALLY NECESSARY
 C        FILTERING STARTS HERE
 213    IF(JCH.EQ."N")GO TO 211
 C        N2=32 IF 64 DATA PER LINE
          AN3=FLOAT(N2)
          RS=AN3-RC
 C        RC=START OF HI-FREQ. FILTER
          IF(RC.LT.AN3)GO TO 180
          RIN=0.0
          RC=AN3+1
          GO TO 181
 180    RIN=1./RS
 181          DEN=1./RL
        NP1=LINE+1
        DO 107 M=1,N2
        RX=FLOAT(M)-1.
        R2=RX*RX
        Z=0
        DO 108 K=1,N2
        Z=Z+1
        R3=Z-1
        R3=R3*R3
        D1=SQRT(R3+R2)
 C        ABOVE CALCS. VECTORS FOR FILTER
          IF(RC.GE.AN3)GO TO 182
        IF(D1.GT.RC+RS)GO TO 408
 182          IF(D1.LE.RL)GO TO 110
        IF(D1.LE.RC)GO TO 108
        D1=RC+RS-D1
        ZA=D1*RIN
          GO TO 111
```

```
110      ZA=DEN*D1
111     ARRAY(M,K)=ARRAY(M,K)*ZA
        ARRAY(M,NP1-K)=ARRAY(M,NP1-K)*ZA
        ARRAY(NP1-M,K)=ARRAY(NP1-M,K)*ZA
        ARRAY(NP1-M,NP1-K)=ARRAY(NP1-M,NP1-K)*ZA
        GO TO 108
408     ARRAY(M,K)=(0.,0.)
        ARRAY(NP1-M,K)=(0.,0.)
        ARRAY(M,NP1-K)=(0.,0.)
        ARRAY(NP1-M,NP1-K)=(0.,0.)
1025    FORMAT(1X,2I6,2E14.4)
108     CONTINUE
107     CONTINUE
C       FILTERED NOW INVERT
211     IF(LCH.EQ."N")GO TO 212
        WRITE(1,1015)
1015    FORMAT(1X,"FREQUENCY DOMAIN")
        FLINE=FLOAT(LINE)
C         ROWS FIRST
        DO 120 LL=1,LINE
        DO 304 MF=1,NCOL
        R(MF)=CONJG(ARRAY(MF,LL))
304     CONTINUE
        CALL FFT(NP,N2,LINE)
        DO 305 MF=1,NCOL
        ARRAY(MF,LL)=R(MF)*FLINE
305     CONTINUE
120     CONTINUE
C       NOW ROTATE FOR COLUMNS
        DO 130 J=1,NCOL
        DO 306 MF=1,LINE
        R(MF)=CONJG(ARRAY(J,MF))
306     CONTINUE
        CALL FFT(NP,N2,NCOL)
        N4=LINE+1
        DO 307 MF=1,LINE
        II=MF
        IF(MF.EQ.1)GO TO 310
        II=N4-MF
310     ARRAY(J,MF)=R(II)*FLINE
307     CONTINUE
130      CONTINUE
C       NOW WRITE AWAY TO LINEAR IDU( )
212     IY=0
        DO 204 KLOOP=LS,LE,NJ
        IY=IY+1
        IX=0
C       THIS REPEATS THE BEGINNING OF PROG
        DO 205 LOOP=KLOOP,KLOOP+NI
        IX=IX+1
        Z=REAL(ARRAY(IY,IX))
        ICON=IFIX(Z)
          NAXOD=MAX0(ICON,NAXOD)
        IF(ICON.GE.0)GO TO 106
        ICON=0
3002    FORMAT(1X,2I6,E14.4)
106     IF(ICON.GT.127)GO TO 224
        IDU(LOOP+9)=ICON
        GO TO 205
```

```
 224    IDIR=ICON.AND.127
        IDU(LOOP+9)=IDIR.OR.128
 205    CONTINUE
 204    CONTINUE
 C      WRITTEN BACK SO DO NEXT BLOCK
        LS=LS+IAR
        LE=LE+IAR
 202    CONTINUE
 C      RESETS FOR NEW BLOCK OF Y
        LS=LE+1-IAR
        LE=LE+(4064*NX)-IAR
 201     CONTINUE
 C        NOW FIXUP THE JOINS
          IF(NDATA.LE.4096)GO TO 257
          NX=IXC/LI
          NY=NX
          NU =(LAP+LAP)
          KST=NX-LAP+9
          KE=NX+LAP+9
          NN=NX*LI
          KY=NY-LAP
          DO 401 K=1,LI
          DO 402 KK=1,KY
 C        NOW GOING DOWN X-INTERFACE
          ISL=IDU(KE)-IDU(KST)
          ISL=ISL/NU
 C        CALC SLOPE
          KL1=KST+1
          KL2=KE-1
          IZ=1
          DO 403 I=KL1,KL2
          IDU(I)=IDU(KST)+(IZ*ISL)
          IZ=IZ+1
 403      CONTINUE
 C        CALC NEXT START & FINISH
          KST=KST+NN
          KE=KE+NN
 402      CONTINUE
 C        NOW TO AVERAGE THE Y-INTERFACE
          IF(K.EQ.LI)GO TO 401
          LST=KST+NX+LAP+1
          LEN=LST+(IXC*5)
 C        GO ALL ALONG X
          DO 404 KX=1,IXC
          ISL=IDU(LEN)-IDU(LST)
          ISL=ISL/NU
          KL1=LST+NN
          KL2=LEN-NN
          IZ=1
          DO 405 I=KL1,KL2,NN
          IDU(I)=IDU(LST)+ISL*IZ
          IZ=IZ+1
 405      CONTINUE
          LST=LST+1
          LEN=LEN+1
 404      CONTINUE
 C        RESTART ON NEXT BLOCK OF X
          KST=LEN+NX-LAP
          KE=KST+NU
```

SSS/SUPERSOFT 8086/88/87 Fortran IV Ver 1.04 cpm PAGE 0007

```
             KY=KY-1
      401    CONTINUE
      C    ALLDONE SO WRITE AWAY
      C    FIRST ADD FILE NAME .FFT
      C      PRINTING LOOP IF REQUIRED
      C      NAMES NEW FILE WITH .FFT
      257  DO 232 ICHAR=1,11
           LCHA=KHAR(ICH,ICHAR)
           IF(LCHA.LT.44.OR.LCHA.EQ.46)GO TO 235
      232  CONTINUE
      235  IZ=0
      3011 FORMAT(1X,I6)
           DO 236 I=ICHAR,ICHAR+3
           IZ=IZ+1
           ICH(I)=IAP(IZ)
      236  CONTINUE
           WRITE(1,3001)ICH
      C      REPLACE OLD MAX O.D. WITH NEW
             ICON=NAXOD
             IF(ICON.GE.0)GO TO 260
             ICON=0
      260    IF(ICON.GT.127)GO TO 262
             IDU(3)=ICON
             GO TO 263
      262    IDIR=ICON.AND.127
             IDU(3)=IDIR.OR.128
      C      FWRIT SUB. WRITES AWAY IN ORIGINAL FORMAT
      263  CALL FWRIT(IER)
      30     CONTINUE
      3001 FORMAT(1X,11A1)
           STOP
           END
```

```
>>> STMT MAP >>>
000999-0620    000990-0626    000889-062C    002002-064A
000020-002F    001000-0632    002010-0638    000021-00DF
000022-00A0    000023-00A3    000024-00BF    002001-063E
000239-00C2    001021-0644    001005-0650    000033-0145
001002-065C    001003-0662    002003-066E    000226-0121
001004-0668    001026-0674    000030-0B0E    000031-01A1
000032-01A4    000237-01C7    000238-01E0    001006-067A
001031-0680    000223-01ED    000002-022B    003000-0686
000001-0228    000010-0249    000011-027E    000255-02AA
000201-0872    000202-084D    001040-068C    000203-0382
000210-037F    000206-0350    000207-035F    000100-03A6
000101-03B1    001020-0692    000104-041D    000300-03F6
003003-0698    000301-041A    001001-069E    000105-046D
000302-0446    000303-046A    000213-0470    000211-0682
000180-04B6    000181-04C1    000107-067F    000108-067C
000182-0561    000408-0634    000110-059A    000111-05A5
001025-06AA    000212-0780    001015-06B6    000120-06F9
000304-06CC    000305-06F6    000130-077D    000306-0727
000307-077A    000310-0762    000204-0834    000205-0831
000106-07F3    003002-06BC    000224-0813    000257-0A4A
000401-0A47    000402-0964    000403-094B    000404-0A1F
000405-0A06    000232-0A74    000235-0A77    003011-06C2
000236-0AA7    003001-06C8    000260-0AD3    000262-0AEF
000263-0B09
```

1

```
 1:  C      ONE DIMENSIONAL FOURIER TAKEN FROM
 2:  C      COOLEY,J.W.,LEWIS,P.A.W., & WELCH,P.D.(1969)
 3:  C      IEEE TRANS. EDUC. VOL E 12,27-34.
 4:         SUBROUTINE FFT(NP,N2,N)
 5:  C      NP=2**NP,N=NO. DATA N2=N/2
 6:         DIMENSION R(64),SI(6),CO(6)
 7:         COMMON /CO/R
 8:         COMPLEX R,U,W,T,CMPLX
 9:  C      DATA IS FOR SINE AND COSINE ASSUMING NP=6(2**NP=NO. DATA)
10:         DATA SI/0.,-1.,-.707106721,-.382683408,-.19509026,-.0978258
11:         DATA CO/-1.,0.,.70710667,.923879428,.98078518,.99518465/
12:         NM1=N-1
13:         FE=FLOAT(N)
14:         FE=1./FE
15:         J=1
16:         DO 3 I=1,NM1
17:         IF(I.GE.J)GO TO 1
18:         T=R(J)
19:         R(J)=R(I)
20:         R(I)=T
21:  1      K=N2
22:  2      IF(K.GE.J)GO TO 3
23:         J=J-K
24:         K=ISHIFT(K,-1)
25:  C      EQUIV. TO K=K/2
26:         GO TO 2
27:  3      J=J+K
28:  C      ABOVE SORTS DATA INSTEAD OF SORTING BITS
29:         DO 5 L=1,NP
30:         LE=2**L
31:         LE1=ISHIFT(LE,-1)
32:  C      EQUIV LE1=LE/2
33:         U=(1.,0.)
34:         A=CO(L)
35:         B=SI(L)
36:  C      LOOK-UPS FOR COSINE AND SINE,INCLUDING SIGN
37:         W=CMPLX(A,B)
38:         DO 5 J=1,LE1
39:         DO 4 I=J,N,LE
40:         IP=I+LE1
41:         T=R(IP)*U
42:         R(IP)=R(I)-T
43:  4      R(I)=R(I)+T
44:  5      U=U*W
45:         DO 6 I=1,N
46:         R(I)=R(I)*FE
47:  6      CONTINUE
48:         RETURN
49:         END
```

Figure 8.5 (*b*) The fast Fourier transform (FFT) called in the program is taken from Cooley *et al* 1969. This program takes approximately 15 min to treat a 128 × 128 image array when run on a 16-bit micro-processer system. This FFT is written for arrays containing 64 × 64 complex numbers and must be changed if other sets are to be transformed. Line 34 must be replaced by a calculation of the cosine of PI/LE1 and that of 35 by its sine, while the data (SI and CO) can be removed.

is reduced to a level equal to the rounding errors or there are no further improvements in the value of Q despite allowing the cycles to continue for a further limited number of cycles.

The procedure is well described in the literature and is relatively slow in execution because it is likely that several iterative cycles (at least five) will be required before converging onto a solution. Thus, for 1000 spots each having 16 pixels (4×4) there will be at least 80 000 evaluations of the exponential. If a linear function were fitted to describe the patterns then $\boldsymbol{A}$ (equation (8.5)) would contain partial derivatives of Q with respect to the parameters of the function, $\boldsymbol{B}$ would hold the parameters and $\boldsymbol{C}$ the residuals. Thus, only one inversion is required after calculating the residuals in order to solve the m simultaneous equations and no iterations or approximations are involved in the calculation.

A3 Image Enhancements

The contrast in an eight-colour display is limited because spots can be hidden within an interval, especially if there is a changing background. Sometimes these effects can be reduced by a procedure called 'histogramming' (Gonzalez and Wintz 1977) in which the frequency distribution of intensities is distorted by calculated probability density functions. The effect is to change the contrast and raise the less intense spots. Naturally this distorts the measured intensities so the transformed value cannot be used for calculating effective masses. Thus, although it is a useful technique to enhance displays, the untainted data should be summed to give the masses.

References

Anderson N G and Anderson N L 1982 *Clin. Chem.* **28** 739–48

Anderson N G, Anderson N L and Tollaksen S L 1979 *Clin. Chem.* **25** 1199–1210

Brown W T and Ezer A 1982 *Clin. Chem.* **28** 1041–4

Dunn M J and Burghes A H M 1983 *Electrophoresis* **4** 173–89

Cooley J W, Lewis P A W and Welch P D 1969 *IEEE Trans. Educ.* **E12** 27–34

Fox S H 1982 *Clin. Chem.* **28** 932–4

Gonzalez R C and Wintz P 1977 *Digital Image Processing* (London: Addison-Wesley)

Lemkin P F, Lipkin L E and Lester E P 1982 *Clin. Chem.* **28** 840–9

McCalla T R 1967 *Introduction to Numerical Methods and* FORTRAN *Programming* (New York: Wiley) pp 256–61

Miller M J, Vo Phong K, Nielsen C, Guiduschek E P and Xuong Nguyen Huu 1982 *Clin. Chem.* **28** 867–75

Milne M J 1949 *Numerical Calculus* (Princeton University Press)

Spragg S P 1973 *New Techniques in Biophysics and Cell Biology* ed. R H Pain and B J Smith Vol. 1 (London: Wiley) pp 209–39

Spragg S P, Jones M I and Hill B J 1983 *Anal. Biochem.* **129** 255–68
Spragg S P, Amess R, Jones M I and Ramasamy R 1985 *Anal. Biochem.* **147** 120–7
Taylor J, Anderson N L, Coulter B P, Scandora A E and Anderson N G 1980 *Electrophoresis 79* ed. B J Radola (Berlin: de Gruyter) pp 329–39
Tracy R P and Young D S 1984 *Clin. Chem.* **30** 462–5

Index